Problem Books in Mathematics

Edited by P. R. Halmos

Problem Books in Mathematics

Series Editor: P.R. Halmos

Unsolved Problems in Intuitive Mathematics, Volume I:
Unsolved Problems in Number Theory
by *Richard K. Guy*
1981. xviii, 161 pages. 17 illus.

Theorems and Problems in Functional Analysis
by *A.A. Kirillov* and *A.D. Gvishiani*
1982. ix, 347 pages. 6 illus.

Problems in Analysis
by *Bernard Gelbaum*
1982. vii, 228 pages. 9 illus.

A Problem Seminar
by *Donald J. Newman*
1982. viii, 113 pages.

Problem-Solving Through Problems
by *Loren C. Larson*
1983. xi, 344 pages. 104 illus.

Demography Through Problems
by *N. Keyfitz* and *J.A. Beekman*
1984. viii, 141 pages. 22 illus.

Problem Book for First Year Calculus
by *George W. Bluman*
1984. xvi, 384 pages. 384 illus.

Exercises in Integration
by *Claude George*
1984. x, 550 pages. 6 illus.

Exercises in Number Theory
by *D.P. Parent*
1984. x, 541 pages.

Problems in Geometry
by *Marcel Berger, Pierre Pansu, Jean-Pic Berry, and Xavier Saint-Raymond*
1984. viii, 266 pages. 244 illus.

Algebraic Logic
by *S.G. Gindikin*
1985. xviii, 360 pages. 93 illus.

S. G. Gindikin

Algebraic Logic

Translated by Robert H. Silverman

With 93 Illustrations

Springer-Verlag
New York Berlin Heidelberg Tokyo

S. G. Gindikin
University of Moscow
Mahmat
Moscow 117234
U.S.S.R.

Robert H. Silverman *(Translator)*
22 Trowbridge Street
Cambridge, Massachusetts 02138
U.S.A.

Editor

Paul R. Halmos
Department of Mathematics
University of Santa Clara
Santa Clara, California 95053
U.S.A.

AMS Subject Classifications: 00A07, 03-01, 06-01

Library of Congress Cataloging-in-Publication Data
Gindikin, S. G. (Semyon Grigor'evich)
Algebraic logic.
Translation of: Algebra logiki v zadachakh.
Bibliography: p.
Includes index.
1. Algebraic logic. I. Title.
QA10.G5613 1985 511.3 85-14847

Original Russian edition: Algebra Logiki v Zadachakh, Nauka, © 1972.

Printed and bound by R.R. Donnelley & Sons, Harrisonburg, Virginia.
Printed in the United States of America.

9 8 7 6 5 4 3 2 1

ISBN 0-387-96179-8 Springer-Verlag New York Berlin Heidelberg Tokyo
ISBN 3-540-96179-8 Springer-Verlag Berlin Heidelberg New York Tokyo

PREFACE

The popular literature on mathematical logic is rather extensive and written for the most varied categories of readers. College students or adults who read it in their free time may find here a vast number of thought-provoking logical problems. The reader who wishes to enrich his mathematical background in the hope that this will help him in his everyday life can discover detailed descriptions of practical (and quite often -- not so practical!) applications of logic.

The large number of popular books on logic has given rise to the hope that by applying mathematical logic, students will finally learn how to distinguish between necessary and sufficient conditions and other points of logic in the college course in mathematics. But the habit of teachers of mathematical analysis, for example, to stick to problems dealing with sequences without limit, uniformly continuous functions, etc. has, unfortunately, led to the writing of textbooks that present prescriptions for the mechanical construction of definitions of negative concepts which seem to obviate the need for any thinking on the reader's part. We are most certainly not able to enumerate everything the reader may draw out of existing books on mathematical logic, however.

Nevertheless, we decided to increase by one the already large number of books on logic, since in most of the elementary books in the field of which we are aware, very little regard is given to the interests of the reader who wishes to study the mathematically *meaningful* theorems and problems. It is to this reader -- primarily students in the early years of the university and pedagogical institutes and students in the upper grades of mathematics high schools -- that the present book is addressed. It seemed like a good idea to familiarize mathematics students with the concepts of finite mathematics, particularly algebraic logic, since these concepts differ qualitatively from the concepts met with in the traditional university course in mathematics.

We will be considering basically three groups of problems: problems of completeness and functionally closed classes; problems of synthesis and evaluation of the complexity of networks; and probability theory in finite Boolean algebras. Everywhere we will strive to introduce the reader to groups of concepts without attempting to derive the most exhaustive results. We would hope that those readers who do not intend to study a particular topic in depth will acquire a sufficient store of knowledge, and that those readers who have shown a special interest in the field will have been prepared for reading the specialized literature.

The term, "algebraic logic" or "algebra of logic," will be understood in the present book in a broad sense. Chapters 1-7 are devoted to mathematical logic proper. Here may be found a discussion of logical operations on propositions (more generally, Boolean operations on regular Boolean algebras; cf. Chapter 2). The remaining chapters deal with related topics. We have not avoided treating traditional topics no matter how often they have been presented before. For example, the reader will find a discussion of the relation between algebraic logic and elementary problems in proof theory and the construction of definitions of negative concepts. We have assumed that the reader will associate what he reads with what he already knows and that the formal techniques set forth in the book will prove useful. However, in no way are these techniques intended to take the place of the ability to reason meaningfully in particular cases. Some of the technical exercises, basically in Chapters 1, 2, and 9, have been included so as to provide necessary material for exercises in the corresponding topics in the course in mathematical analysis at pedagogical institutes. Only a small portion of the book makes formal use of results outside the scope of the university course in mathematics. Certain facts from mathematical analysis are occasionally applied, primarily in isolated problems and examples. These problems and examples may be omitted (though not without some loss).

Some remarks on the structure of the book are in order. Virtually any mathematics textbook may be thought of as a home-study course, often modeled on the efforts of students who are actually taking special courses. It is known that solving groups of problems into which theorems have been divided is one of the better ways of learning certain branches of mathematics. Many university seminars are constructed in this way. Unfortunately, however, it is the rare textbook that is modeled on this type of seminar. The present book originated in seminars conducted by the author at the Lenin State Pedagogical Institute in Moscow; some parts of the book were also used in a series of lectures and seminar sessions in the evening mathematics institute of Moscow State University for seven- and eight-year students. The instructor in this type of seminar is responsible for a variety of functions. He gives definitions and states problems, complementing the definitions and problems with whatever explanations are necessary. (It is this role which is served by the main text of the present book.) If the students in the seminar are unable to find some solution on their own, he gives hints in several stages, answers sometimes serving as such hints. (Our section, *Hints*, which follows the main text of each chapter, plays this role here.) Finally, he may himself have to solve the problem if the hints do not suffice; here, again, an answer itself may serve as such a solution. (Our section, *Solutions*, corresponds to this role.)

Even if the reader is forced to peek ahead to the *Solutions* section without having obtained a final solution, the effort he will have expended in reflecting on the problem will most certainly not have been in vain. And, if he has found the solution to some problem on his own, he should review our hints and solution, since they may contain additional information. (At the same time, we do not, of course, wish to insist in any way on precisely the version of the solution presented in the book to the exclusion of all other solutions.) While there will be those who prefer to immediately read the solution of a particular problem, with this group of readers we are concerned only minimally.

There is no detailed bibliography in the book; the only references are to literature where supplementary material may be found. Nor are the more important hints systematized.

I would like to take this occasion to express my appreciation to P. S. Novikov, my first teacher of mathematical logic, whose suggestions I have made regular use of in the course of teaching and with whom I discussed the plan for this book. I would also like to thank I. M. Yaglom for his unwavering interest in the present book, which has helped to bring my efforts to fruition, and Yu. A. Gastev and V. V. Donchenko, the editors of the book, for their counsel and criticism.

GUIDELINES AND SUGGESTIONS

The book consists of twelve chapters. Each chapter has its own numbering scheme for the definitions and problems, the number preceding the dot representing the chapter number (for example, Problem 3.12, Definition 8.3).

Chapter 1 serves as an informal introduction to the rest of the book. Here we introduce operations on propositions, discuss the statements of the basic problems of algebraic logic in the terminology of operations with propositions, and show how algebraic logic is related to elementary proof techniques.

We are relying on intuitive concepts of what constitutes a "proposition" and what constitutes an "instance". In particular cases, it will be clear which instances may be encountered in deciding on the truth or falsity of one or more (finitely many) propositions. However, only in special cases does the set of possible instances for infinite collections of propositions admit a natural description.

Formally speaking, there are no essential references from Chapter 1 onward. Therefore, the reader who, for one reason or another, is not interested in learning about propositional algebra proper need give this chapter only a cursory review. However, we suggest that the reader be careful in deciding this question since we believe that a fluent mastery of the interpretations of the basic facts of algebraic logic in the language of propositional algebra is extremely important for the study of algebraic logic. Some readers, by contrast, may be interested in a more unhurried introduction with a large number of examples and analogies to well-known results. Any one of the numerous books on mathematical logic may be used for this purpose, depending upon the particular reader's "taste". Some of these books are presented in the list of literature for Chapter 1.

Thus, strictly speaking, the reader may begin the book with Chapter 2. Here we introduce the central subject of our study: the logical functions (Definition 2.1); at this point, it is not essential to understand that the two values which the variables may assume may be interpreted as the "truth" values of the propositions. A whole series of facts from Chapters 1 and 2 may be restated, and often proved anew, in the new language. In sec. 2.2, we introduce the basic operation defined on logical functions -- composition, or the formation of a composite function. Its analog is probably well known to most readers from their study of the functions of a real variable. However, there are a number of difficulties that may arise in studying the intuitive definition of the composition when it is broken up into a series of steps. Since this definition is only required for rigorous proofs beginning with Chapter 3 (Problem 3.7), until that point we will understand composition to refer simply to an operation that consists of multiple substitutions of certain functions for the arguments of other functions and the "equating" of some of these arguments. Nevertheless, it is important to study Definition 2.3, and even more important to think through the scheme of intuitive proofs upon which it is based. Later on in the book, the reader will frequently come across analogous proofs in other contexts (cf. Chapters 8 and 9). In sec. 2.2 the basic equivalences of algebraic logic and a number of problems that involve the transformation of formulas may be found. In general, Chapter 2 contains a rather large number of practical exercises, though rather cumbersome manipulations are needed for their solution. We leave it to the reader to decide which problems to solve, and how detailed the algebraic manipulations ought to be. It is nevertheless important for what follows that the reader learn how to perform these manipulations without any difficulty. Boolean algebras are discussed in sec. 2.3 and 2.4. The results achieved here lead to an interpretation of results dealing with the logical functions applicable not only to propositional algebra, but also to the algebra of sets, for example.

In sec. 2.4 we introduce an important class of Boolean algebras, called regular Boolean algebras, for which we define Boolean operations, which are operations that may be applied to any of the logical functions in a natural way. These topics are considerably more difficult than the main part of the book, and may be omitted on a first reading. However, we suggest that the reader familiar with the foundations of set theory think through the general definition of a set-theoretic operation (cf. page 36) and the relationship of a set-theoretic operation to the logical functions. Subsequent references in the text to sec. 2.3 and 2.4 may be thought of as supplementary in nature. In Sections 2.5-2.7 normal forms are discussed. These are a special way of representing the logical functions in terms of the basic operations of conjunction, disjunction, and negation. These operations are considered in part for the case of propositions at the end of Chapter 1. There are many technical exercises in these sections.

The brief, though extremely important, Chapter 3 is devoted to the law of duality in algebraic logic. The reader who has not read sec. 2.4 will have to omit the first several paragraphs, and start with the paragraph preceding Definition 3.1. Problems 3.10 and 3.11 are supplementary in nature, and are needed in Chapter 6 to prove Post's theorem.

Chapter 4 discusses the representation of the logical function by means of arithmetic polynomials modulo 2 (Zhegalkin polynomials). The basic material may be found in Problems 4.1-4.7. Problems 4.10-4.12 prepare the reader for Chapter 6.

In Chapter 5, which is devoted to monotone logical functions, a first reading may be limited to Problems 5.1-5.11; Problems 5.12 and 5.13 are needed in Chapter 7, and Problems 5.14-5.16 in Chapter 6.

The central theorem of algebraic logic -- Post's theorem -- and topics related to it are discussed in Chapter 6. Post's theorem provides necessary and sufficient conditions on a system of functions according to which all the logical functions may be represented by compositions of functions occurring in the system. The basic material may be found in Problems 6.1-6.21; Problems 6.22-6.27 supply supplementary material. The concept of a functionally closed class (Definition 6.2), or class of functions closed relative to composition, is essential here.

Five functionally closed classes, referred to as pre-complete (they are, in some sense, maximal) occur in Post's theorem. The next chapter (Chapter 7) is devoted to the set of all the functionally closed classes. In sec. 7.1 a conceptual description of the structure of this set (Post schema) may be found, along with a description of the kinds of problems which may be solved by means of functionally closed classes. In Section 7.2 we consider the simplest problem of functionally closed classes containing constants. These classes are related to problems that involve not only composition, but also substitution of constants (extended composition). By means of this expansion, we are able to eliminate the basic difficulties of the general problem. In addition, by means of the general answer, which is derived comparatively easily, the situation that obtains in these problems may be handled without any difficulty. In concluding this section, we present as an example a problem that may be solved by means of functionally closed classes (for extended composition); the problem of self-dual completeness. This problem is considered in sec. 7.3 under the conditions of ordinary composition. The section is not made use of subsequently, and therefore may be omitted if desired, with the reader limiting himself to Definition 7.8, which will be needed later on. In sec. 7.4 we construct a fragment of the Post schema (third level), by means of which the problem of bases in pre-complete classes is solved. Finally, the Post schema is presented without proof in sec. 7.5, together with several problems that illustrate its application. References to Chapter 7 are found only in sec. 8.5 and 10.10.

Chapters 8 and 9 are devoted to network theory. Chapter 8 considers networks formed from functional elements; that is, devices that realize the logical functions. The reasoning of sec. 8.1 assumes that these elements produce an instantaneous response. Connections of elements corresponding to compositions of logical functions are studied in detail, and the concept of feedback is introduced. Under the assumptions of sec. 8.1, Post's theorem tells us how many basic functional elements must be supplied to realize an arbitrary logical function by means of a network formed from these elements.

In sec. 8.2 we drop the assumption that the functional elements produce an instantaneous response. This contracts the class of networks which realize logical functions. At the same time, a new type of element is used, called delay elements. Delay elements are needed to equalize the arrival time of signals at the inputs of functional elements. Everywhere below, it is assumed that the response time is the same for all basic elements. This rather burdensome constraint markedly simplifies our discussion. Once the response time of elements is taken into account, Post's theorem no longer is of any help in determining completeness conditions for a system of functional elements. Instead, Problems 8.7-8.11 are used for this purpose (under the above assumptions). Note that sets which are no longer functionally closed classes relative to ordinary composition occur in the answer. In Problems 8.12-8.14 several other examples of completeness problems that do not involve functionally closed Post classes may be found. These problems are not made use of subsequently, and may be omitted. In sec. 8.3 and 8.4 we study networks that do not realize any of the logical functions. A description of how these networks operate leads to the important concept of a finite automaton. A number of completeness problems arise here; their solution reduces to the completeness problem solved in sec. 8.2. In sec. 8.5 we consider von Neumann's method of realizing the logical functions by means of functional elements. The completeness problem is solved for this class of networks by means of the results of sec. 7.2 and 7.3.

A second class of networks, what are known as relay-contact networks, are considered in Chapter 9; sec. 9.1 discusses the general operating principle of such networks. We wish to emphasize here the second part of Problem 9.1; this part implies that relays with both negative and positive contacts are needed to realize the logical functions. The remaining sections consider only networks lacking any connections between the relay coils and the contacts, or contact networks. Section 9.2 is devoted to the general theory of such networks, and sec. 9.3 to the minimization of the networks, i.e., the construction of realization of functions that contain the least number of contacts. Examples of minimal networks are presented. It is extremely important to think through the different methods of proving minimality. The proof of Shannon's estimate of the number of contacts necessary for realizing a function of n variables (for large n) is of central importance here. In sec. 9.4 we construct realizations of linear logical functions (cf. Chapter 4) and, by means of these realizations, construct networks for summing numbers in the binary system. In sec. 9.5 we consider a realization of arithmetic operations by means of networks consisting of functional elements. The main part of Chapter 9 uses only Chapter 2; in sec. 9.3 we use elementary results from analysis in the proof of Shannon's theorem.

In Chapter 10 we construct probability theory for finite Boolean algebras. Accordingly, it is necessary to know the definitions of a Boolean algebra and of a regular Boolean algebra (Definitions 2.5 and 2.7); incidentally, the latter definition may be replaced by Definition 2.5, complemented with the axioms presented in the remark on page 238. Basic results from probability theory are presented in sec. 10.1-10.6. In sec. 10.7 we introduce Bernshtein polynomials and use them to prove Weierstrass' theorem for the uniform approximation of continuous functions on a segment by means of polynomials. In sec. 10.8 we introduce special classes of Bernshtein polynomials which relate to the logical functions. In this section, the classes are used to solve certain problems from automaton theory and, in the succeeding section, to study the reliability of contact networks. The last section is devoted to the synthesis of reliable networks from functional elements with unreliable realization, i.e., the completeness problem for such elements. In sec. 10.1-10.7, we use only Chapter 2, though in sec. 10.8 and 10.9 we will also require the definitions of self-dual and monotone functions from Chapters 3 and 5, and definitions from Chapters 8 and 9; sec. 10.10 relies on the results of Chapter 7. Results on the approximations of functions by means of Bernshtein polynomials are attained using simple assertions from analysis.

Chapter 11 considers a generalization of 2-valued algebraic logic to the finite-valued case; this chapter draws on results from Chapters 1-6.

Chapter 12 presents a brief survey of predicate logic, using Chapters 1 and 2. Besides examples from analysis, there are several problems in the chapter that were already solved in the proof of Post's theorem, and these may be omitted here. This chapter is of especial importance. All the foregoing discussions had been related in one way or another to the concept of a logical operation, and, accordingly, the propositions were treated relative to a fixed instance; in

Chapter 12 we study the relation between proposition and instance.

There are a number of ways of studying the book. An elementary course may consist of Chapters 1 and 2, with the abridgements indicated above in sec. 2.3 and 2.4, sec. 2.1 and 2.2, the survey portion of sec. 9.3, and Chapter 12. Such a reading may be extended if desired through the addition of the basic parts of Chapters 3-5, the statement of Post's theorem and examples illustrating the application of the theorem from Chapter 6, and the survey portion of Chapter 8. Thus, it is possible to construct the basic portion of the curriculum in mathematical logic as presented in pedagogical institutes (to this it would be necessary to add somehow the foundations of the theory of algorithms, for example, Turing machines). Further on, Chapters 6 and 9 may be studied. Additional sequences may consist of Chapters 7 and 8 (functionally closed classes, networks created from functional elements, and the completeness problem), and Chapter 10 (probabilistic logic). These sequences intersect in sec. 10.8 and 10.10 (completeness of systems of unreliable functional elements); Chapter 10 may be used to study probability theory at the point where it comes into contact with mathematical logic, though in this case it is best to expand upon the number of examples and informal explanations.

In the appendix may be found reference material helpful in the solution of the problems.

CONTENTS

Chapter 1
OPERATIONS ON PROPOSITIONS

The greater portion of the present book is devoted to the least advanced topic in mathematical logic, that of algebraic logic. Elsewhere the name of propositional algebra, which emphasizes the extremely important interpretation of this theory with which we shall begin our discussion, has often been used. Propositional algebra considers topics related to the composition of propositions. If we have several propositions, new and different propositions may be formed by means of the logical connectives and negations of these connectives. The initial propositions are then said to be simple propositions, and the newly formed propositions, composite propositions. These designations are not fixed for all time, and so a proposition that in one situation is "simple", elsewhere might be a "composite" proposition, and conversely. We will speak of this in more detail below. Note that in grammar the logical connectives simply refer to different ways of forming compound sentences from simple sentences. Here are some examples.

Suppose we are given the following two propositions: "The sun is shining outside" and "A lesson is under way in the classroom." Starting with these simple propositions, we may construct composite propositions in different ways:

(1) The sun is shining outside, and a lesson is under way in the classroom.

(2) The sun is not shining outside.
(3) The sun is shining outside, though a lesson is under way in the classroom.
(4) A lesson is under way in the classroom, though the sun is shining outside.
(5) The sun is shining outside or a lesson is under way in the classroom.
(6) Either the sun is shining outside or a lesson is under way in the classroom.
(7) If the sun is shining outside, a lesson is under way in the classroom.
(8) If a lesson is under way in the classroom, the sun is shining outside.
(9) The sun is not shining outside or a lesson is under way in the classroom.
(10) The sun is shining outside if and only if a lesson is under way in the classroom.

The reader should have no trouble extending this list of composite propositions, and we have no doubt that propositions even more absurd than (8) or (10) may be obtained. This may be accomplished most easily by taking as the simple propositions pairs of sentences that are semantically least related to each other, for example, "The 17th page of the book begins with the letter 'v'" and "Our city is on the shore of the Volga River." Here is the first special feature of propositional algebra. It admits any grammatically correct method of formation of composite propositions, ignoring entirely the *semantic* content of the resulting sentence. From the standpoint of the propositional algebra, any of the above ten compound sentences is permissible. In the propositional algebra, we are interested only in the *truth* or *falsity* (*truth value*) of propositions. More precisely, in the propositional algebra we study the truth of a compound proposition as a function of the truth values of the simple propositions occurring in it, and from this standpoint, investigate the different logical connectives.

We have not yet made any distinction between "propositions" and "sentences". From the standpoint of our subject, however, such a distinction is indeed necessary. We cannot assert of just any sentence that it is true or false. As examples, consider sentences that express a command ("Come to me!"), regret ("If only I had known earlier..."), a question

("Will you be home?"), congratulations ("Happy New Year!"), etc. Without dwelling on the description of this class of sentences, we agree that *a proposition is understood to mean a sentence for which it is meaningful to speak of truth or falsity.*

In general, selecting a particular proposition ("The sun is shining outside") scarcely helps us decide on its truth value. It is still necessary to determine the situation or instance (the precise time and place). Therefore, it is more accurate to speak of a proposition being true (false) *in a particular instance.* The instance may be determined uniquely in the proposition itself. ("The second correction of the spaceship's course was performed on February 4, 1971 at 5:01 Moscow time.") Sometimes in propositional algebra the discussion is limited exclusively to these propositions, but we will not do so here. There are propositions that are true (correspondingly, false) in all possible instances. Such propositions are said to be *absolutely true* (correspondingly, *absolutely false*). The propositions, "The Volga flows into the Caspian Sea" and "A unique line passes through two distinct points of the Euclidean plane," are absolutely true; the proposition, "$2a = a$, where a is a positive number," is absolutely false. Propositions that are absolutely true or absolutely false are called *logical constants.* If the instance is uniquely determined in the proposition itself, it is a logical constant. Now we can be more rigorous in speaking of the logical connectives. In studying the logical connectives we correlate the truth values of the simple propositions and the resulting composite proposition in the same instance. One of the most important questions we shall be interested in is to decide under what conditions (on the simple propositions) a logical connective yields an absolutely true proposition. Let us start by considering examples of logical connectives (operations).

1. The logical connective corresponding to the word "and" is referred to as *conjunction* and denoted by &.*

The proposition A & B, called the conjunciton of A and B, is true if and only if both A and B are true (cf. Example 1). This fact may be expressed by means of what is known as the

*The sign $\vee$ is often used in place of &. Note also that conjunction is referred to as logical multiplication.

A	B	$A\&B$
T	T	T
T	F	F
F	T	F
F	F	F

truth table for conjunction. In this table the symbol T denotes a true proposition, and the symbol F, a false proposition. The truth table enumerates all possible *truth values* (T or F) of the propositions A and B and correlates them with the truth values of the proposition $A \& B$. From the standpoint of propositional algebra, the logical connective (operation) of conjunction is completely *defined* by this table. The proposition $A \& B$ is absolutely true if and only if both A and B are absolutely true.

In constructing the table, we started with the conjunction "and". But we could have arrived at the same truth table by starting with any of the words "while", "but", "however", "although", and so on (cf. Examples 3 and 4). Thus, though the logical connectives corresponding to these words have different connotations, from the standpoint of propositional algebra they are indistinguishable. In addition, the connective corresponding to the operation of conjunction is sometimes realized without any words at all, as in the example, "The sun is shining outside; a lesson is under way in the classroom."). A similar abbreviation will be applied below (page 26) for the conjunction sign itself.

2. The *negation* $\bar{A}$ of a proposition A (cf. Example 2) is specified by the accompanying table. Negation is a *unary*

A	$\bar{A}$
T	F
F	T

operation in the sense that the new proposition is constructed from a single simple proposition A. By contrast, conjunction is a *binary* operation, as there the composite proposition is constructed from two simple propositions. The negation of an absolutely true proposition is absolutely false and conversely.

3. The binary logical operation corresponding to the connective "or" is called the *disjunction** (denoted $\vee$). But in ordinary speech the connective "or" is used in at least two different senses, that of the inclusive "or" and the exclusive "or". In the first type of composite proposition, the truth of *at least* one of its simple propositions is asserted, while in the second type we are concerned with the truth of one and only one of its simple propositions (sometimes using the more precise phrase, "either...or..."). Disjunction corresponds to the inclusive "or"; it is specified by the accompanying truth table.

A	B	$A \vee B$
T	T	T
T	F	T
F	T	T
F	F	F

The proposition $A \vee B$ is said to be absolutely true if in every instance at least one of the propositions A and B is true.

1.1. The logical operation corresponding to the exclusive "or" is called the *exclusive disjunction* (denoted $A \mathbin{\Delta} B$). Compile its truth table and express the operation in terms of the logical operations you are already familiar with.

4. We say that a logical operation is *identically true* (*identically false*) if a composite proposition obtained from any simple proposition by means of the operation is absolutely true (absolutely false). This is equivalent to the assertion that the composite proposition is true (false) for any truth values of the simple propositions. Do not, however, confuse the two concepts of identically true and absolutely true. The former refers to logical operations, and the latter,

*Disjunction is sometimes referred to as logical addition (this is not the same operation as arithmetic addition modulo 2, which we shall discuss in Chapter 2, page 65).

to individual propositions. In the next problem, we present examples of unary operations that are identically true (identically false).

1.2. Prove that the operation $A \vee \bar{A}$ is identically true, and that the operation $A \mathbin{\&} \bar{A}$ is identically false.

In the binary logical operations we have considered so far, the simple propositions A and B occurred symmetrically (i.e. commutatively: $\omega(A,B) = \omega(B,A)$). Now we wish to consider operations where the simple propositions occur non-symmetrically. (This is analogous to the transition from compound (coordinated) sentences to complex sentences.)

5. The binary logical operation corresponding to the phrase, "if...,then...," used in the formation of conditional sentences is called a *material implication* (or simply *implication*). The composite proposition, "if A, then B", is written in the form, $A \to B$. The simple proposition A is called the *premise* of the implication, and B its *conclusion*. We first present the truth table for the implication, and then provide some explanatory remarks.

A	B	$A \to B$
T	T	T
T	F	F
F	T	T
F	F	T

The discrepancy between the ordinary understanding of the truth of a composite proposition and the idealized standpoint adopted in propositional algebra is far more marked in the case of implication than in the case of the other logical operations. The principal difference is that, according to the semantic content of propositions (and not just their truth value), the phrase, "if..., then...", comprehends a *causal relation* between a premise and a conclusion. From the standpoint of propositional algebra, the truth of an implication in any one instance asserts only that if in this instance the premise is true, then so is the conclusion. As a result, such implications as "If there are five floors in the house, Ivanov lives in apartment 3," "If it is raining in Voronezh, the book is gray," and even more "remarkable" propositions can be true. A more careful analysis shows that in our ordinary speech, we do not question the truth of the proposition, "B follows from A" for any particular instance, and instead are interested only in the absolute truth of $A \to B$; only in this case do we say that B is a consequence of A. This is perfectly natural from the standpoint of the truth table for implication. In fact, by our intuitive conception, the proposition $A \to B$ is absolutely true if and only if in every instance in which A is true, so is B, i.e. there exist no instances in which A is true and B false. On the other hand, the truth table must assert that $A \to B$ is true in every possible instance. By comparing these two standpoints, it follows that a truth table must have the form given above. There is one fine terminological point worth noting. We usually say that "B does not follow from A" whenever the proposition $A \to B$ is not absolutely true, i.e. whenever $A \to B$ is false in some instance; this must not be confused with the absolute falsity of $A \to B$.

As an illustration, note that in proving a theorem it is necessary to decide whether or not some implication is absolutely true. In fact, every theorem has the form of an implication $A \to B$ in which the premise (A) is "what is given" and the conclusion (B), "what is to be proved." With the converse theorem we associate the implication $B \to A$, and with the contrapositive theorem, the implication $\bar{A} \to \bar{B}$.

Digression on Implication. We often assume a more general definition of the converse theorem: If $(A \,\&\, C) \to B$ is the direct theorem, the converse is not only the theorem $B \to (A \,\&\, C)$, but also the theorems $(B \,\&\, \mathrm{A}) \to C$ and $(B \,\&\, C) \to A$, i.e. from "part" of the premise and the conclusion follows some other "part" of the premise. As an example, we may consider the "theorem of three perpendiculars": "A line ℓ that belongs to the plane P and is perpendicular to the projection m in the plane P of a line n is perpendicular to n." Let us consider the structure of this theorem. We let A denote the proposition, "The line ℓ belongs to the plane P," B the proposition, "The line m is the projection in the plane P of the line n," C the proposition, "The line ℓ is perpendicular to m," and D the proposition, "The line ℓ is perpendicular to n". The theorem then takes the form of the proposition $((A \,\&\, B) \,\&\, C) \to D$. It is clear that the implication $D \to ((A \,\&\, B) \,\&\, C)$, is, in general false, i.e. it is not absolutely true. However, the proposition $((A \& B) \& D) \to C$ is absolutely true; this proposition asserts that "The line ℓ, which belongs to the plane P and is perpendicular to the line n, is perpendicular to the projection m of n in P." This theorem is usually referred to as the theorem converse to the three perpendiculars theorem.

In proving a theorem, we assume the truth of the premise and establish the truth of the conclusion, i.e. we exclude the unique case (cf. truth table for $A \to B$) in which the implication may be false, and thereby prove the absolute truth of the particular implication $A \to B$ corresponding to the theorem. Here we must distinguish between the truth of an implication that holds always (independent of the truth of the premise) if the theorem can be proved, and the truth of the conclusion, which in this case may be asserted only if the premise is true. In the case of a false premise, however, the validity of the theorem tells us nothing about the truth or falsity of the conclusion. Conversely, if B is false, so is A, but if B is true, nothing may be said as to the truth of A.

A misunderstanding of these facts may lead to an incorrect application of some theorem. Suppose that an absolutely true implication $A \to B$ corresponds to a theorem that has been already proved. A correct application of the theorem asserts that if in some instance A is true, we may conclude that B is true. As an example, consider the theorem, "If a quadrangle is a rectangle, its diagonals are equal."* The premise is true for every square: the diagonals of a square are equal. A mistake common in mathematical reasoning is to apply the correct theorem $A \to B$ to a proposition A whose truth has not been established, and, by virtue of the truth of B, to conclude that A is true. That such reasoning is mistaken follows from the third row of the truth table for $A \to B$ (a true implication with a true conclusion may contain a false premise). Here is an example of this kind of incorrect reasoning. As proposition A we have "The number 6 is divisible by 4" and $A \to B$ is "If a number is divisible by 4, it is divisible by 2." Now $A \to B$ is a true theorem, and B is a true proposition, however we cannot conclude that A is true. In this case, we

*Perhaps this formulation of the theorem is somewhat unusual; we present it in this form as it is then an implication.

may further state that "true results may be obtained from a false assertion by means of valid methods." (Of course, it is not possible to obtain a false result from a true assertion by means of valid methods!) Yet another example showing how mistakes can result from a misunderstanding of the properties of the implication can be taken from the study of the solutions of equations (and is even more glaring in the case of inequalities). Suppose that we are given the equation $P(x) = 0$, and let the equation $Q(x) = 0$ be its consequence, i.e., the following implication is true: "If x_0 is a root of the equation $P(x) = 0$, it is also a root of the equation $Q(x) = 0$" ($A \rightarrow B$). Thus, once we have found the roots of $Q(x) = 0$, we may then find all the roots of $P(x) = 0$ among them (the truth of B follows from the truth of A). However, there may be extraneous roots among the roots of the second equation, i.e., those which are not roots of the first equation (the truth of A does not, in general, follow from the truth of B). Thus, each root of $Q(x) = 0$ must be checked before it may be concluded that it is a root of $P(x) = 0$. Conversely, if we know all the roots of the first equation, this does not mean that we know all the roots of the second equation. Some roots may be lost as we go from $Q(x) = 0$ to $P(x) = 0$. There are no extraneous roots in the first case -- and neither is there any loss of roots in the second case -- if and only if not only is the above implication $A \rightarrow B$ true, but so is the implication $B \rightarrow A$. The equations are then said to be equivalent; the set of roots of such equations coincide.

1.3. Prove that the logical connectives $\bar{B} \rightarrow \bar{A}$, $(A \ \& \ \bar{B}) \rightarrow \bar{A}$, $(A \ \& \ \bar{B}) \rightarrow B$, and $(A \ \& \ \bar{B}) \rightarrow \wedge$, where $\wedge$ is a fixed, absolutely false proposition, all have the same truth table as the implication $A \rightarrow B$.

By virtue of this problem, instead of proving the absolute truth of $A \rightarrow B$, we need only prove the absolute truth of any one of these four propositions. As an example, consider the implication $\bar{B} \rightarrow \bar{A}$. We wish to prove the truth of $\bar{A}$ under the assumption that $\bar{B}$ is true. But this is the simplest method of proving the theorem $A \rightarrow B$ "by contradiction", i.e., we start by assuming the contrary to what is to be proved, and obtain a contradiction to the assumptions. In carrying out a proof by contradiction by means of the other three connectives, we assume not only the contrary to what we wish to prove, but also assume that we have been given the proposition A & $\bar{B}$. Then to prove the theorem $A \to B$, it is sufficient to either arrive at a contradiction to what we have assumed ($\bar{A}$), derive what we wish to prove (B), or, finally, derive any absolutely false proposition ($\wedge$). By Problem 1.2, we need only derive any proposition C along with its negation $\bar{C}$ as our $\wedge$, since the compound proposition C & $\bar{C}$ must then be true. This last method of proof is in some sense the most general method of proof by contradiction.

1.4. Prove that the absolute truth of the implication $A \to B$ is equivalent to the assertion that the truth of B is a necessary test for the truth of A, and that the truth of A is a sufficient test for the truth of B.

As a result, the absolute truth of the implication $A \to B$ may be understood as asserting that B is necessary for A (and that A is sufficient for B).

6. As a final example of a logical operation, consider the connective, called the material equivalence [*ekvivalentnost'*] (denoted ~). It corresponds to such phrases as "if and only if...," "in order that...," "it is necessary and sufficient that...," and so on. We present the truth table for material equivalence below.

A	B	$A \sim B$
T	T	T
T	F	F
F	T	F
F	F	T

In the same way as with implication, the use of material equivalence in the propositional algebra disregards the semantic content of propositions. Here our intuitive understanding of equivalence relates to the case in which a composite proposition $A \sim B$ is absolutely true (only in this case are we accustomed to asserting "A and B are equivalent"). Henceforth, we will sometimes say that "A and B are absolutely equivalent" if the proposition $A \sim B$ is absolutely true.

1.5. Prove that the logical operation $(A \to B)$ & $(B \to A)$ has the same truth table as the material equivalence.

From Problem 1.5 and the truth table for conjunction, it follows that the absolute truth of a material equivalence is logically the same as the absolute truth of both the implications $A \to B$ and $B \to A$, i.e. in this case the direct theorem $A \to B$ and the converse theorem $B \to A$ are together valid. To prove the absolute truth of $A \sim B$, it is necessary to prove that the truth of A implies the truth of B, and conversely, the truth of B implies the truth of A. Thus in all instances A *and* B are either simultaneously true or simultaneously false. If the material equivalence $A \sim B$ is absolutely true, A is a necessary and sufficient condition for B, and conversely.

Note that the errors in reasoning we spoke of in sec. 1.5 do not arise if our theorems are not just absolutely true implications, but absolutely true material equivalences. When we speak of equal equations, we are also basically dealing with material equivalences.

1.6. How many distinct (i.e., in terms of truth tables) binary logical relations are there?

1.7. How many distinct commutative binary logical relations are there? Express them in terms of the operations we have already introduced.

We now have several "basic" logical operations (negation, disjunction, conjunction, implication, material equivalence) by means of which composite propositions may be derived from simple propositions. Note that composite propositions that have already been constructed may be used as simple propositions. As a result, it is possible to apply multi-step procedures in the construction of composite propositions, making repeated use of the logical operations that have already been introduced. The term *formula* refers to a logical operation obtained by combining a finite number of the basic logical operations. For every formula, it is possible to construct a truth table that successively uses the truth tables for the basic operations. Formulas with identical truth tables may be considered equivalent. We give an exact definition.

Definition 1.1.* Suppose that $\mathfrak{A}$ and $\mathfrak{B}$ are two formulas of the propositional algebra, and let $A_1, \dots, A_n$ be a set of simple propositions that occur in at least one of the formulas. The formulas $\mathfrak{A}$ and $\mathfrak{B}$ are said to be (logically) *equivalent* [*ravnosil'nost'*] if their truth values coincide for all truth values of $A_1, \dots, A_n$. The logical equivalence of $\mathfrak{A}$ and $\mathfrak{B}$ is denoted by the ordinary equality sign, thus: $\mathfrak{A} = \mathfrak{B}$.

1.8. Prove the following equivalences: (a) $A \vee A = A$; (b) $(A \vee \overline{A})\ \&\ B = B$; (c) $A \vee \overline{A} = B \vee \overline{B}$; (d) $A \vee (A\ \&\ B) = A$.

From the definition it is clear that the two sets of simple propositions occurring in the equivalent formulas $\mathfrak{A}$ and $\mathfrak{B}$ may be distinct. In this case, the simple propositions occurring in only one of the formulas will occur in the other one only formally, i.e., the truth of the other formula is independent of the truth of these simple propositions.

1.9. Attempt to define rigorously what it means to say that a simple proposition B occurs in a formula as an actual variable (not formally).

In the next problem, we attempt to give as effective a definition of the above concept as possible. Until now, we have been unable to give a formal description of everything we might wish (we will do so in Chapter 12), and have limited ourselves to requiring that only simple propositions may be negated. Such a definition of a dummy variable cannot be obtained by the simple negation of the definition of an actual variable.

1.10. In what case does a simple proposition B occur in a formula as a dummy variable (i.e., formally)?

Now we may give a rigorous proof of the above assertion.

---

*Definition 1.1 basically contains the definition of logical equivalence for any logical operation whatsoever (cf. Chapter 2).

1.11. If $\mathfrak{A} = \mathfrak{B}$ and if the simple proposition A occurs in the formula $\mathfrak{A}$ but not in the formula $\mathfrak{B}$, it occurs in $\mathfrak{A}$ only formally.

1.12. Prove that the formulas $\mathfrak{A}$ and $\mathfrak{B}$ are equivalent if and only if the formula $\mathfrak{B} \sim \mathfrak{A}$ is identically true (cf. sec. 1.4).

Note that the concept of logical equivalence relates to formulas, i.e., to logical operations. By contrast, the rather analogous concept of an absolutely true material equivalence applies to individual propositions.

We had in fact already considered formulas and implicitly spoken of the logical equivalence of formulas in our study of the basic logical operations. In particular, we saw that these basic operations are not independent, in the sense that any one of them may be expressed in terms of the others, for example, material equivalence in terms of implication and conjunction (cf. Problem 1.5). More precisely, there exists a formula with only implications and conjunctions that is logically equivalent to a material equivalence. We will sometimes express this fact by saying that a material equivalence may be expressed (to within logical equivalence) in terms of conjunctions and implications. Let us present several such examples.

1.13. Express (i.e., find the logically equivalent formula) all the basic operations in terms of disjunction, conjunction, and negation.

1.14. Express all the basic operations in terms of conjunction and negation; in terms of disjunction and negation.

1.15. Express all the basic operations in terms of implication and negation.

1.16. The following operations are called Sheffer operations:

$$\omega(A,B) = \overline{A \vee B} = \overline{A} \mathbin{\&} \overline{B}; \quad \psi(A,B) = \overline{A \mathbin{\&} B} = \overline{A} \vee \overline{B}.$$

Prove that all the basic operations may be expressed in terms of any one of the Sheffer operations.

1.17. Express $A \vee B$ in terms of $A \rightarrow B$.

There are paradoxical constructions of composite propositions to which no truth value may be assigned (though, unlike the examples presented earlier in the section, they are expressed by means of narrative sentences). Here is one example: "What I am now saying is false." It is clear that we arrive at a contradiction whether we consider the proposition true or consider it false. Another example constructed by the same principle: "On one side of a sheet of paper it is written that what is written on the other side is true, while on the other side it is written that on the first side what is written is false." Such riddles are quite common. The reader is surely familiar with the story of the tsar who would behead a stranger who spoke the truth, and hang him if he lied. A traveler who claimed he would be hanged placed the tsar in quite a difficult position indeed. The paradox is due to the fact that the proposition as a whole is implicitly 'contained within itself' as a simple proposition, and that to determine the truth value of the proposition it is necessary to know this truth value itself in advance. Situations of this kind are impossible in the case of composite propositions, which are formulas constructed inductively (by our interpretation) by means of the basic logical operations.

Composite propositions that are formulas are constructed by means of the five basic operations we have introduced. Another list of basic operations may sometimes be used. In this case, we might ask how large is the set of logical operations* that may be constructed by combining these basic operations. It turns out that any logical operation may be expressed in terms of the basic operations (to within logical equivalence); moreover, to construct any logical operation, disjunction, conjunction, and negation are all that is needed. And by Problem 1.14, we may limit ourselves to disjunction and negation, or conjunction and negation.

---

*As is clear from the foregoing, by a <u>logical connective</u> (or <u>logical operation</u>) we understand an arbitrary method of forming a composite proposition from simple propositions in which the truth value of the composite proposition is uniquely determined for every set of truth values of the simple propositions.

Suppose that we are given a logical operation $\mathfrak{U}$ defined over the simple propositions $A_1, \ldots, A_n$. To within logical equivalence, this operation may be characterized by a truth table. For the sake of definiteness, we consider a particular

A_1	A_2	A_3	$\mathfrak{U}$
T	T	F	T
T	F	T	T
F	F	T	T
F	F	F	T
T	T	T	F
F	T	T	F
F	T	F	F
T	F	F	F

operation defined on three simple propositions. The table for $\mathfrak{U}$ has been written so that the rows for which $\mathfrak{U}$ is true come first. Then an operation with a given truth table may be written out by simply enumerating the instances in which $\mathfrak{U}$ is true (saying to oneself the names of the logical connectives "and", "or", and "not" in place of the symbols &, $\vee$, and $\overline{\ }$):

$$(A_1 \,\&\, A_2 \,\&\, \overline{A}_3) \vee (A_1 \,\&\, \overline{A}_2 \,\&\, A_3) \vee (\overline{A}_1 \,\&\, \overline{A}_2 \,\&\, A_3)$$
$$\vee\, (\overline{A}_1 \,\&\, \overline{A}_2 \,\&\, \overline{A}_3)$$

1.18. Check that the operation we have constructed in fact has the given truth table.

1.19. Generalize the construction to the case of arbitrary logical operations $\mathfrak{U}$.

The construction obtained by solving Problem 1.19 is called a *principal disjunctive normal form* (abbreviated PDNF). We will speak about it in more detail in the next chapter.

A logical operation may be constructed from a truth table starting not only with those rows in which $\mathfrak{U}$ is true, but also with those in which it is false. In this case we exclude those possibilities for which $\mathfrak{U}$ is false:

$$(\bar{A}_1 \vee \bar{A}_2 \vee \bar{A}_3) \mathbin{\&} (A_1 \vee \bar{A}_2 \vee \bar{A}_3) \mathbin{\&} (A_1 \vee \bar{A}_2 \vee A_3)$$
$$\mathbin{\&} (\bar{A}_1 \vee A_2 \vee A_3)$$

(Here again we read the names of the corresponding connectives whenever we see &, $\vee$, and $\bar{\ }$).

1.20. Check that the operation we have constructed has the given truth table.

1.21. Generalize this construction to the case of arbitrary connectives $\mathfrak{U}$.

The construction obtained by solving Problem 1.21 is called a *principal conjunction normal form* (abbreviated PCNF).

1.22. Find the PDNF and PCNF for the basic logical operations.

The simplest facts from the propositional algebra are often helpful when solving elementary logical problems. We present here just one example of such a problem.

1.23. A traveler finds herself in city *A* or city *B*, but in which one precisely, that she doesn't know. She can ask one question only, which may be answered with a "yes" or "no"; her respondent's answer may be either true or false (precisely which, this too the traveler does not know). Try to think of a question whose answer would correctly tell the traveler which city she is in.

Hints

1.1. Here is the truth table for $A \vartriangle B$:

A	B	$A \vartriangle B$
T	T	F
T	F	T
F	T	T
F	F	F

The operation $(A \,\&\, \bar{B}) \vee (\bar{A} \,\&\, B)$ and $(A \vee B) \,\&\, (\bar{A} \vee \bar{B})$ have the same truth table.

1.4. We need only recall the definitions of necessary and sufficient condtions, after which the assertion is a tautology.

1.6. There are 16 operations, each specified by its own truth table, so that the problem reduces to finding out how many distinct truth tables there are in all.

1.7. There are eight operations (the number is found as in the preceding problem): identically true, identically false, disjunction, conjunction, material equivalence, and their negations. Note that the negation of a material equivalence is the same as the exclusive disjunction (cf. Problem 1.1).

1.9. A simple proposition B occurs in $\mathfrak{U}$ *actually* if there exists a set of truth values of the other simple propositions $A_1, \dots, A_n$ in which the truth value of $\mathfrak{U}$ depends on the truth value of B (i.e., for this set the truth values of $\mathfrak{U}$ are distinct for distinct truth values of B).

1.10. The simple proposition B occurs formally in $\mathfrak{U}$ if for every set of truth values of the other simple propositions $A_1, \dots, A_n$, $\mathfrak{U}$ has the same truth value for both truth values of B.

1.11. The proof follows directly from the definition of a formal occurrence (Problem 1.10) and the definition of logical equivalence of formulas.

1.13. We need the implication $A \to B$, since, by Problem 1.5, a material equivalence is expressed in terms of implication and conjunction:

$$A \sim B = (A \to B) \,\&\, (B \to A).$$

We may also use the equivalence $A \sim B = \overline{A \,\Delta\, B}$ and Problem 1.1. To express $A \to B$, it is best to compare its truth table with the truth tables for disjunction and use the fact that in both cases we have a unique set in which the result of the operation is false.

1.14. It is necessary to express disjunction in terms of conjunction and negation, and conjunction in terms of disjunction and negation. Use the fact that the truth tables for disjunction and conjunction may be transformed into each other by interchanging F and T.

1.15. We need only express conjunction and negation or disjunction and negation in terms of implication and negation.

1.19. Compile the truth table for $\mathfrak{U}$, beginning with those rows in which $\mathfrak{U}$ is true. With each such row, associate the conjunction of those simple propositions that are true in this row, and the negations of the others. We then need only take the disjunction of these conjunctions.

1.21. With each row from the table in which $\mathfrak{U}$ is false, associate the disjunction of those simple propositions whose truth values in this row are false, and the negations of the other simple propositions. Then take the conjunction of these disjunctions.

1.23. For the sake of definiteness, suppose that the answer "yes" means that the traveler is in city A, and "no" that she is in city B. A question with a two-valued answer may be interpreted as a question about the truth or falsity of some proposition. Suppose that this proposition is the composite proposition formed from the simple propositions

X: *The traveler is in city A.*
Y: *Her respondent is telling the truth.*

Then:

$\overline{X}$: *The traveler is in city B.*
$\overline{Y}$: *Her respondent is telling a lie.*

Thus the composite proposition $\mathfrak{U}(X,Y)$ must be constructed, with the understanding that what the respondent says is true if X is true, and false if X is false.

Solutions

1.6. The truth tables have a form that may be schematically depicted as in the accompanying table:

A	B	$\mathfrak{U}(A,B)$
T	T	*
T	F	*
F	T	*
F	F	*

Tables for distinct operations may be distinguished by means of their last columns (here the elements have been replaced by asterisks. Either T or F may occur in place of an asterisk, and there may be any combination of T's and F's whatsoever; there are $2^4 = 16$ distinct combinations. (We are dealing with what is known as a permutation of two elements repeated four times.)

1.7. Again it is necessary to compute the number of truth tables, but unlike the preceding problem, here the second and third elements of the last column must coincide. Therefore, we select three elements arbitrarily, resulting in $2^3 = 8$ distinct possibilities. The particular form of the operations may be checked directly.

1.13. $A \to B = \overline{A} \vee B$; $A \sim B = (\overline{A} \vee B) \; \& \; (\overline{B} \vee A)$
$= (A \; \& \; B) \vee (\overline{A} \; \& \; \overline{B})$.

1.14. $A \vee B = \overline{\overline{A} \; \& \; \overline{B}}$; $A \; \& \; B = \overline{\overline{A} \vee \overline{B}}$.

1.15. $A \vee B = \overline{A} \to B$.

1.16. $\overline{A} = \omega(A, A)$; $A \vee B = \overline{\overline{A \vee B}} = \overline{\omega(A, B)} = \omega(\omega(A,B), \omega(A,B))$ or $A \; \& \; B = \omega(\overline{A}, \overline{B}) = \omega(\omega(A,A), \omega(B,B))$. The operation $\psi(A,B)$ may be considered analogously.

1.17. $A \vee B = (A \to B) \to B$.

1.19. Each of the conjunctions we have constructed is true

only for those truth values of the simple propositions found in the row corresponding to it. Since we have taken a disjunction over all sets of truth values of the simple propositions in which $\mathfrak{U}$ is true, the resulting proposition is true over all these sets and only over these sets. (Clearly, a logical operation is determined by the collection of sets over which the resulting composite proposition is true and false over the other sets.)

1.21. Each of the disjunctions constructed in the hint to the problem is false only over the set of truth values of the simple propositions corresponding to it. Therefore, the conjunction of these disjunctions is false over the same sets as is $\mathfrak{U}$.

1.22. PDNF: $\overline{X} = \overline{X}$; $X \,\&\, Y = X \,\&\, Y$;

$$X \vee Y = (X \,\&\, Y) \vee (\overline{X} \,\&\, Y) \vee (X \,\&\, \overline{Y}).$$
$$X \to Y = (X \,\&\, Y) \vee (\overline{X} \,\&\, Y) \vee (\overline{X} \,\&\, \overline{Y});$$
$$X \sim Y = (X \,\&\, Y) \vee (\overline{X} \,\&\, \overline{Y}).$$

PCNF: $\overline{X} = \overline{X}$; $X \,\&\, Y = (\overline{X} \vee Y) \,\&\, (X \vee \overline{Y}) \,\&\, (X \vee Y)$;

$$X \vee Y = X \vee Y; \quad X \to Y = \overline{X} \vee Y;$$
$$X \sim Y = (\overline{X} \vee Y) \,\&\, (X \vee \overline{Y}).$$

1.23. We construct the truth table for the desired proposition. It is clear that if Y is true, $\mathfrak{U}(X,Y)$ is logically equivalent to X; otherwise, it is equivalent to $\overline{X}$, i.e.,

$$\mathfrak{U}(X,Y) = (X \,\&\, Y) \vee (\overline{X} \,\&\, \overline{Y}).$$

The form of $\mathfrak{U}(X,Y)$ may be obtained from the accompanying table by means of the construction in Problem 1.19. Thus,

X	Y	$\mathfrak{U}(X,Y)$
T	T	T
T	F	F
F	T	F
F	F	T

our traveler might ask herself, "Is it true that I am in city A

and is my respondent telling the truth, or am I in city B and is my respondent speaking falsely?" Note that

$$\mathfrak{U}(X,Y) = X \sim Y;$$

therefore, the question may be restated thus: "Is the assertion that the traveler is in city A materially equivalent to the assertion that her respondent is telling the truth?"

Of course, problems such as 1.23 may be solved without using the propositional algebra; the propositional algebra only makes it possible to solve them at a more formal level.

Chapter 2
LOGICAL FUNCTIONS. NORMAL FORMS

1. *Logical Functions. Logical Equivalence of Formulas.* In the preceding chapter, we explained that from the standpoint of the propositional algebra, logical operations are completely determined by their truth tables. We could even forget that we are considering operations over propositions and deal only with the truth tables themselves. In this way we arrive at the concept of a logical function, which we shall study below. However, we are not suggesting that the reader forget the interpretations of the logical connectives we have given above, as these interpretations supply an explanation of quite a number of relations in algebraic logic. Instead of the symbols T and F we used in Chapter 1, here we will use the symbol 1 and 0.*

Definition 2.1. A function $f(x_1, \dots, x_n)$ of n variables $x_1, \dots, x_n$ that takes the values 1 and 0 and whose arguments also take the values 1 and 0 will be called a *logical function* [literally, *function of algebraic logic* -- trans.].

The function $f(x_1, \dots, x_n)$ is specified by a truth table of its own (see page 23). In each row of the table, we first give

*There is no particular "meaning" assigned to these symbols (not even the ordinary arithmetic meaning), unless otherwise stipulated (cf. Chapter 4).

the sequence of truth values of the variables $(\alpha_1, \dots, \alpha_n)$, and then the truth value of the function on this sequence. It can be easily seen that there is a finite number of distinct binary sequences of length n (ordered sequences $(\alpha_1, \dots, \alpha_n)$ consisting of zeros and units).*

x_1	x_2	x_3	...	x_{n-1}	x_n	$f(x_1, \dots, x_n)$
0	0	0	...	0	0	$f(0,0, \dots, 0,0)$
1	0	0	...	0	0	$f(1,0, \dots, 0,0)$
...	...	...	...	...	...	....
1	1	1	...	1	0	$f(1,1, \dots, 1,0)$
1	1	1	...	1	1	$f(1,1, \dots, 1,1)$

2.1. How many distinct binary sequences $(\alpha_1, \dots, \alpha_n)$ of length n are there?

2.2. How many distinct logical functions of n variables are there?

2.3. How many distinct logical functions of n variables that preserve zero (i.e., functions that vanish on the null set: $f(0, \dots, 0) = 0$) are there?

We may similarly consider functions that preserve the unit. Here are truth tables in the new notation for the basic logical operations.

x	y	$\bar{x}$	$x \,\&\, y$	$x \vee y$	$x \to y$	$x \sim y$
0	0	1	0	0	1	1
0	1	1	0	1	1	0
1	0	0	0	1	0	0
1	1	0	1	1	1	1

*These sequences are often referred to as n-letter words in the two-word alphabet {0,1}.

In addition to these basic functions, we also encounter the constants 0 and 1 and functions that *coincide* with their own argument.

Let us now stipulate which logical functions are to be considered identical (cf. Definition 1.1). Note that specifying a function means also deciding on the notation for the arguments. We will abbreviate $f(x_1, \dots, x_n)$ by f whenever this will not cause any misunderstanding, however.

Definition 2.2. Suppose that f and g are logical functions and let $x_1, \dots, x_n$ be the set of those arguments that occur in at least one of the two functions. We will say that f and g are (*logically*) *equivalent* (and then write $f = g$)* if the values of f and g coincide for all values of $x_1, \dots, x_n$.

By analogy with Chapter 1 (Problems 1.9 and 1.10), we may introduce the concepts of dummy [*fiktivnyi*] and actual [*sushchestvennyi*] variables. We will sometimes use the fact that it is always possible to formally [*fiktivno*] add to the arguments of a function some new (dummy) argument, after which a function equivalent to the initial one is obtained that depends formally on the newly introduced arguments.

Note that by Definition 2.2, functions that have identical truth tables, but which differ by virtue of the names of their variables, are not equivalent.

2. *Composition of Logical Functions.* We now define the basic operation that may be performed on logical functions: composition, or the operation of forming a composite function. Intuitively, this concept may be understood to mean that new functions are substituted for the arguments of some given function, certain variables are "equated", and, finally, the procedure itself may be repeated. A rigorous definition may be given by induction.

Definition 2.3. Suppose that $\Phi = \{\omega_1(x_{11}, \dots, x_{1k_1}), \omega_2(x_{21}, \dots, x_{2k_2}), \dots, \omega_m(x_{m1}, \dots, x_{mk_m})\}$ is a finite system of logical functions.

---

*In particular, $f(x) = 1$ will mean that f is identically equal to 1.

The function ψ is called an elementary composition or *composition* of rank ℓ (denoted $\psi \in \Phi^{(1)}$) if it may be obtained by any one of the following methods:

(a) by renaming some of the variables x_{jk} of a function $\omega_j \in \Phi$, i.e., we have the form

$$\omega_j(x_{j1}, \dots, x_{ji-1}, y, x_{ji+1}, \dots, x_{jk_j}),$$

where y, in particular, may coincide with any one of the variables $x_{\ell n}$;

(b) by substituting some function $\omega_\ell \in \Phi$ for an argument x_{ji} of one of the functions $\omega_j \in \Phi$:

$$\omega_j(x_{j1}, \dots, x_{ji-1}, \omega_\ell(x_{\ell 1}, \dots, x_{\hat{\ell} k_\ell}), x_{ji+1}, \dots, x_{jk_j}).$$

The function ψ obtained as a result depends on the arguments

$$(x_{j1}, \dots, x_{ji-1}, x_{\ell 1}, \dots, x_{\ell k_\ell}, x_{ji+1}, \dots, x_{jk_j}),$$

i.e., on all the variables of the functions ω_j and ω_ℓ, except possibly for x_{ji}.

Once we have described the class $\Phi^{(r)}$ of functions that are compositions of rank r of functions from Φ, the class $\Phi^{(r+1)}$ is then seen to consist of elementary compositions of functions from $\Phi^{(r)}$ (i.e., $\Phi^{(r+1)} = (\Phi^{(r)})^{(1)}$).

Functions that occur in any of the classes $\Phi^{(r)}$ are referred to as compositions of functions from Φ.

Remark 1. If two functions ω and ψ have the same truth tables and differ only in the names of the variables, then by Definition 2.3(a), each is a composition of the other.

Remark 2. By Definition 2.3(a), $\Phi \quad \Phi^{(1)}$, that is, $\Phi^{(r)} \subset \Phi^{(r+1)}$, and in general, $\Phi^{(r)} \subset \Phi^{(s)}$ for $r \leqslant s$.

Remark 3. If in the process of renaming the variables by Definition 2.3(a) we replace x_{ji} by some $x_{j\ell}$ $(\ell \neq i)$, we then obtain a function of a lesser number of variables. In this case, we say that the variables x_{ji} and $x_{j\ell}$ in ω_j have been *equated* (*identified*). For example, equating the variables x and y turns the two functions $x \vee y$ and $x \,\&\, \overline{y}$ into $x \vee x = x$ and $x \,\&\, \overline{x} = 0$, respectively. Any number of variables may be equated by repeating this procedure.

Remark 4. In precisely the same way, by repeating the substitution procedure of Definition 2.3(b) we may substitute arbitrary functions from Φ for any number of variables of any function from Φ.

Definition 2.4. A composition of the basic functions $\bar{x}$, $x \,\&\, y$, $x \vee y$, $x \to y$, and $x \sim y$ is called a *formula.*

The concept of (logical) equivalence of formulas introduced in Chapter 1 (Definition 1.1) is consistent with Definition 2.2.

Below, we will often have to make implicit use of the following nearly self-evident, though important, property of composition.

2.4. Suppose that $\Phi = \{\omega_j\}$ is a finite system of functions, and let $\tilde{\Phi}$ consist of functions ψ_j each of which is (logically) equivalent to some function from Φ. With each function $\tilde{f}$ which is a composition of functions from $\tilde{\Phi}$, we associate the function f obtained by replacing all functions in $\tilde{\Phi}$ from which $\tilde{f}$ was obtained by means of composition, by certain equivalent functions in Φ (a rigorous definition of f may be given by induction). Prove that $\tilde{f}$ and f are equivalent.

Despite the self-evident nature of this assertion, we advise the reader to carefully prove it so as to become accustomed to the technique of this type of inductive proof. The assertion should be taken to mean that forming the composition of the functions $\{\omega_i\}$ involves replacing each occurrence of these functions by equivalent functions. We may similarly prove that in any given composition, compositions of lesser rank occurring in it may be replaced by equivalent functions.

Let us simplify the notation a bit. We will omit the conjunction symbol $\&$, i.e., in place of $x \,\&\, y$ we will write simply xy.* We then reduce the number of parentheses by establishing a "hierarchy" of operations, with conjunction understood as the "first" operation in the hierarchy, followed by disjunction and, finally, implication and material equivalence (the latter two operations are not ordered). Our stipulation is that the leading operation is executed first if the parentheses do not prescribe otherwise. For example, the

*This notation recalls the formation of a compound sentence without the use of connectives (cf. page 4).

formula $(x \mathbin{\&} y) \vee (z \to t)$ may be rewritten thus: $xy \vee (z \to t)$, and the formula $(x \vee y) \to (y \mathbin{\&} z)$ may be rewritten $x \vee y \to yz$.

Here is a list of the most important logical equivalences of algebraic logic:

$$\overline{\overline{x}} = x; \tag{2.1}$$

$$xy = yx; \tag{2.2}$$

$$(xy)z = x(yz); \tag{2.3}$$

$$x \vee y = y \vee x; \tag{2.4}$$

$$(x \vee y) \vee z = x \vee (y \vee z); \tag{2.5}$$

$$x(y \vee z) = xy \vee xz; \tag{2.6}$$

$$x \vee yz = (x \vee y)(x \vee z); \tag{2.7}$$

$$\overline{x \vee y} = \overline{x}\,\overline{y}; \tag{2.8}$$

$$\overline{xy} = \overline{x} \vee \overline{y}; \tag{2.9}$$

$$x \vee x = x; \tag{2.10}$$

$$xx = x; \tag{2.11}$$

$$1x = x; \tag{2.12}$$

$$0 \vee x = x. \tag{2.13}$$

2.5. Prove the equivalences (2.1) - (2.13).

The equivalences (2.2), (2.3), (2.4), and (2.5) show that conjunction and disjunction are commutative and associative. Since the operations are associative, we may omit the parentheses in conjunctions and disjunctions of several variables, and since they are commutative, we may arrange conjunctions and disjunctions in any order. The equivalences (2.6) and (2.7) are the distributive laws of conjunction and disjunction (unlike the arithmetic case, here there are two such laws). By means of (2.6), we are able to remove parentheses; and in general, both (2.6) and (2.7) may be used to transform expressions in such a way that the operations occurring in them may be executed in reverse order. For example, if disjunction is to be executed first in some initial expression followed by conjunction, an equivalent formula may be obtained in which conjunction is executed first and then disjunction. We will again use this fact in our discussion of normal forms. Here we require the following:

2.6. Prove that

$$(x \vee y)(z \vee t) = xz \vee yz \vee xt \vee yt.$$

2.7. Prove that

$$xy \vee zt = (x \vee z)(y \vee z)(x \vee t)(y \vee t).$$

The equivalences (2.8) and (2.9) (called de Morgan's laws) have already been mentioned in Chapter 1 in our discussion of how to express a conjunction in terms of disjunctions and negations or a disjunction in terms of conjunctions and negations. These relations are also used to transform negations applied to composite propositions into their constituent simple propositions.

2.8. Transform (find an equivalent formula) the following to formulas in which only the arguments are negated:
(a) $\overline{xy \vee \overline{z}}$; (b) $x(xy \vee \overline{yz}) \quad \overline{(y \vee \overline{t}z)}$.

2.9. Express the negation of an implication in terms of the basic operations in such a way that only the arguments are negated.

The procedure of transforming negations to simple propositions is used to construct definitions of negative concepts. We have already touched on this in Chapter 1 (Problem 1.10) and will return to it repeatedly in what follows.

2.10. A boy decided to finish reading his book on Sunday, then go to a museum or the movies, and if the weather was nice, go swimming in the river. In which case may it be said that the boy's decision has not been fulfilled? (In your answer, only the simple propositions may be negated.)

In answering this question, you should not have to resort to the propositional algebra. In any case, try to given an answer directly. However, then you must solve the problem by means of algebraic logic.

2.11. Solve Problem 2.10 using propositional algebra.

2.12. Think of the most complicated composite proposition containing the different connectives, and construct its negation directly and by means of algebraic logic.

3. *Boolean Algebras.*

Definition 2.5. A Boolean algebra is understood to refer to a set $\mathfrak{M}$ on which we define two binary operations, $x, y \Rightarrow xy$ and $x, y \Rightarrow x \vee y$, and a single unary operation $x \Rightarrow \bar{x}$, and in which we distinguish two elements 0 and $1 \in \mathfrak{M}$. The axioms (2.1)-(2.12) are satisfied by these operations and elements, with equality understood here to mean that two elements of $\mathfrak{M}$ coincide.

In defining a Boolean algebra, we could have started with a single binary operation (for example, xy) and completed the definition with some other operation by means of (2.1) and de Morgan's laws (2.8) and (2.9). We will not bother discussing the construction of a system of independent axioms for a Boolean algebra, however.

In studying the logical operations using the truth tables, we were essentially dealing with a Boolean algebra consisting of two elements {0,1}, where the operations were defined in accordance with the truth table on page 23.

Digression on Axiomatic Structures. Boolean algebras constitute an example of an axiomatically defined algebraic structure (other examples are groups, rings, and fields, a discussion of which may be found in any textbook on higher algebra). These structures constitute sets in which certain elements are distinguished (in a Boolean algebra, the elements 0 and 1; in a group, the identity element), and certain operations defined. These elements and operations must satisfy a finite set of relations, called axioms, but otherwise are defined arbitrarily. It may be proved that this use of the term, "axiom," contradicts our ordinary understanding of the term as given in the university course in geometry. A slight difference is that here we are speaking of an axiomatization of operations, whereas in Euclidean geometry it is a matter of axiomatizing certain relations between the fundamental structures ("belongs to," "lies between," etc.). A more significant difference is that, roughly speaking, there is one and only one Euclidean plane, whereas there are many distinct groups or Boolean algebras. A precise formulation of this assertion may be given in terms of isomorphism (for Boolean algebras, see Definition 2.6 below): All Euclidean planes are isomorphic, though there exist non-isomorphic Boolean algebras (the latter follows from Problem 2.19, for example). A tacit agreement as to what constitutes an isomorphism strongly influences the presentation of axiomatic theories in the university course in geometry and often becomes a major psychological stumbling block to the introduction of axiomatic structures that admit non-isomorphic realizations. In the case of such axiomatic structures, besides proving theorems -- which may be thought of as assertions that are consequences of the axioms -- it is also important to study topics related to the existence of non-isomorphic structures, for example, special classes of structures that may be distinguished by means of additional axioms, such as commutative groups and regular Boolean algebras (cf. sec. 2.4). In Euclidean geometry, such questions do not arise for the reason just given. But do not think that axiomatic theories that lead to non-isomorphic structures are the exclusive reserve of algebra; these types of axiomatic structures may also be found in non-Euclidean geometry.

Finally, in the university course in geometry it is explained that the axioms are not proved (it is sometimes even said that they "do not require proof"). On the contrary, we will verify (prove!) the axioms of Boolean algebra for each of the examples we shall consider (cf. below). In fact, as we have already mentioned, in elementary geometry we are concerned only with obtaining corollaries from the axioms (theorems), and thus limit our discussion to the "internal" topics of the theory. As for the validity of the axioms for the only structure we are interested in -- the idealized model of our real space -- this sort of question falls outside our theory and is answered on the basis of our intuitive representations. In the case of the theory of Boolean algebras, for example, these "external" topics arise when studying various special structures. Rigorous mathematical techniques must be applied in each case to verify the axioms, though such verification is outside the scope of the theory. In other words, the concept of an axiomatic system aquires a novel nuance. Here we are no longer dealing with a set of truths that for various reasons is assumed without proof, but rather a set of tests that, if satisfied by some structure, automatically implies the validity of the entire set of theorems forming the given axiomatic theory. This evolution of our standpoint with regard to the axioms is the reason for the widespread use of the axiomatic method in modern mathematics.

Note that we must verify (again, prove!) the axioms when considering models of Euclidean geometry. In the models, the basic structures (points, line, etc.) are created starting with previously developed mathematical theories (for example, the theory of real numbers). Thus, in treating the analytic interpretation of Euclidean plane geometry, pairs of real numbers (x,y) are declared to be points and all the basic relations ar described in terms of real numbers. The validity of all the axioms is then verified by means of proofs based on the properties of the real numbers. You probably know that the consistency of Lobachevskian plane geometry is proved by constructing its interpretation in the Euclidean plane. In this interpretation, the axioms of Lobachevskian geometry become theorems of Euclidean geometry.

To avoid any misunderstanding, note that the word "algebra" is used in two different senses, either as the name of some mathematical theory (for example, algebraic logic, i.e., algebra of logic) or to denote some class of axiomatic structures (for example, Boolean algebra). There is no reason for this, other than tradition.

Let us now consider some examples of Boolean algebras.

(1) Suppose that M is some set that, for the sake of simplicity, may be considered the set of points on the line or set of natural numbers, and let $\mathfrak{M}_M$ be the set of its subsets. By xy $(x,y \in \mathfrak{M}_M)$ we denote the intersection of x and y; by $x \vee y$, their union; by $\overline{x}$, the complement of x with respect to the entire set M; by 0, the empty set; and by 1, the set M itself.

2.13. Verify that the set $\mathfrak{M}_M$ of subsets of M with the operations and elements 0 and 1 thus defined is a Boolean algebra.

(2) Suppose that $\mathfrak{M}$ consists of numbers x such that $0 \leqslant x \leqslant A$, with $A \neq 0$. We set $\bar{x} = A - x$, $x \vee y = max\{x, y\}$, and $x \ \& \ y = min\{x, y\}$. The number 0 will serve as our "0" and the number A as our "1".

2.14. Verify that the set $\mathfrak{M}$ with the operations and elements 0 and 1 thus defined is a Boolean algebra.

(3) If there is some subset in a Boolean algebra that is closed relative to the operations defined in this algebra (i.e., the application of the operations to the elements of the subset results again in elements of the same subset) that contains 0 and 1, it is a Boolean algebra relative to these operations and elements (a Boolean subalgebra). Thus, if in Example (2) A is a natural number, we may consider a subset of the natural numbers. In general, in this example any subset symmetric relative to the mid-point of the segment $[0, A]$ and containing its endpoints forms a Boolean subalgebra. (Prove this!)

(4) Suppose that N is a natural number and let $\mathfrak{M}_N$ be the set of its integral positive divisors. For $x, y \in \mathfrak{M}_N$, we set $\bar{x} = N/x$, and let xy be the greatest common divisor of x and y, $x \vee y$ the least common multiple of x and y, the number 1 our "0" and the number N our "1".

2.15. Prove that we have transformed the set of divisors of a number N into a Boolean algebra.

(5) Since the corresponding operations have been defined on the set of propositions, it is natural to suppose that it is a Boolean algebra. For this purpose, however, some of the propositions in the set of propositions must be equated. We will say that the propositions A and B are identical ($A \equiv B$) if the material equivalence $A \sim B$ is absolutely true (i.e., true in all instances; cf. page 3). The set of propositions, considered to within identification as just defined (more precisely, the classes of identical propositions) forms a Boolean algebra. The "0" is then the class of absolutely false propositions (they are all identical to each other), and the "1" the class of absolutely true propositions. Clearly, if the logical operations are applied to certain propositions, a replacement of some of the

propositions by identical propositions induces a replacement of the resulting proposition by an identical one. Thus, in this case equalities (2.1)-(2.13) denote an identity between propositions; the algebra we have constructed is called a *Boolean propositional algebra.**

(6) Let us consider certain subalgebras in the Boolean propositional algebra $\mathfrak{M}$. Suppose that M is an arbitrary set. With each fixed subset $x \subset M$ ($x \in \mathfrak{M}_M$) we associate the proposition $\tilde{x}$: "This element of m is in the subset x."

Obviously, if x and y do not coincide, we will always have $\tilde{x} \not\equiv \tilde{y}$. (We need only consider an element that belongs to one of the two sets x and y, but not to the other.)

2.16. Prove that the set of propositions $\tilde{x}$ ($x \in \mathfrak{M}_M$) is a subalgebra in the Boolean algebra $\mathfrak{M}$.

Deciding on an instance for the propositions $\tilde{x}$ of Example (6) means finding an element of M. Two instances that correspond to distinct elements of M are essentially distinct; that is, there exists a proposition $\tilde{x}$ that is true in one instance and false in the other. It is therefore natural to identify m with the set of instances of $\tilde{\mathfrak{M}}_M$.

For an arbitrary set of propositions, the concept of an instance is fairly undefined; it is not even certain that there is really an instance for the set of all propositions in general. Suppose, however, that $\mathfrak{M}$ is some subalgebra in the propositional algebra $\mathfrak{M}$ for which it is possible to describe the set of distinct instances of M. Two instances will be considered distinct in a natural way if there exists a proposition that is true in one instance, and false in the other. Then with each proposition $\tilde{x} \in \tilde{\mathfrak{M}}$ we may associate a subset $x \subset M$ of instances in which x is true. Two subsets x and y coincide if and only if $\tilde{x} \equiv \tilde{y}$, i.e., if and only if the material equivalence $\tilde{x} \sim \tilde{y}$ is absolutely true. The set $\mathfrak{M}$ of subsets of M obtained as a result is a subalgebra in the Boolean algebra $\mathfrak{M}_M$. If a proposition is constructed from $x \in \mathfrak{M}$ by the method indicated in (6), a proposition is obtained that is identical to $\tilde{x} \in \tilde{\mathfrak{M}}$.

---

*We recommend that the reader carefully think through why it is necessary to equate absolutely materially equivalent propositions in order to turn a set of propositions into a Boolean algebra.

We have thus obtained examples of *isomorphic* Boolean algebras. (Another example is discussed in the hint to Problem 2.15.)

Definition 2.6. Suppose that $\mathfrak{M}$ and $\mathfrak{N}$ are two Boolean algebras. The mapping $x \to \tilde{x}$ which associates with each $x \in \mathfrak{M}$ an element $\tilde{x} \in \mathfrak{N}$ is called a *homomorphism* of the Boolean algebra $\mathfrak{M}$ into the Boolean algebra $\mathfrak{N}$ if

$$\tilde{0}_{\mathfrak{M}} = 0_{\mathfrak{N}};$$

$$\tilde{1}_{\mathfrak{M}} = 1_{\mathfrak{N}};$$

$$\tilde{\bar{x}} = \bar{\tilde{x}};$$

$$\tilde{x} \vee \tilde{y} = \widetilde{x \vee y};$$

$$\tilde{x}\tilde{y} = \widetilde{xy}$$

for all $x, y \in \mathfrak{M}$; the subscripts 0 and 1 indicate from which algebra the particular elements have been taken.

If the homomorphism $x \to \tilde{x}$ establishes a one-to-one correspondence between $\mathfrak{M}$ and $\mathfrak{N}$, it is called an *isomorphism* between the Boolean algebras $\mathfrak{M}$ and $\mathfrak{N}$. In this case, the Boolean algebras $\mathfrak{M}$ and $\mathfrak{N}$ are said to be *isomorphic*.

We proved earlier that the Boolean algebra $\mathfrak{M}_M$ of subsets of M is isomorphic to some Boolean subalgebra $\tilde{\mathfrak{M}}_M$ of propositions and that the Boolean algebra $\tilde{\mathfrak{M}}$ is isomorphic to some Boolean algebra of subsets $\mathfrak{M}$ of the set of instances of M. We have thereby established a natural correspondence between Boolean propositional algebras and sets.

2.17. Suppose that N is a natural number whose decomposition into prime factors contains only distinct factors, and let $\hat{N}$ denote the set of these factors. Prove that the Boolean algebra of divisors of $\mathfrak{M}_N$ (Example 4) is isomorphic to the Boolean algebra $\mathfrak{M}_{\hat{N}}$ of subsets of $\hat{N}$.

There is another type of isomorphism that is of extraordinary importance. According to the axioms of Boolean algebra, if we interchange 0 and 1 and disjunction and conjunction everywhere, we obtain the same axiomatic structure as before. This fact may be expressed in the following way. Suppose that $\mathfrak{M}$ is a Boolean algebra. We construct a new Boolean algebra $\mathfrak{M}^+$ to consist of the same elements as $\mathfrak{M}$, but with different operations. The elements x, y, ... of $\mathfrak{M}$, understood as elements of $\mathfrak{M}^+$, are denoted x^+, y^+, ... ; the elements 0 and 1, understood as elements of $\mathfrak{M}^+$, are denoted 0^+ and 1^+. We then set

$0^+ = 1$ (i.e., the zero in $\mathfrak{M}^+$ is the unit in $\mathfrak{M}$);

$1^+ = 0$;

$x^+ \vee y^+ = (xy)^+$

(i.e., the disjunction of elements in $\mathfrak{M}^+$ coincides with their conjunction in $\mathfrak{M}$);

$x^+y^+ = (x \vee y)$.

The Boolean algebra $\mathfrak{M}^+$ is said to be *dual* to $\mathfrak{M}$. Note that $(\mathfrak{M}^+)^+ = \mathfrak{M}$.

2.18. Prove that the mapping $x \Rightarrow \bar{x}$ realizes an isomorphism of $\mathfrak{M}$ onto $\mathfrak{M}^+$.

4. *Regular Boolean Algebras and Boolean Operations.* In the case of the two-element Boolean algebra {0,1}, the operations xy, $x \vee y$, and $\bar{x}$ are special cases of the general operations $f(x_1, \ldots, x_n)$, that is, of the logical functions. We might then ask whether analogous operations may be introduced in certain other Boolean algebras. Naturally, we could consider functions in which the arguments and value of the function itself belong to a Boolean algebra. But on the other hand, in the case of the algebra of Proposition 2.14 we are considering arbitrary functions of real variables without taking into account the structure of the Boolean algebra. We might then attempt to associate with the logical functions operations defined over Boolean algebras that satisfy all the relations that hold in the algebra {0,1}. The most natural path would seem to involve associating with each of the logical functions an operation defined over a Boolean algebra, starting with a representation of the logical functions in normal form (cf. sec. 2.5, as well as the end of Chapter 1), i.e., in terms of

operations already existing in any Boolean algebra. However, this path does not always lead to our goal, since there are equivalences even between the basic operations xy, $x \vee y$, and $\bar{x}$ (considered as logical functions) that do not hold in every Boolean algebra.

2.19. Give an example of a Boolean algebra in which the equivalences $x \vee \bar{x} = 1$ and $x\bar{x} = 0$ do not hold.

Operations corresponding to the logical functions may be introduced in an entirely natural way for certain special Boolean algebras. Let us consider the Boolean algebra $\mathfrak{M}_M$ of subsets of M. A *set-theoretic operation* will be understood to refer to an arbitrary mapping $F(x_1, \dots, x_n)$ that associates with every sequence of n subsets $x_1, \dots, x_n \in \mathfrak{M}_M$ some subset $y = F(x_1, \dots, x_n) \in \mathfrak{M}_M$ such that some element $m \in M$ will belong to y exclusively as a function of which set $x_1, \dots, x_n$ it occurs in.

There is a one-to-one correspondence between the set-theoretic operations and the logical functions. In fact, suppose that $F(x_1, \dots, x_n)$ is a set-theoretic operation. With every binary sequence $\alpha = (\alpha_1, \dots, \alpha_n)$ we associate the number 1 if the element m of M that belongs to all subsets x_i (and to no other x_i) when $\alpha_i = 1$ also belongs to $y = F(x_1, \dots, x_n)$. But if such an element does not belong to y, we associate with the given sequence α the number 0. The logical functions $f_F(\alpha_1, \dots, \alpha_n)$ are thereby created. Clearly, this correspondence is one-to-one.

All the operations over sets usually considered in set theory satisfy our definition of a set-theoretic operation. Let us consider some examples. Recall that the difference between two sets, denoted $x \setminus y$, is understood to refer to the complement of their intersection with respect to x, and the symmetric difference $x \Delta y$, to the complement of their intersection with respect to their union. (In Figure 1 the results of these operations are shaded.)

2.20. (a) Which of the logical functions correspond to the difference and symmetric difference of two sets?

(b) Prove that the symmetric difference is associative.

2.21. Which set-theoretic operations correspond to the

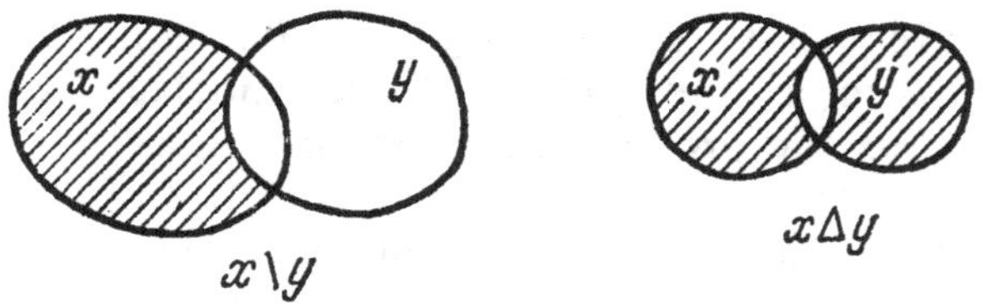

Figure 1

implication, material equivalence, and Sheffer functions (Problem 1.16)?

An interpretation of logical functions in the Boolean propositional algebra may be found at the end of Chapter 1 and beginning of the present chapter. Recall that a logical opertion in the propositional algebra is understood to refer to a mapping $F(x_1, \dots, x_n)$ that associates with each sequence of n propositions the proposition $y = F(x_1, \dots, x_n)$; further, the truth of y depends solely on the truth or falsity of the propositions $x_1, \dots, x_n$.

With each logical operation we have associated a logical function (sec. 2.1). Note that our correspondence between the logical and set-theoretic operations and the logical function is in agreement with the isomorphism established above and between Boolean algebras of sets and propositions.

Our reasoning may be given a somewhat more general form. Suppose that $\mathfrak{M}$ is a Boolean algebra. We wish to consider the homomorphisms of $\mathfrak{M}$ into the two-element Boolean algebra $\{0,1\}$, denoting by $M(\mathfrak{M})$ the set of these homomorphisms.

Definition 2.7. We will say that a Boolean algebra $\mathfrak{M}$ is *regular* if for every pair of elements $x, y \in \mathfrak{M}$, a separable homomorphism $\Phi \in M(\mathfrak{M})$ of x and y may be found, i.e., such that $\Phi(x) \neq \Phi(y)$. (The regularity constraint asserts that the set $M(\mathfrak{M})$ is, in some sense, large enough.)

2.22. Give an example of a non-regular Boolean algebra.

2.23. Prove that the algebras of propositions and subsets are regular.

2.24. Prove that a subalgebra of a regular Boolean algebra is regular.

2.25. Prove that every regular Boolean algebra is isomorphic to some subalgebra of the Boolean algebra of subsets (and thus the propositional algebra as well).

Definition 2.8. In a regular Boolean algebra $\mathfrak{M}$, the operations $F(x_1, \dots, x_n)$, where $x_i \in \mathfrak{M}$, are called *Boolean operations* if for every $\Phi \in M(\mathfrak{M})$, the value of $\Phi(y)$, $y = F(x_1, \dots, x_n)$, is determined by the values of $\Phi(x_1), \dots, \Phi(x_n)$.

2.26. Prove that the Boolean operations are in a one-to-one correspondence with the logical functions.

In the course of proving the regularity of algebras of subsets and propositions, we have constructed sets of separable homomorphisms in the algebra {0,1}. However, in none of these cases did we determine the intersection of all the homomorphisms in {0,1}, (i.e., find the entire set $M(\mathfrak{M})$). In fact, in our examples we did find all such homomorphisms. For example, in the algebra of subsets we found two elements and associated 1 with the subset containing both these elements, and 0 with the subset not containing at least one of the elements.

The concept of a Boolean operation may be associated with every subset $M' \subset M(\mathfrak{M})$ containing a separable homomorphism for every pair of elements of $\mathfrak{M}$. In the definition given above, we need only replace $M(\mathfrak{M})$ by M'.

2.27. Prove that the concept of a Boolean operation is independent of the set $M'' \subset M(\mathfrak{M})$ (which contain a separable homomorphism for every pair of elements of $\mathfrak{M}$) with which the concept is associated.

Hence, it follows that Boolean operations are set-theoretic operations when defined over the algebra of subsets, but logical operations when defined over the propositional algebra.

Now let us return to the study of the logical functions, i.e., the Boolean algebra {0,1}. However, bear in mind that all the results obtained so far may be correspondingly interpreted for Boolean operations over arbitrary Boolean algebras.

5. *Disjunctive Normal Form (DNF).* We have already spoken of complete disjunctive and complete conjunctive

normal forms at the end of Chapter 1 in connection with the representation of arbitrary logical operations in terms of the basic operations. We now wish to return to this problem.

We introduce the notation

$$x^{\sigma} = \begin{cases} x & \text{if } \sigma = 1 \\ \bar{x} & \text{if } \sigma = 0 \end{cases}$$

Note that $\sigma^{\sigma} = 1$.

Definition 2.9. The formula $x_1^{\sigma_1}, \dots, x_n^{\sigma_n}$, where $\sigma = (\sigma_1, \dots, \sigma_n)$ is some binary sequence and where some of the x_i may coincide, is called an *elementary conjunction.*

Definition 2.10. Every disjunction of elementary conjunctions is called a *disjunctive normal form* (abbreviated DNF).

2.28. Prove that a disjunctive normal form vanishes if and only if in every one of its elementary conjunctions there is some variable (for each elementary conjunction, this, in general, is simply the variable of the conjunction, itself) which occurs together with its negation.

Definition 2.11. An elementary conjunction is said to be *proper* if each variable occurs in it at most once (including any negated occurrences).

2.29. Which of the following elementary conjunctions is proper: (a) $x_1\bar{x}_2x_3$; (b) $x_1\bar{x}_3x_1$; (c) $x_2\bar{x}_2x_2x_1$; (d) $x_1\bar{x}_2x_3\bar{x}_4$?

Definition 2.12. A proper elementary conjunction is said to be *complete* relative to the variables $x_1, \dots, x_n$ if each of these variables occurs in it once and only once (possibly negated).

For example, the conjunction of Problem 2.29(a) is totally relative to the variables x_1, x_2, and x_3, while the conjunction of Problem 2.29(b) is totally relative to the variables x_1, x_2, x_3, and x_4.

Definition 2.13. A disjunctive normal form relative to the variables $x_1, \dots, x_n$ in which there are no isolated conjunctions and in which all the elementary conjunctions are proper and complete relative to the variables $x_1, \dots, x_n$ is called a *principal disjunctive normal form* (PDNF).

Since the variables of a PDNF are usually clear from its form, we will say simply PDNF, omitting the words, "relative to the variables $x_1, \dots, x_n$."

From Problem 1.19 it follows that every logical function $f(x_1, \dots, x_n)$ that is not identically zero may be represented by a principal disjunctive normal form:

$$f(x_1, \dots, x_n) = \bigvee_{f(\sigma_1,\dots,\sigma_n)=1} x_1^{\sigma_1} \dots x_n^{\sigma_n}, \tag{2.14}$$

where the symbol denotes that the disjunction is taken over all the sequences indicated beneath the symbol (in this case, over all sequences on which f is equal to 1). If $f = 0$, the set of conjunctions on the right side is empty.

2.30. Prove that a representation of a function in PDNF is unique.

Equality (2.14) may also be written thus:

$$f(x_1, \dots, x_n) = \bigvee_{(\sigma_1,\dots,\sigma_n)} f(\sigma_1, \dots, \sigma_n) x_1^{\sigma_1} \dots x_n^{\sigma_n}, \tag{2.15}$$

where the disjunction is taken over all binary sequences $(\sigma_1, \dots, \sigma_n)$. However, it is clear that only terms with coefficients of 1 remain in the disjunction, i.e., only those conjunctions for which $f(\sigma_1, \dots, \sigma_n) = 1$.

We will also require a decomposition of a function in PDNF relative to some of its variables, i.e., the formula

$$f(x_1, \dots, x_n; y_1, \dots, y_m) = \bigvee_{(\sigma_1,\dots,\sigma_n)} f(\sigma_1, \dots, \sigma_n; y_1, \dots, y_m) x_1^{\sigma_1} \dots x_n^{\sigma_n}. \tag{2.16}$$

Here f is decomposed in PDNF relative to its first n variables.

2.31. Prove the equivalence (2.16).

2.32. Find the PDNF for:
(a) a function of three variables, equal to 1 if most of its arguments are equal to 1;
(b) a function of four variables, equal to 1 if an even number of its arguments is equal to 1.

6. *Algorithm for Transforming a Formula Into PDNF.* By means of (2.14), it is possible to find the PDNF of a function from its truth table. However, this method is often unsuitable if the function is specified by a formula. Let us now discuss how to proceed in this case. We divide the construction of the PDNF into two stages. We first construct the DNF by means of the formulas, and then construct the PDNF from the DNF. The procedure described below constitutes an algorithm for the transformation of a formula into its PDNF.

Digression on Algorithms. In general, we understand by the term *algorithm* a method of solving some class of problems. An algorithm consists in a sequence of operations (one and the same sequence for all problems in a particular class) that for various reasons (e.g., each of them can actually be carried out) are thought of as elementary. Thus, by means of Euclid's algorithm for finding the greatest common divisor of two natural numbers or the algorithm for finding the square root of a natural number, it is possible to solve the given problems by executing elementary arithmetic operations in some definite sequence. The algorithm for solving a quadratic equation in fact uses the operation of extracting a square root as an elementary operation. It is known that there is no analogous algorithm that uses arithmetic operations and the operation of extracting square roots as its elementary operations for algebraic equations of degree greater than four (formulas, i.e., algorithms, are known for third-degree equations). In computer programming, the elementary operations are those operations which may be implemented on a computer (and sometimes problems for which algorithms, known as "routines," are already known). The branch of mathematical logic known as theory of algorithms is concerned with the study of different problems related to algorithms. (The central problem is to derive a rigorous definition of the concept of an algorithm.) Proofs that there is no algorithm for solving a particular problem occupy an important place in the theory. Transformations of formulas that use the axioms (2.1) - (2.13) and their simplest corollaries are the elementary operations here.

(1) Transform a formula in such a way that there are only disjunctions, conjunctions, and negations left in it, with negations applying only to arguments.

We have already performed such a transformation. In this type of transformation, Problem 1.13 is used to remove both implications and material equivalences, and equalities (2.8) and (2.9), to transpose negations (cf. Problem 2.8).

Note that formulas that are in DNF may be described as formulas that contain only disjunctions, conjunctions, and negations, and in which negations apply only to arguments, further, where all conjunctions are performed first, followed by disjunctions. Thus, we must also:

(2) Transform a formula in such a way that all conjunctions are performed before any disjunction.

2.33. Transform the following formulas into DNF: (a) $\overline{x \vee y}$; (b) $\overline{(x \vee z)(x \to y)}$; (c) $(x \sim y)\overline{(z \to t)}$.

Now let us transform DNF into PDNF.

(3) If there are seveal identical elementary conjunctions in DNF, we leave only one.

By virtue of the equivalence (2.10), this transformation leads to an equivalent formula.

(4) Turn all elementary conjunctions into proper conjunctions by means of the following two transformations:

(a) if some variable occurs in an elementary conjunction together with its negation, eliminate this conjunction from the DNF;

(b) if some variable occurs several times in an elementary conjunction, and is either nowhere negated or is everywhere negated, leave only one occurrence of the variable.

Transformation (a) leads to an equivalent formula, since $x\overline{x} = 0$; transformation (b) likewise, by virtue of (2.11).

Now we must derive the complete conjunctions.

(5) If some conjunction $x_1^{\sigma_1} \ldots x_k^{\sigma_k}$ lacks the variable y, we

have to consider the equivalent expression $x_1^{\sigma_1} \dots x_k^{\sigma_k}(y \vee \bar{y})$ and again apply transformation (2). If there are several missing variables, several conjunctive terms of the form $(y \vee \bar{y})$ must be added.

Recall that $y \vee \bar{y} = 1$ and $1x = x$. Identical conjunctions are again obtained after applying transformation (5). Therefore,

(6) Again apply transformation (3).

Thus concludes the process of transforming a formula into PDNF. We have not mentioned that it is of course necessary to everywhere use the fact that conjunction and disjunction are commutative and associative operations.

2.34. Find the PDNF for the formulas of Problem 2.33, and also for the formulas (d) $x \vee yz$; (e) $xy\bar{x}z \vee xt$; (f) $\bar{x}y \vee yzt \vee \bar{x}yzt$.

7. *Conjunctive Normal Forms.* Now let us present analogous reasoning for conjunctive normal forms.

Definition 2.14. A formula of the form $x_1^{\sigma_1} \vee \dots \vee x_n^{\sigma_n}$ is called an *elementary disjunction.*

Definition 2.15. Every conjunction of elementary disjunctions is called a *conjunctive normal form.*

2.35. Prove that a CNF will be identically equal to 1 if and only if every variable that occurs in each elementary disjunction is accompanied by an occurrence of its negation.

Definition 2.16. An elementary disjunction is said to be *proper* if each variable in it occurs at most once, including occurrences to which negation applies.

Definition 2.17. A proper elementary conjunction is said to be *complete* relative to the variables $x_1, \dots, x_n$ if each of these variables occurs in it once and only once (possibly negated).

Definition 2.18. A *principal conjunctive normal form* (PDNF) relative to the variables $x_1, \dots, x_n$ is understood to refer to a conjunctive normal form in which there are no isolated elementary disjunctions and in which all elementary disjunctions are proper and complete relative to the variables $x_1, \dots, x_n$.

By Problem 1.21, every function $f(x_1, \dots, x_n)$ that is not identically equal to 1 may be represented in PCNF:

$$f(x_1, \dots, x_n) = \prod_{f(\bar{\sigma}_1, \dots, \bar{\sigma}_n)} x_1^{\sigma_1} \vee \dots \vee x_n^{\sigma_n}, \tag{2.17}$$

where the symbol $\prod$ denotes that the conjunction is to be taken over all the sequences indicated beneath it.

2.36. Prove that the representation of a function in PCNF is unique.

Equality (2.17) may be rewritten thus:

$$f(x_1, \dots, x_n) \tag{2.18}$$

$$= \prod_{(\sigma_1, \dots, \sigma_n)} f(\bar{\sigma}_1, \dots, \bar{\sigma}_n) \quad x_1^{\sigma_1} \quad \dots \quad x_n^{\sigma_n},$$

where the conjunction is taken over all binary sequences $(\sigma_1, \dots, \sigma_n)$.

A function may be put in PCNF also relative to only some of its variables, thus:

$$f(x_1, \dots, x_n; y_1, \dots, y_m)$$

$$= \prod_{(\sigma_1, \dots, \sigma_n)} f(\bar{\sigma}_1, \dots, \bar{\sigma}_n; y_1, \dots, y_m) \tag{2.19}$$

$$\vee\, x_1^{\sigma_1} \quad \dots \vee x_n^{\sigma_n}.$$

2.37. Prove the equivalence (2.19).

2.38. Find the PCNF for the functions of Problem 2.32.

2.39. Describe an algorithm for transforming a formula into PCNF (similar to sec. 2.6).

2.40. Find the PCNF for formulas (a) and (b) of Problem

2.33, and also for the formulas: (c) $(x \vee y)(y \vee z)(z \vee t)$; (d) $x(y \vee \overline{z})(x \vee y \vee z)$.

8. *Simplifying Normal Forms.* The following equivalences may be used to simplify DNF or CNF:

$$x \vee xy = x; \tag{2.20}$$

$$x(x \vee y) = x; \tag{2.21}$$

$$x \vee \overline{x}y = x \vee y; \tag{2.22}$$

$$\overline{x} \vee xy = \overline{x} \vee y; \tag{2.23}$$

$$x(\overline{x} \vee y) = xy; \tag{2.24}$$

$$\overline{x}(x \vee y) = \overline{x}y. \tag{2.25}$$

Equivalences (2.20) and (2.21) are called *absorption laws.*

2.41. Prove the equivalences (2.20)-(2.25).

2.42. Simplify the formulas: (a) $xyz \vee xy\overline{z} \vee x\overline{yz} \vee \overline{x}yz$; (b) $x \vee xy \vee yz \vee \overline{xz}$; (c) $(x \vee y)$ & $(\overline{xy} \vee z) \vee \overline{z} \vee (x \vee y)(u \vee v)$.

Hints

2.1. The number of sequences $p(n) = 2^n$.

2.2. This number of equal to the number of binary sequences of length 2^n, i.e., 2^{2^n}.

2.3. $2^{2^{n-1}}$.

2.4. Determine the function f by induction on the rank of the composition f.

2.5. It is sufficient to prove that the truth tables of the formulas on the left and right sides are identical. However, it is not necessary to compile complete truth tables, as we need only check that the sets of sequences on which these formulas are equal to 1 (or 0) coincide.

2.6. Use (2.6) and (2.2).

2.7. Use (2.7) and (2.4).

2.8. Use the equivalences (2.8) and (2.9). (a) $(\bar{x} \vee \bar{y})z$; (b) $x(xy \vee \bar{y} \vee \bar{z} \vee \bar{y}(t \vee \bar{z}))$.

2.9. $\overline{A \to B} = A\bar{B}$. Use Problem 1.13 ($A \to B = \bar{A} \vee B$) and equivalences (2.8) and (2.1).

2.11. Identify the simple propositions and then write this proposition in the form of a logical formula; then construct the negation on the basis of equivalences (2.8) and (2.9).

2.15. Well-known properties of common divisors and multiples may be used or these properties may be directly derived. It is also possible to reduce the problem to the preceding one in the following way.

Suppose that $N = p_1^{A_1}p_2^{A_2} \dots p_k^{A_k}$ is a decomposition of a number N into prime factors, with $A_i > 0$ and $1 \leqslant i \leqslant k$. Every divisor x of N has the form $x = p_1^{x_1}p_2^{x_2} \dots p_k^{x_k}$, where $0 \leqslant x_i \leqslant A_i$. For every $i \leqslant k$, we consider the Boolean algebra $\mathfrak{N}_i$ of natural numbers $0 \leqslant x_i \leqslant A_i$ discussed in Example (3). It can be easily seen that if one of the operations we have defined ($\bar{x}$, xy, or $x \vee y$) is applied to the divisors $x = p_1^{x_1}p_2^{x_2} \dots p_k^{x_k}$, and $y = p_1^{y_1}p_2^{y_2} \dots p_k^{y_k}$, the identical operation is applied to the corresponding indices x_i and y_i in the Boolean algebra ${}_i$, $1 \leqslant i \leqslant k$. Hence, it follows at once that conditions (2.1) - (2.13) are satisfied by the set $\mathfrak{M}_N$ of divisors of N and that $\mathfrak{M}_N$ is a Boolean algebra.

We have proved here that the Boolean algebra $\mathfrak{M}_N$ is isomorphic to the direct product of the Boolean algebras $\mathfrak{N}_1, \dots, \mathfrak{N}_k$ (cf. Definition 2.6). If we are dealing with several Boolean algebras $\mathfrak{N}_1, \dots, \mathfrak{N}_k$, their *direct product* is the set $\mathfrak{N} = \mathfrak{N}_1 \times \dots \times \mathfrak{N}_k$ whose elements are sequences $(x_1, \dots, x_k)$, each element of which is taken from a Boolean algebra $\mathfrak{N}_i$. The operations over the sequences are performed element by element:

$$\overline{(x_1, \ldots, x_k)} = (\overline{x}_1, \ldots, \overline{x}_k);$$

$$(x_1, \ldots, x_k)(y_1, \ldots, y_k) = (x_1y_1, \ldots, x_ky_k);$$

$$(x_1, \ldots, x_k) \vee (y_1, \ldots, y_k)$$

$$= (x_1 \vee y_1, \ldots, x_k \vee y_k).$$

Sequences of the correspondingly identified elements (0, ..., 0), and (1, ..., 1) are the elements to be equated in $\mathfrak{N}$. Obviously, $\mathfrak{N}$ thus becomes a Boolean algebra.

2.16. We have

$$\overline{\tilde{x}} = \tilde{\overline{x}},\ \tilde{x}\tilde{y} = \widetilde{xy},\ \tilde{x} \vee \tilde{y} = \widetilde{x \vee y},\ x,y \in \mathfrak{M}_M.$$

2.17. With each divisor of N, associate the set of its prime divisors.

2.18. Use axiom (2.1) and de Morgan's laws (2.8) and (2.9).

2.19. Consider the Boolean algebra of Example (3) or some subalgebra of it.

2.20. Truth tables, for example, could be constructed.

2.22. Use Problem 2.19.

2.23. In the first case, we need only consider for each instance the homomorphism that associates with the proposition its truth value in this instance. (Propositions that are not separable are, by definition, identical.) In the second case, with each element $m \in M$ we associate the homomorphism $x \Rightarrow 1$ if m belongs to the subset, and otherwise the homomorphism $x \Rightarrow 0$.

2.24. Consider restrictions of homomorphisms of the algebra to its subalgebra.

2.30. (1) Take the PDNF in general form and find sequences on which it is equal to 1.

(2) Though this idea is quite simple, we advise the reader to consider another method of solving the problem. Thus,

find the total number of distinct PDNF of n variables and compare this number to the number of functions of n variables. Such "cardinality" considerations are often useful.

2.31. Use the equivalence (2.15).

2.36. The elementary (proper) disjunction $x_1^{\sigma_1} \vee \ldots \vee x_n^{\sigma_n}$ is equal to 0 only on the sequence $(\bar{\sigma}_1, \ldots, \bar{\sigma}_n)$. A "cardinality" proof may also be derived (cf. solution to Problem 2.30).

2.41. All the equivalences (2.21)-(2.25) may be obtained by means of the distributive laws (2.6) and (2.7).

2.42. Use the equivalences (2.21)-(2.25). Answers: (a) $xy \vee xz \vee yz$; (b) $x \vee z$; (c) $x \vee y \vee \bar{z}$.

Solutions

2.1. The answer may be easily guessed at in the following way. There are two possibilities for the first element α_1 of the sequence: 0 and 1; in each of these possibilities, α_2 may be selected in either of two ways. Thus the first two elements α_1 and α_2 may be selected in $2^2 = 4$ ways. In each of these four cases, there are two possibilities for selecting α_3, i.e., the elements α_1, α_2, and α_3 may be selected in $2^3 = 8$ ways, and so on, and n elements $(\alpha_1, \ldots, \alpha_n)$ may be selected in 2^n ways. A rigorous proof may be conducted by means of mathematical induction.

2.2. In the truth table for the functions $f(x_1, \ldots, x_n)$, we establish a method of ordering the rows in some way, i.e., an order of the sequences $(\alpha_1, \ldots, \alpha_n)$ of values of the arguments $x_1, \ldots, x_n$ is established. Then the functions may be uniquely determined by their last columns, i.e., by sequences of 2^n zeros and units (by Problem 2.1, there are 2^n rows in the table). There are as many distinct functions as there are distinct sequences of length 2^n, i.e., 2^{2^n}.

2.3. The values of the function on all sequences other than the null sequence, i.e., on $(2^n - 1)$ sequences, may be selected arbitrarily. Such a selection may be performed in $2^{2^n - 1}$ ways.

Incidentally, it can be easily seen that precisely half the functions preserve zero, since it is possible to establish a one-to-one correspondence between functions that preserve zero and functions that do not (simply take functions that coincide on all sequences, except for the null sequence).

2.4. We give a rigorous definition of the mapping $\tilde{f} \Rightarrow f$.* We define f so that $f \in \Phi^{(k)}$ if $\tilde{f} \in \tilde{\Phi}^{(k)}$. If $\tilde{f} \in \tilde{\Phi}$, the definition of f is self-evident. Suppose that the functions $f \in \Phi^{(k)}$ have been defined for all $\tilde{f} \in \tilde{\Phi}^{(k)}$; we define f for $\tilde{f} \in \tilde{\Phi}^{(k+1)}$. Suppose that $\tilde{f} \in \tilde{\Phi}^{(k+1)}$. There are two cases. If $\tilde{f}$ is obtained from some function $\tilde{g} \in \tilde{\Phi}^{(k)}$ by renaming the variable in accordance with Definition 2.3(a), we take the function $g \in \Phi^{(k)}$ corresponding to $\tilde{g}$ by the induction hypothesis, and rename the same variable in it. (Of course, every function may be assumed to depend on an arbitrary variable, at least formally.) Analogously, if $\tilde{f} \in \tilde{\Phi}^{(k+1)}$ is obtained by substituting the functions $\tilde{g} \in \tilde{\Phi}^{(k)}$ for some variable x of the function $\tilde{h} \in \tilde{\Phi}^{(k)}$ in accordance with Definition 2.3(b), we take the functions $g, h \in \Phi^{(k)}$ constructed by the induction hypothesis, and in h substitute g for x. The resulting function is also denoted f; $f \in \Phi^{(k+1)}$. It is also clear that $\tilde{f} = f$ for $\tilde{f} \in \Phi$, and if we suppose that $g = \tilde{g}$ and $h = \tilde{h}$, we obtain $f = \tilde{f}$ in both of the cases considered.

2.5. As an example, consider the equivalences (2.3) and (2.7).

Let us prove that $(xy)z = x(yz)$. We find values of the variables for which the left side is equal to 1. Since the final operation is a conjunction, it is only necessary that $xy = 1$ and $z = 1$, that is (1, 1, 1) is the only sequence on which $(xy)z$ is equal to 1. It may be similarly proved that the right side is also equal to 1 only on this sequence.

Let us now show that $x \vee yz = (x \vee y)(x \vee z)$. We find sequences of variables on which these formulas are zero. Let us consider the left side. Since the last operation is a disjunction, it is necessary that $x = 0$ and $yz = 0$. The latter will hold if $y = 0$ or $z = 0$. On the right side, the final operation is a conjunction; therefore, $x \vee y$ or $x \vee z$ must

*We use the symbol $\Rightarrow$ for mappings, in contradistinction to the symbol $\rightarrow$ for implications in algebraic logic.

equal zero. In the first case, x and y must both equal zero, and in the second case, x and z are both zero. Thus, the conditions under which the left and right sides vanish are the same.

2.6. By equation (2.6), $(x \vee y)(z \vee t) = (x \vee y)z \vee (x \vee y)t$. By the commutativity of the conjunction, we may again apply equation (2.6) to $(x \vee y)z$ and $(x \vee t)t$. By the associativity of the disjunction, the parentheses may be omitted from the resulting formula.

2.7. The solution is analogous to that of Problem 2.6. The sole difference is that the law (2.7) is used in place of the distributive law (2.6).

2.8. (a) By (2.8) and (2.1), we have $\overline{xy \vee \bar{z}} = \overline{xy}z = (\bar{x} \vee \bar{y})z$.

2.10. The boy's decision will not hold if he does not finish reading his book or goes neither to the museum or the movies and, though the weather was nice, does not go swimming in the river.

2.11. The simple propositions are as follows:

x: *The boy finished reading his book on Sunday.*
y: *The boy went to the museum on Sunday.*
z: *The boy went to the movies on Sunday.*
t: *The boy went swimming in the river on Sunday.*
u: *The weather was nice*

The entire sentence has the form $x(y \vee z)(u \vee t)$; its negation is $\bar{x} \vee \overline{yz} \vee \overline{ut}$. Substituting the simple propositions in place of the variables, we arrive at the answer to Problem 2.10 given above.

2.19. In Example (3), let us consider the following subalgebra of three elements: $\{0, \frac{1}{2}, 1\}$. Then $\overline{\tfrac{1}{2}} = \frac{1}{2}$ and $\frac{1}{2} \cdot \overline{\tfrac{1}{2}} = \frac{1}{2}$ and $\frac{1}{2} \quad \overline{\tfrac{1}{2}} = \frac{1}{2}$. We have simultaneously proved that a Boolean algebra may contain an element that coincides with its own negation.

2.20. (a) $x \setminus y$ corresponds to $\overline{xy}$; $x \,\Delta\, y$ corresponds to $x\bar{y} \vee$

$\bar{x}y$ (i.e., the exclusive disjunction $x \,\Delta\, y = \overline{x \sim y}$; cf. Problem 1.1).

(b) $(x \,\Delta\, y) \,\Delta\, z$ is equal to 1 if and only if one and only one of the arguments x, y, and z is equal to 1. Hence follows the associativity of the exclusive disjunction, that is, the symmetricity of the difference.

2.21. The complement of the difference $x \,\Delta\, y$ with respect to the entire set or the union of the complement of x (with respect to the entire set) and the subset y corresponds to the implication; the complement of the symmetric difference or the union of the intersections of the sets themselves (xy) and their complements ($\bar{x}\bar{y}$) correspond to the material equivalence; the complement of the intersection or the union of the complements ($\overline{xy} = \bar{x} \vee \bar{y}$) and the complement of the union of the intersection of the complements ($\overline{x \vee y} = \bar{x}\bar{y}$), respectively, correspond to the two Sheffer operations.

2.22. In the algebra considered in the course of solving Problem 2.19, there is no homomorphism into the algebra $\{0,1\}$, since $x \neq \bar{x}$ in this algebra.

2.25. With each element m of a proper algebra $\mathfrak{M}$, we associate the subset of the set $M(\mathfrak{M})$ of its homomorphisms into $\{0,1\}$ under which the element m becomes 1. It may be directly verified that we thus have derived a homomorphism into the algebra of subset of $M(\mathfrak{M})$. The subalgebra in which $\mathfrak{M}$ thus turns into a homomorphism is isomorphic to $\mathfrak{M}$, since distinct subsets correspond to distinct elements of $\mathfrak{M}$ by virtue of the existence of separable homomorphisms into $\{0,1\}$.

2.27. It is sufficient to prove that for each subset M' of homomorphisms, the Boolean operations are the same as for $M(\mathfrak{M})$. This follows from the fact that the value of any homomorphism $\Phi \in M(\mathfrak{M})$ on an arbitrary element $x \in \mathfrak{M}$ is uniquely determined by the values of the homomorphisms $\psi \in M'$ on this element. (The element x itself may be uniquely described by these values; cf. solution to Problem 2.25.)

2.28. That the condition is sufficient follows from the fact that the function $x\bar{x}$ is identically zero; therefore, all the elementary conjunctions, thus the entire DNF, are identically zero.

Let us prove necessity. Suppose that $x_1^{\sigma_1}, \ldots, x_n^{\sigma_n}$ is an elementary conjunction in which no variable occurs together with its negation. We may then consider the sequence of values of the variables $(\sigma_1, \ldots, \sigma_n)$, since though some of the variables may coincide, the same variable either occurs everywhere itself or its negation occurs everywhere. On this set, our elementary conjunction is equal to 1, that is, the entire DNF is 1. (If one term of a disjunction is equal to 1, the entire disjunction is equal to 1.)

2.29. (a) and (d); in (b) the variable x_1 occurs twice, once negated; in (c) the variable x_2 occurs three times, once negated.

2.30. First solution. A complete proper elementary conjunction $x_1^{\sigma_1} \ldots x_n^{\sigma_n}$ is equal to 1 on the isolated sequence $(\sigma_1, \ldots, \sigma_n)$. A disjunction of several such conjunctions is equal to 1 on the sequences $(\sigma_1, \ldots, \sigma_n)$, and only on these sequences, which are sequences of indices of the elementary conjunctions occurring in it. Therefore, the decomposition (2.14) is the only one possible and distinct functions correspond to distinct PDNF.

Second solution. We will find the number of PDNF of n variables $x_1, \ldots, x_n$. We enumerate the complete proper elementary conjunctions $x_1^{\sigma_1} \ldots x_n^{\sigma_n}$ in some way. There are as many such conjunctions as there are binary sequences of n elements, i.e., 2^n (Problem 2.1). With each PDNF of the variables $x_1, \ldots, x_n$, we may associate in a one-to-one fashion a sequence of 2^n zeros and units not equal to the null sequence. Units are put at each entry where there is a number corresponding to a conjunction occurring in the PDNF, and zeros elsewhere. A null seqquence is not obtained here, since it would correspond to an empty PDNF. Thus, there are as many distinct PDNF as there are sequences of length 2^n not equal to the null sequence, i.e., $2^{2^n} - 1$.

There are also $2^{2^n} - 1$ functions of the variables $x_1, \ldots, x_n$ which are not identically zero, and since each such function may be represented in PDNF, the representation is unique. This simple combinatorial consideration may be reasoned through by means of the following scheme. Suppose there are N boxes and N balls. In each box there is at least one ball. Then in each box there is precisely one ball. The proof of this assertion is perfectly obvious. In our case, the logical functions serve as boxes, and the PDNF as the balls. We will use this idea below.

2.31. To prove the equivalence (2.16), it is necessary to prove that the values of the left and right sides coincide for any values of the variables $x_1, \dots, x_n$ and $y_1, \dots, y_m$. We first establish the values of $y_1, \dots, y_m$. Then the values will coincide once the remaining variables are established as a consequence of the equivalence (2.15).

We could also expand f with respect to all the variables, and then regroup terms using the distributive law.

2.32. (a) $xyz \vee xy\bar{z} \vee x\bar{y}z \vee \bar{x}yz$; (b) $xyzt \vee xy\overline{zt} \vee x\overline{yzt} \vee \overline{xy}zt \vee \bar{x}yzt \vee \bar{x}yz\bar{t} \vee x\bar{y}z\bar{t} \vee \overline{xyzt}$.

2.33. (a) $x \vee y$ is in DNF; (b) Applying the procedure of (1), we obtain $\overline{xz}(\bar{x} \vee y)$. Now, applying the equivalence (2.6), we obtain $\overline{xzx} \vee \overline{xz}y$. (c) On the first step, we obtain $(\bar{x} \vee y)(x \vee \bar{y})z\bar{t}$. Now, applying (2.6) and Problem 2.6, we obtain

$$\bar{x}xz\bar{t} \vee \overline{xy}z\bar{t} \vee yxz\bar{t} \vee y\bar{y}z\bar{t}.$$

2.34. (a) In the DNF $x \quad y$, all the elementary conjunctions are proper, so that we may immediately apply rule (5): $x(y \vee \bar{y}) \vee y(x \vee \bar{x})$. We then obtain $xy \vee x\bar{y} \vee xy \vee \bar{x}y$. Of the two occurrences of the conjunction xy, we are now left with one: $xy \vee x\bar{y} \vee \bar{x}y$.

(b) Begin with rule (4): $\bar{x}z \vee \bar{x}y\bar{z}$. The first conjunction is not proper: $\bar{x}z(y \vee \bar{y}) \vee \overline{xy}z = \bar{x}yz \vee \overline{xy}z \vee \bar{x}y\bar{z}$. This formula is in PDNF.

(c) By rule (4), obtain the DNF $\overline{xy}z\bar{t} \vee xyz\bar{t}$. This formula is already in PDNF.

(d) $x \vee yz = x(y \vee \bar{y})(z \vee \bar{z}) \vee (x \vee \bar{x})yz = xyz \vee x\bar{y}z \vee xy\bar{z} \vee x\overline{yz} \vee xyz \vee \bar{x}yz = xyz \vee x\bar{y}z \vee xy\bar{z} \vee x\overline{yz} \vee \bar{x}yz$.

(e) $xy\bar{x}z \vee xt = xt = x(y \vee \bar{y})(z \vee \bar{z})t = xyzt \vee x\bar{y}zt \vee xy\bar{z}t \vee x\overline{yz}t$.

(f) $x\bar{y} \vee yzt \vee \bar{x}yzt = x\bar{y}(z \vee \bar{z})(t \vee \bar{t}) \vee (x \vee \bar{x})yzt \vee \bar{x}yzt = x\bar{y}zt \vee x\overline{yz}t \vee x\bar{y}z\bar{t} \vee x\overline{yzt} \vee xyzt \vee \bar{x}yzt$.

2.38. (a) $(x \vee y \vee z)(x \vee y \vee \bar{z})(x \vee \bar{y} \vee z)(\bar{x} \vee y \vee z)$.

(b) $(\bar{x} \vee y \vee z \vee t)(x \vee \bar{y} \vee z \vee t)(x \vee y \vee z \vee t)(x \vee y \vee z \vee \bar{t})(x \vee \bar{y} \vee \bar{z} \vee t)(\bar{x} \vee y \vee \bar{z} \vee \bar{t})(\bar{x} \vee \bar{y} \vee z \vee \bar{t})(\bar{x} \vee \bar{y} \vee \bar{z} \vee t)$.

2.39. (1) Make use of an equivalent formula in which there are only disjunctions, conjunctions, and negations, with negations applying only to arguments.

(2) Transform the resulting formula in such a way that all disjunctions are performed first, using the distributive law (2.7) and Problem 2.7. As a result, we obtain a CNF.

(3) If there are several isolated elementary disjunctions, leave only one. By (2.11), an equivalent formula is thus obtained.

(4) Transform all the elementary disjunctions into proper disjunctions as follows:

(a) Remove elementary disjunctions in which some variable occurs together with its negation.

(b) If some variable occurs several times in the same form (i.e., either always negated or never negated), we leave only one occurrence.

In (a), we are using the relation $x \vee \bar{x} = 1$, and in (b), the relation of (2.10).

(5) If some variable y does not occur in the elementary conjunction $x_1^{\sigma_1} \vee \dots \vee x_k^{\sigma_k}$, we add the term $y\bar{y}$ to it by means of a disjunction, $x_1^{\sigma_1} \vee \dots \vee x_k^{\sigma_k} \vee y\bar{y}$, and then again apply the transformation (2). We have used here the fact that $\bar{y} = 0$, and relation (2.13).

(6) We again apply the transformation (2.13).

2.40. (a) $x \vee y$ is in PCNF.

(b) As in the solution of Problem 2.33, we first obtain $\overline{xz}(\bar{x}$ $y)$. This is already in CNF. We now construct the PCNF, thus:

$$\overline{xz}(\bar{x} \vee y) = (\bar{x} \vee y\bar{y} \vee z\bar{z})(x\bar{x} \vee y\bar{y} \vee \bar{z})(\bar{x} \vee y \vee z\bar{z})$$

$$= (\bar{x} \vee y \vee z)(\bar{x} \vee \bar{y} \vee z)(\bar{x} \vee y \vee \bar{z})$$

$$\& \ (\bar{x} \vee \bar{y} \vee \bar{z})(x \vee y \vee \bar{z})(\bar{x} \vee y \vee \bar{z})$$

$$\& \ (x \vee \bar{y} \vee \bar{z})(\bar{x} \vee \bar{y} \vee \bar{z})(\bar{x} \vee y \vee z)(\bar{x} \vee y \vee \bar{z})$$

$$= (\bar{x} \vee y \vee z)(\bar{x} \vee \bar{y} \vee z)(\bar{x} \vee y \vee \bar{z})(\bar{x} \vee \bar{y} \vee \bar{z})$$

$$\& \ (x \vee y \vee \bar{z})(x \vee \bar{y} \vee \bar{z}).$$

(c)

$$(x \vee y)(y \vee z)(z \vee t)$$

$$= (x \vee y \vee z\bar{z} \vee t\bar{t})(x\bar{x} \vee y \vee z \vee t\bar{t})$$

$$\& (x\bar{x} \vee y\bar{y} \vee z \vee t) = (x \vee y \vee z \vee t)(x \vee y$$

$$\vee \bar{z} \vee t)(x \vee y \vee z \vee \bar{t})$$

$$\& (x \vee y \vee \bar{z} \vee \bar{t})(x \vee y \vee z \vee t)(\bar{x} \vee y \vee z \vee t)$$

$$\& (x \vee y \vee z \vee \bar{t})$$

$$\& (\bar{x} \vee y \vee z \vee \bar{t})(x \vee y \vee z \vee t)(\bar{x} \vee y \vee z \vee t)$$

$$\& (x \vee \bar{y} \vee z \vee t)(\bar{x} \vee \bar{y} \vee z \vee t)$$

$$= (x \vee y \vee z \vee t)(x \vee y \vee \bar{z} \vee t)(x \vee y \vee z \vee \bar{t})$$

$$\& (x \vee y \vee \bar{z} \vee \bar{t})(\bar{x} \vee y \vee z \vee t)(\bar{x} \vee y \vee z \vee \bar{t})$$

$$\& (x \vee \bar{y} \vee z \vee t)(\bar{x} \vee \bar{y} \vee z \vee t).$$

(d)

$$x(y \vee z)(x \vee y \vee z)$$

$$= (x \vee y\bar{y} \vee z\bar{z})(x\bar{x} \vee y \vee \bar{z})(x \vee y \vee z)$$

$$= (x \vee y \vee z)(x \vee \bar{y} \vee z)(x \vee y \vee \bar{z})(x \vee \bar{y} \vee \bar{z})$$

$$\& (x \vee y \vee \bar{z})(\bar{x} \vee y \vee \bar{z})(x \vee y \vee z)$$

$$= (x \vee y \vee z)(x \vee \bar{y} \vee z)(x \vee y \vee \bar{z})$$

$$\& (x \vee \bar{y} \vee \bar{z})(\bar{x} \vee y \vee \bar{z}).$$

2.41. (2.20): $x \vee xy = 1$ if and only if $x = 1$, since if $xy = 1$, then $x = 1$.

2.42. (a)

$$xyz \vee xy\bar{z} \vee x\bar{y}z \vee \bar{x}yz = xy(z \vee \bar{z}) \vee x\bar{y}z \vee \bar{x}yz$$

$$= xy \vee x\bar{y}z \vee \bar{x}yz = x(y \vee \bar{y}z) \vee \bar{x}yz$$

$$= xy \vee xz \vee \bar{x}yz = xy \vee z(x \vee \bar{x}y)$$

$$= xy \vee xz \vee yz.$$

(b)

$$x \vee xy \vee yz \vee \bar{x}z = x \vee yz \vee \bar{x}z$$

$$= x \vee z \vee yz = x \vee z.$$

(c)

$$(x \vee y)(\overline{xy} \vee z) \vee \bar{z} \vee (x \vee y)(u \vee v)$$

$$= (x \vee y)(\overline{\bar{x} \vee \bar{y}} \vee z) \vee \bar{z} \vee (x \vee y)(u \vee v)$$

$$= (x \vee y)z \vee \bar{z} \vee (x \vee y)(u \vee v)$$

$$= (x \vee y) \vee \bar{z} \vee (x \vee y)(u \vee v) = x \vee y \vee \bar{z}.$$

Chapter 3
LAW OF DUALITY IN ALGEBRAIC LOGIC

At the end of sec. 2.3, we spoke of the property of duality present in the axiomatic structures of Boolean algebras. Because of this property, the structure of the dual Boolean algebra $\mathfrak{M}^+$ can be introduced on a set $\mathfrak{M}$ equipped with the structure of a Boolean algebra. Between $\mathfrak{M}$ and $\mathfrak{M}^+$ there is what is known as a canonical isomorphism, the zero and unit in the one algebra corresponding to the unit and zero in the other, likewise, disjunction (conjunction) in the one algebra corresponding to conjunction (disjunction) in the other. In this chapter, we shall explain how Boolean operations defined on regular Boolean algebras behave under the isomorphism $\mathfrak{M} \rightarrow \mathfrak{M}^+$ (sec. 2.4).

3.1. Prove that a Boolean algebra $\mathfrak{M}^+$ dual to a regular Boolean algebra $\mathfrak{M}$ is itself regular.

Using the correspondence between Boolean operations and the logical functions, we may limit our discussion to the latter, i.e., the two-element algebra {0,1}.

The concept of dual functions will be basic in this chapter. A dual function is obtained from some initial function by replacing the values of all the variables by their opposite values, i.e., everywhere in the truth table 0 is replaced by 1 and 1 by 0.

Definition 3.1. The function $f^+(x_1, \dots, x_n)$ dual to the function $f(x_1, \dots, x_n)$ is determined by the equality

$$f^+(x_1, \dots, x_n) = \overline{f(\overline{x}_1, \dots, \overline{x}_n)}.$$

Definition 3.2. A function equivalent to its dual function, i.e., a function such that

$$f(x_1, \dots, x_n) = f^+(x_1, x_2, \dots, x_n) = \overline{f(\overline{x}_1, \dots, \overline{x}_n)},$$

is said to be *self-dual.*

Thus, a self-dual function takes opposite values on opposite sequences $(\alpha_1, \dots, \alpha_n)$ and $(\overline{\alpha}_1, \dots, \overline{\alpha}_n)$.

3.2. Construct a function dual to the following functions:

(a) the basic logical operations and constants 0 and 1;

(b) a function of five variables, equal to 1 if an even number of the variables is equal to 1;

(c) the analogous function of six variables.

Which of these functions are self-dual?

3.3. Prove that the function $xy \vee xz \vee yz$ is self-dual.

3.4. Find all self-dual functions of two variables.

3.5. How many self-dual functions of n variables are there?

3.6. Give a definition of a non-self-dual function.*

We now formulate the assertion

Law of Duality. A function dual to a composition of designated functions is equivalent to the corresponding** composition of dual functions (cf. Definition 2.3).

---

*In this problem, as in previous problems of this type, we are assuming a definition in which negations occur only in the simple propositions.

**A precise definition of a "corresponding composition" may be derived by means of induction (cf. solution to Problem 3.7).

3.7. Prove the law of duality.

From the law of duality, it follows that a composition of self-dual functions is self-dual.

The law of duality is useful for finding functions dual to functions that may be represented by means of formulas. Of course, we could use the definition of a dual function and apply the negation to arguments and to the entire formula. However, we would then obtain from a formula containing only disjunctions, conjunctions, and negations -- in which only the arguments are negated -- a formula that does not possess this property. But if we construct a dual function by means of the law of duality (replacing operations by their duals, that is, conjunctions by disjunctions, and conversely*), this property is preserved.

3.8. For the functions dual to those indicated below, construct representations by means of formulas in which only the arguments are negated: (a) $(\overline{x} \vee y\overline{z})(xy \vee x\overline{z})$; (b) $(x \vee \overline{y})z\overline{t} \vee \overline{x}t$.

Thus, if we have a formula that contains only conjunctions, disjunctions, and negations, then, by replacing the conjunctions everywhere by disjunctions and conversely, we arrive at a formula dual to the initial one. By applying this procedure to equivalent formulas, we obtain equivalent formulas. (It follows at once from the definition that functions dual to equivalent functions are equivalent.) By the law of duality, we often understand precisely this corollary of the law. It is in fact an extremely important corollary, as it may be used to derive assertions of algebraic logic from assertions previously derived. For example, from the equivalences (2.2), (2.3), (2.6), (2.8), (2.10), (2.20), (2.22), and (2.23), we may obtain (2.4), (2.5), (2.7), (2.9), (2.11), (2.21), (2.24), and (2.25); similarly, the assertion of Problem 2.7 may be obtained from Problem 2.6. Thus, the properties of CNF and PCNF may be derived from those of DNF and PDNF.

3.9. Based on the possibility of decomposing any function

---

*Cf. solution to Problem 3.2 (recall that $(\overline{x})^{+} = \overline{x}$).

f other then the additive identity into PDNF (2.14), prove that any function ω other then the multiplicative identity may be decomposed into PCNF (2.18). Analogously, derive (2.19) from (2.15).

We will require below certain results on non-self-dual functions.

3.10. Suppose we are given a non-self-dual function. Equate its variables in such a way so as to obtain a non-self-dual function with the least number of variables. What might be this number?

3.11. Prove that the constants 0 and 1 may be obtained from an arbitrary non-self-dual function and a negation of compositions.

Hints

3.1. The homomorphisms of $\mathfrak{M}^+$ into $\{0,1\}$ may be obtained by a composition of a canonical isomorphism of $\mathfrak{M}^+$ into $\mathfrak{M}$ and homomorphisms of $\mathfrak{M}$ into $\{0,1\}$ (i.e., $\mathfrak{M}^+$ is first mapped isomorphically into $\mathfrak{M}$, and then $\mathfrak{M}$ is mapped homomorphically into $\{0,1\}$). It may be directly verified that these mappings equate the elements of $\mathfrak{M}^+$ if the homomorphisms of $\mathfrak{M}$ into $\{0,1\}$ possess this property.

3.5. A self-dual function is determined by its value on a set of sequences containing one element from each pair of dual sequences. (The function will take its opposite value on an opposite sequence.)

3.10. We may always obtain a non-self-dual function of two variables. A further decrease in the number of variables is, in general, impossible.

3.11. We may limit our discussion to the case of two variables.

Solutions

3.2. (a) $0^+ = 1$ (since the function does not depend actually on any argument, it is necessary to apply negation only to the value of the function); $1^+ = 0$; $(\bar{x})^+ = \bar{\bar{\bar{x}}} = \bar{x}$, i.e., $\bar{x}$ is a self-dual function; $(xy)^+ = x \vee y$ (cf. (2.9)); $(x \vee y)^+ = xy$ (cf. (2.10)); $(x \to y)^+ = \overline{\bar{x} \vee \bar{\bar{y}}} = \overline{\bar{y} \to \bar{x}}$; $(x \sim y)^+ = \overline{\bar{x} \sim \bar{y}}$.

(b), (c) The functions are equal to 0 if an even number of variables are equal to 0. In example (b), this is the same function as the initial one, i.e., the given function is self-dual.

3.3.

$$(xy \vee yz \vee xz)^+ = \overline{\bar{x}\bar{y} \vee \bar{y}\bar{z} \vee \bar{x}\bar{z}} = \overline{\bar{x}\bar{y}}\;\overline{\bar{y}\bar{z}}\;\overline{\bar{x}\bar{z}}$$

$$= (x \vee y)(y \vee z)(x \vee z)$$

$$= xy \vee yz \vee xz.$$

We have used equality (2.21).

3.4. x, y, $\bar{x}$, and $\bar{y}$, i.e., all the self-dual functions of two variables depend actually on a single variable.

3.5. $2^{2^{n-1}}$. The set of sequences, the values on which determine the self-dual function, contains 2^{n-1} elements (half of all the sequences). We further apply the method used above to compute the number of funtions (cf. Problem 2.3).

3.6. A function is non-self-dual if there exists a sequence $(\alpha_1, \dots, \alpha_n)$ such that $f(\alpha_1, \dots, \alpha_n) = f(\bar{\alpha}_1, \dots, \bar{\alpha}_n)$.

3.7. Besides solving the problem, we will also assign an exact meaning to the words, "corresponding composition of dual functions." Suppose that we are considering a composition of functions from the system Φ. We associate with the function $\omega \in \Phi$ the dual functions ω^+ (denoting the set of such functions Φ^+). The assertion of the law of duality is a tautology for these functions. Suppose that we have determined for compositions of rank k from Φ, the corresponding compositions $\tilde{\omega} \in \Phi^{+(k)}$ of the system of dual functions Φ^+ and that we have proved that the law of duality is satisfied by these functions, i.e., $\tilde{\omega} = \omega^+$ if $\omega \in \Phi^{(k)}$. Then we may associate with the compositions

$$F_1(x_1, \dots, x_{i-1}, y, x_{i+1}, \dots, x_n)$$

$$= \omega(x_1, \dots, x_{i-1}, y, x_{i+1}, \dots, x_n);$$

$$F_2(x_1, \dots, x_{i-1}, x_{i+1}, \dots, x_n; y_1 \dots, y_\ell)$$

$$= \omega(x_1, \dots, x_{i-1}), \psi(y_1, \dots, y_\ell), x_{i+1}, \dots, x_n),$$

where $\omega(x_1, \dots, x_{i-1}, x_i, x_{i+1}, \dots, x_n), \psi(y_1, \dots, y_\ell) \in \Phi^{(k)}$ the compositions

$$\widetilde{F}_1(x_1, \dots, x_{i-1}, y, x_{i+1}, \dots, x_n)$$

$$= \widetilde{\omega}(x_1, \dots, x_{i-1}, y, x_{i+1}, \dots, x_n);$$

$$\widetilde{F}_2(x_1, \dots, x_{i-1}, x_{i+1}, \dots, x_n; y_1, \dots, y_\ell)$$

$$= \widetilde{\omega}(x_1, \dots, x_{i-1}, \widetilde{\psi}(y_1, \dots, y_\ell), x_{i+1}, \dots, x_n),$$

where $\widetilde{\omega}, \widetilde{\psi} \in \Phi^{+(k)}$ are compositions of functions from Φ^+ already constructed by the induction hypothesis and corresponding to ω and ψ. Now that the required definition has been given, it remains for us to verify that the law of duality is valid. By the induction hypothesis, $\widetilde{\omega} = \omega^+$ and $\widetilde{\psi} = \psi^+$. We must prove that $F_1^+ = \widetilde{F}_1$ and $F_2^+ = \widetilde{F}_2$. By Definition 3.1 and the induction hypothesis,

$$F_1^+(x_1, \dots, x_{i-1}, y, x_{i+1}, \dots, x_n)$$

$$= \overline{\omega(\overline{x}_1, \dots, \overline{x_{i-1}}, \overline{y}, \overline{x_{i+1}}, \dots, \overline{x}_n)}$$

$$= \omega^+(x_1, \dots, x_{i-1}, y, x_{i+1}, \dots, x_n)$$

$$= \widetilde{F}_1(x_1, \dots, x_{i-1}, y, x_{i+1}, \dots, x_n);$$

$$F_2^+(x_1, \dots, x_{i-1}, x_{i+1}, \dots, x_n; y_1, \dots, y_\ell)$$

$$= \overline{\omega(\overline{x}_1, \dots, \overline{x_{i-1}}, \psi(\overline{y}_1, \dots, \overline{y}_\ell), \overline{x_{i+1}}, \dots, \overline{y}_n)}$$

$$= \overline{\omega(\bar{x}_1, \dots, \bar{x}_{i-1}, \overline{\psi(\bar{y}_1, \dots, \bar{y}_\ell)}, \bar{x}_{i+1}, \dots, \bar{x}_n)}$$

$$= \omega^+(x_1, \dots, x_{i-1}, \psi^+(y_1, \dots, y_\ell), x_{i+1}, \dots, x_n)$$

$$= \tilde{\omega}(x_1, \dots, x_{i-1}, \tilde{\psi}(y_1, \dots, y_\ell), x_{i+1}, \dots, x_n)$$

$$= \tilde{F}_2(x_1, \dots, x_{i-1}, x_{i+1}, \dots, x_n; y_1, \dots, y_\ell).$$

3.8. We replace conjunctions by disjunctions, and conversely. (a) $\bar{x}(y \vee \bar{z}) \vee (x \vee y)(x \vee \bar{z})$; (b) $(x\bar{y} \vee z \vee \bar{t})(\bar{x} \vee t)$.

3.9. Suppose that $\omega(x_1, \dots, x_n)$ is not the multiplicative identity. We consider $f(x_1, \dots, x_n) = \omega^+(x_1, \dots, x_n)$; it turns out that f is not the additive identity. We represent $f(x_1, \dots, x_n)$ in PDNF (2.14). On the left side, we introduce a dual function, and on the right side we replace all conjunctions by disjunctions, and conversely, obtaining

$$f^+(x_1, \dots, x_n) = \omega(x_1, \dots, x_n)$$

$$= \prod_{f(\sigma_1, \dots, \sigma_n)=1} x_1^{\sigma_1} \vee \dots \vee x_n^{\sigma_n}$$

$$= \prod_{\omega(\sigma_1, \dots, \sigma_n)=0} x_1^{\sigma_1} \vee \dots \vee x_n^{\sigma_n}.$$

We have found a representation of ω in PCNF. The transition from (2.15) to (2.19) may be performed in an analogous fashion.

3.10. In solving Problem 3.6, we gave a definition of a non-self-dual function. Suppose that $f(x_1, \dots, x_n)$ is such a function and, for the sequence $(\alpha_1, \dots, \alpha_n)$, let

$$f(\alpha_1, \dots, \alpha_n) = f(\bar{\alpha}_1, \dots, \bar{\alpha}_n).$$

We divide the variables $x_1, \dots, x_n$ into two groups, in the first of which are those variables x_i for which $\alpha_i = 1$, and the second, all the other variables (for which $\alpha_i = 0$). Equating all the variables of the first group to each other (we rename them y_1), and then equating all the variables of the second group (substituting y_2 for them), we obtain as a result a function of two variables $\omega(y_1,y_2)$. (It may turn out to be a function of just one variable if $(\alpha_1, \dots, \alpha_n)$ is the unit or null sequence.)

It is clear that

$$\omega(1,0) = f(\alpha_1, \dots, a_n);\ \omega(0,1) = f(\overline{\alpha}_1, \dots, \overline{\alpha}_n).$$

Therefore,

$$\omega(0,1) = \omega(1,0),$$

that is, $\omega(y_1, y_2)$ is a non-self-dual function.

It may not be possible to equate any other variables without violating the property of non-self-duality. For example, xy is a non-self-dual function, and the only way of equating variables leads to the self-dual function x.

3.11. Note that the non-self-dual functions of a single variable are constants. By the preceding problem, we may limit our discussion to the case of a function $\omega(x,y)$ of two variables.

Suppose that

$$\omega(\alpha,\beta) = \omega(\overline{\alpha},\overline{\beta}).$$

Then we may consider the function

$$\psi(x,y) = \omega(x^{\alpha},y^{\beta}).$$

This function is a composition of $\omega(x,y)$ and $\overline{x}$. Since $\alpha^{\alpha} = \beta^{\beta} = 1$,

$$\psi(1,1) = \psi(0,0).$$

In $\psi(x,y)$, equate the variables x and y: $\tau(x) = \psi(x,x)$. We have $\tau(1) = \tau(0)$; i.e., $\tau(x)$ is a constant. Substituting this constant in $\overline{x}$, we obtain another constant. Thus, we have represented both constants 0 and 1 in the form of a composition of $\omega(x,y)$ and $\overline{x}$. Recall that constants are non-self-dual functions.

Chapter 4
ARITHMETIC OPERATIONS IN ALGEBRAIC LOGIC

In the Boolean algebra {0,1} the conjunction xy coincides with the arithmetic operation of multiplication over the numbers 0 and 1. Ordinary arithmetic addition extends beyond the set {0,1}, though we could consider addition modulo 2. The logical function which we will denote $x + y$ (without stipulating that addition is performed modulo 2, since we will only encounter this type of addition) thus arises; we define it by the accompanying table.

x	y	$x+y$
1	1	0
0	1	1
1	0	1
1	1	0

Note that $x + y = \overline{x \sim y} = x \,\Delta\, y$.* All the basic arithmetic laws (commutativity, associativity, distributivity of

*In set theory, the symmetric difference (Problem 2.20) corresponds to arithmetic addition. We have already noted the associativity of the symmetric difference (Problem 2.20b); that it is commutative is self-evident. We suggest that the reader think through the proof of the distributivity of intersection relative to the symmetric difference.

multiplication with respect to addition) are satisfied by the addition and multiplication (conjunction) operations we have introduced. Therefore, without making any special stipulations, we will use all the ordinary simplifications to write out the arithmetic expressions.

4.1. Represent $x + y$ in PDNF and PCNF; find $(x + y)^{+}$.

The compositions of the functions xy and $x + y$ and the constants may be considered "polynomials" by virtue of the above remark.

4.2. Prove that every logical function may be represented by arithmetic polynomials (modulo 2).

Let us find the canonical form of a polynomial. By (2.11), $x^n = x$ if $n \geqslant 1$; further, $x + x = 0$.

Definition 4.1. A *Zhegalkin polynomial** is understood to refer to a polynomial which is the sum of a constant and distinct monomials in which all the variables occur linearly only:

$$\Sigma x_{i_1} x_{i_2} \ldots x_{i_k} + a.$$

In each sequence $(i_1, \ldots, i_k)$, all the i_j are distinct, while the summation is taken over some set of distinct sequences. It is convenient to consider the constants as, formally speaking, monomials.

That an arbitrary arithmetic polynomial may be transformed into a Zhegalkin polynomial follows from the remarks presented just before the definition.

4.3. Represent by means of Zhegalkin polynomials: (a) the basic logical operations; (b) $x \vee y \vee z$; (c) $xy \vee yz \vee xz$; (d) $xy\bar{z} \vee x\bar{y}z \vee \bar{x}yz \vee \overline{xyz}$.

4.4. Prove that a representation of functions by means of Zhegalkin polynomials is unique.

Definition 4.2. Functions of the form $x_{i_1} + x_{i_2} + \ldots + x_{i_k} + a$,

*I. I. Zhegalkin. Matem. Sb., Vol. 34, No. 1, pp. 9-28 (1927) (added in translation).

where a is a constant, are said to be *linear.* Linear functions may be written in the form

$$\sum_{i=1}^{n} a_i x_i + a_0, \tag{4.1}$$

where a_i and a_0 are equal to 0 or 1.

4.5. How many linear functions of n variables are there?

Note that every function of one variable is linear.

4.6. Which linear functions are self-dual?

Remark. To prove that some function is linear, it is sufficient to prove that there is a term of degree higher than one in the Zhegalkin representation of the function. The uniqueness of the representation of functions by means of Zhegalkin polynomials is essential here. That is, if the negation of the proposition presented in Definition 4.2 is constructed directly, to prove that the function given in the definition is a linear function it is first necessary to prove that it can be represented in the form (4.1). But this follows from the nonlinearity of its Zhegalkin polynomial, by virtue of its uniqueness (since (4.1) is a Zhegalkin polynomial).

4.7. Which of the functions of Problem 4.3 are linear?

4.8. Prove that a function that can be represented by a Zhegalkin polynomial depends actually on all the variables occurring in it.

4.9. Prove that the following definition of a linear function is equivalent to the one adopted above: A function is linear if and only if it will still depend actually on all its unknown variables after its variables have been arbitrarily equated (cf. Problems 1.9 and 1.10).

4.10. Suppose that we are given an arbitrary nonlinear function. What is the least number of variables possible in the nonlinear function obtained by equating the variables of the original function?

4.11. What is the least number of variables possible in the nonlinear function constructed from the composition of an arbitrary nonlinear function and one of the constants?

4.12. Suppose that we are given some nonlinear function of two variables and negation. Prove that all nonlinear functions of two variables may be obtained from compositions of negation and this function.

Hints

4.3. In (b), (c), and (d), use (a); in (c), simplify the algebra by using (b); in (d) it is a good idea to use the representations $x + y$ and $x + y + 1$ in PDNF.

4.4. First method. It is sufficient to prove that a Zhegalkin polynomial containing at least one monomial other than the constant 0 is not identically 0 (i.e., is not the additive identity). Prove that no linear terms may occur in this polynomial; then by substituting constants for some of the variables so that the polynomial becomes a conjunction or its negation, we arrive at a contradiction.

Second method. As in the proof of this uniqueness of the representation of a function in PDNF (Problem 2.30), find the number of distinct Zhegalkin polynomials of n variables and compare this number with the number of functions of n variables.

4.5. 2^{n+1}.

4.6. Functions in which an odd number of coefficients a_i ($i \neq 0$) is equal to 1.

4.10. It is possible to obtain a nonlinear function of three variables but in general, never of two variables (give an example). In the proof, it is best to use a grouping of terms of the Zhegalkin polynomial relative to some variable already used to solve the preceding problems (Problems 4.8 and 4.9).

4.11. A nonlinear function of two variables may be obtained.

4.12. Write down the general form of a nonlinear function of two variables. Derive conjunction, for example, from this form by means of negation.

Solutions

4.1. $x + y = x\bar{y} \vee \bar{x}y = (x \vee y)(\bar{x} \vee \bar{y})$; $(x + y)^{+} = x+y+1$.

4.2. It is sufficient to express conjunction and negation by means of the arithmetic operations (Problem 1.14). But xy is itself an arithmetic operation, and $\bar{x} = x + 1$.

4.3. (a)

$$xy = xy; \quad x \vee y = \overline{\bar{x}\bar{y}} = (x+1)(y+1) + 1$$

$$= xy + x + y + 1 + 1 = xy + x + y,$$

$$\bar{x} = x + 1;$$

$$x \to y = \bar{x} \vee y = (x+1) + y + (x+1) + y$$

$$= xy + y + x + 1 + y = xy + x + 1;$$

$$x \sim y = x + y + 1.$$

(b)
$$x \vee y \vee z = (xy+x+y) \vee z$$

$$= (xy+x+y)z + xy + x + y + z$$

$$= xyz + xy + xz + yz + x + y + z.$$

(c) Use the preceding example:

$$xy \vee yz \vee xz$$

$$= x^2y^2z^2 + xy^2z + x^2yz + xyz^2 + xy + yz \vee xz$$

$$= 4xyz + xy + yz + xz = xy + yz + xz.$$

(d) Note that $x\bar{y} \vee \bar{x}y = x + y$. (Problem 4.1). Analogously, $xy \vee \bar{x}\bar{y} = x + y + 1 = \overline{x + y}$. Then

$$xy\bar{z} \vee x\bar{y}z \vee \bar{x}yz \vee \bar{x}\bar{y}\bar{z} = z(x\bar{y} \vee \bar{x}y) \vee z(xy \vee \bar{x}\bar{y})$$

$$= z(x+y) \vee \bar{z}\,\overline{(x+y)} = x + y + z + 1.$$

4.4. *First Proof.* If there were two distinct ways of representing some function by means of Zhegalkin polynomials, then, by equating these polynomials and transposing all the monomials to the same side (bearing in mind that $-x = x$, since $x + x = 0$), we would obtain a nontrivial Zhegalkin polynomial that is identically zero. Thus, the problem devolves to deciding on the uniqueness of the representation of the additive identity by means of a Zhegalkin polynomial.

Suppose that some Zhegalkin polynomial is identically zero. If a function is identically zero, it preserves this property under any substitution of constants for the variables. Suppose that there is a linear term x_1 in the polynomial. We substitute zeros for all the variables other than x_1, obtaining $x_i + a$, where a is a constant. This function is not identically zero, so that the initial polynomial is not identically zero. Thus, there are no linear terms in our polynomial. Let us consider some arbitrary monomial of least degree (there may be several such monomials), for the sake of definiteness the variables x_1, x_2, ... , x_k (this may be achieved by renaming the variables) and substitute zeros in place of all the other variables. All the monomials other than the one we have picked out contain at least one of these variables and therefore vanish. As a result, we obtain the polynomial $x_1 \ldots x_k + a$ which is not identically zero. Thus, there cannot be any monomials of degree higher than zero in the initial polynomial.

Second Proof. Let us find the number of Zhegalkin polynomials of n variables, first computing the number of monomials. With each monomial we associate a binary sequence of length n in which units occur at those entries which correspond to the number of variables that occur in the monomial; the null sequence is associated with the constant 1; the resulting correspondence is one-to-one. Therefore, there are 2^n distinct monomials (including unit). We then number the monomials in some way. We associate with the Zhegalkin polynomials sequences of length 2^n in which the numbers of monomials occurring in the polynomial are labeled by units. Therefore, there are 2^{2^n} distinct Zhegalkin polynomials, and since there are 2^{2^n} distinct functions (Problem 2.2), the representation is unique (cf. second solution of Problem 2.30).

4.5. A linear function is determined by the values of the coefficients a_i and the free term a_0, i.e., by a binary sequence of length $(n+1)$. Therefore, there are 2^{n+1} linear functions.

4.7. $\bar{x}$, $x \sim y$, $xy\bar{z} \vee x\bar{y}z \vee \bar{x}yz \vee \overline{xyz}$.

4.8. Suppose that x_1 is such a variable. We group together terms in which x_1 occurs, and extract x_1. We obtain*

$$f(x_1, \ldots, x_n) = x_1\omega(x_2, \ldots, x_n) + \psi(x_2, \ldots, x_n)$$

where the function ω is not identically zero, since otherwise x_1 would not occur in the polynomial for f (by virtue of the uniqueness of the Zhegalkin polynomial). We take the values of the variables $x_2, \ldots, x_n$ on which ω is equal to 1. Then the value of f will depend on the value of x_1.

4.9. Suppose that we are given the linear function (4.1). Variables occurring in (4.1) with coefficients equal to 1 are (by virtue of Problem 4.8) actual variables. If we equate some of the variables, we obtain a function in which variables which have not been equated will occur with the same coefficients as in the initial function. That is, the actual variables remain actual, and necessity is proved.

Let us now prove sufficiency. Suppose that f is a nonlinear function; we consider its Zhegalkin polynomial, and let x_1 be a variable that occurs in some term with exponent greater than 1. Let us consider the same representation as in the preceding problem. In this representation, the function ω is not identically 0 or 1 (otherwise, x_1 would not occur in terms with exponent greater than 1). In place of $x_1, \ldots, x_n$, we substitute some set on which ω is zero. As a result, we obtain a constant which depends formally on x_1, even though f depends actually on x_1 (Problem 4.8).

4.10. Suppose that $f(x_1, \ldots, x_n)$ is a nonlinear function and, for the sake of definiteness, suppose that x_1 occurs in some monomial (of degree greater than 1) of its Zhegalkin polynomial. As in Problems 4.8 and 4.9, we consider the decomposition

---

*This decomposition may be considered a Zhegalkin polynomial in the variable x_1 (with coefficients that depend on the remaining variables).

$$f(x_1, \ldots, x_n) = x_1\omega(x_2, \ldots, x_n) + \psi(x_2, \ldots, x_n).$$

The function ω is not identically equal to zero or unit (cf. solution to Problem 4.9). We let a denote the value of ω on the null set: $\omega(0, \ldots, 0) = a$. By what we have said above, we have found a sequence $(\alpha_2, \ldots, \alpha_n)$ on which ω takes some other value: $\omega(\alpha_2, \ldots, \alpha_n) = \bar{a}$. We divide the variables $x_2, \ldots, x_n$ into two groups. The first group consists of those variables for which $\alpha_i = 0$, and the second group, those variables for which $\alpha_i = 1$. We equate the variables in each of these groups, obtaining a function $\tilde{\omega}(y_1, y_2)$ such that $\tilde{\omega}(0,0) = a$ and $\tilde{\omega}(0,1) = \bar{a}$. An analogous process of equating is performed for f.

Further,

$$\tilde{f}(x_1,y_1,y_2) = x_1\tilde{\omega}(y_1,y_2) + \tilde{\psi}(y_1,y_2).$$

The function $\tilde{f}$ will be nonlinear, since $\tilde{\omega}$ is not identically constant, so that x_1 will occur in some term of the Zhegalkin polynomial for $\tilde{f}$ with exponent greater than 1 (the Zhegalkin polynomial for $\tilde{f}$ may be obtained by substituting the polynomials for $\tilde{\omega}$ and $\tilde{\psi}$). Thus, we have obtained a nonlinear function of three variables.

Remark 1. In our discussion of the function ω, we could have started with a unit set rather than a null set; it is only important that all the elements of the set be equal.

Remark 2. We have in fact proved that a necessary and sufficient condition for $f(x_1, \ldots, x_n)$ to be nonlinear is the existence of a variable (say, x_1) such that

$$f(x_1, \ldots, x_n) = x_1\omega(x_2, \ldots, x_n) + \psi(x_2, \ldots, x_n),$$

where the function ω is not identically constant.

Now let us prove that, in general, it is impossible to obtain a nonlinear function of two variables by equating variables. Consider the following nonlinear function of three variables: $xy \vee yz \vee xz = xy + yz + xz$ (Problem 4.3c)). It is symmetric in all its variables. As an example, equate the variables x and y. We obtain $x \vee xz \vee xz = x$ (cf. (2.20), i.e., a nonlinear function. Equating any other two variables leads to the same result. Equating all three variables does not lead to our goal.

4.11. Using the same notation as in our solution of the preceding problem, suppose that we have one more constant 0 (if we have the unit, we will have to begin with the unit set in 4.10; cf. Remark 1 following the solution of this problem).

In place of y_1, we substitute zero:

$$\tilde{f}(x_1, 0, y_2) = x_1\tilde{\omega}(0, y_2) + \tilde{\psi}(0, y_2).$$

Suppose that $\hat{\omega}(y_2) = \tilde{\omega}(0, y_2)$. Then $\hat{\omega}(0) = a$ and $\hat{\omega}(1) = \bar{a}$, i.e., $\hat{\omega}$ is not a constant, that is, $\hat{f}(x_1, y_2) = \tilde{f}(x_1, 0, y_2)$ is a nonlinear function of two variables (Remark 2 following the solution of Problem 4.10). A further reduction in the number of variables is not possible, since there are no nonlinear functions with fewer variables.

4.12. The general form of a nonlinear function of two variables is as follows:

$$\omega(x,y) = xy + \alpha x + \beta y + \gamma.$$

We first obtain the conjunction xy. Let us eliminate linear terms. Suppose that $\alpha = 1$. Then substitute $\bar{y}$ for y:

$$\omega(x,\bar{y}) = x(y+1) + \alpha x + \beta(y+1) + \gamma$$

$$= xy + (\alpha+1)x + \beta y + (\beta+\gamma).$$

Since $\alpha = 1$, $\alpha + 1 = 0$ and the term containing x vanishes. It is important here that the coefficient of y does not change. If $\beta = 1$, the term y is analogously eliminated. The free term changes after these substitutions. If it is then equal to 1, negation must be applied to the entire function, and as a result we are left with the term xy alone.

Passing from one nonlinear function of two variables to another is carried out in the general case in a similar fashion. The case we have considered will be sufficient for our purposes below.

Chapter 5
MONOTONE LOGICAL FUNCTIONS

We order the set $\{0,1\}$ by letting $0 < 1$. Since it is necessary to deal with functions of several variables, we introduce a partial ordering of binary sequences of the same length.

Definition 5.1. Suppose that $\alpha = (\alpha_1, \ldots, \alpha_n)$ and $\beta = (\beta_1, \ldots, \beta_n)$ are two binary sequences. We will say that α *precedes* ("is smaller than") β (denoted $\alpha \prec \beta$) if $\alpha_i \leq \beta_i$ for all i and if strict inequality holds for at least one i. We will write $\alpha \preceq \beta$ if $\alpha \prec \beta$ or if the sequences α and β coincide.

This ordering is only a partial ordering as it does not allow us to compare arbitrary sequences.

Definition 5.2. A logical function $f(x_1, \ldots, x_n)$ is said to be *monotone* if for any pair of sequences $\alpha = (\alpha_1, \ldots, \alpha_n)$ and $\beta = (\beta_1, \ldots, \beta_n)$ such that $\alpha \prec \beta$, we have

$$f(\alpha_1, \ldots, \alpha_n) \leq f(\beta_1, \ldots, \beta_n).$$

Note that the requirement $f(\alpha) \preceq f(\beta)$ if $\alpha \preceq \beta$ is logically equivalent to the above assertion.

It would make sense to refer to logical functions that satisfy this condition as "nondecreasing"; however, since we will be concerned with "nonincreasing" functions (with two exceptions; cf. Problems 5.8 and 6.5), we will say that these functions are "monotone".

5.1. Give a definition of a nonmonotone function.

5.2. (a) Are constants monotone functions?
(b) Which of the basic logical operations are monotone?

5.3. Which linear functions are monotone?

5.4. Prove that a monotone function that does not preserve zero (unit) is identically equal to unit (zero) (cf. Problem 2.3).

5.5. List all the monotone functions of two variables.

In all the preceding cases, we were able to find in a relatively simple manner how many functions there were in a particular class. It is, however, quite difficult to determine how many monotone functions of n variables there are, and in fact, the problem remains unsolved; this number has been determined only for special cases of small n. Comparatively good estimates for the number of monotone functions have recently been obtained in the general case [1].

5.6. Which of the following functions are monotone: (a) $xy \vee xz \vee \bar{x}z$; (b) $x \to (x \to y)$; (c) $\overline{x \vee y} \sim \bar{x} \vee \bar{y}$; (d) $\overline{x \vee y} \sim \overline{xy}$; (e) $xy \vee x \vee \bar{x}z$; (f) $xy \vee yz \vee xz$?

We now attempt to formulate the basic property of monotone functions.

5.7. Prove that a composition of monotone functions is monotone.

5.8. Prove that a composition of monotonically decreasing functions may be neither decreasing nor even increasing.

Problems 5.7 and 5.8 show why we do not consider decreasing or nonincreasing functions in algebraic logic. In fact, in algebraic logic we are usually interested in "hereditary" properties of functions, i.e., properties that remain true after application of a composition.

By virtue of Problem 5.7, we may readily verify the monotonicity of a broad class of functions that may be represented by formulas. To see how simple a proof of monotonicity often is, the reader need only look at Problem 5.6 once again.

From Problem 5.7 (as well as Problem 5.2), it follows in particular that a composition of conjunctions and disjunctions is always a monotone function. The converse assertion is also true.

5.9. Prove that a function that is not identically constant is monotone if and only if it may be represented in the form of a composition of conjunctions and disjunctions.

5.10. Prove that a function dual to a monotone function is itself monotone.

5.11. Prove that those functions (and only those functions) that are either constants or admit a representation in CNF or DNF without any negations are monotone.

The representation of a monotone function in DNF or CNF is not unique. Let us now construct canonical representations.

Definition 5.3. A disjunctive normal form that does not contain any negations is said to be *proper* if:

(a) it does not contain isolated elementary conjunctions;

(b) all the elementary conjunctions occurring in it are proper;

(c) no one elementary conjunction absorbs another, i.e., entirely contains variables that occur in another conjunction.*

5.12. Prove that a monotone function may be represented in the form of a proper DNF, and that this representation is unique.

5.13. By analogy with DNF, give a definition of a proper CNF and prove the uniqueness of the representation of a monotone function in the form of a proper CNF.

Now let us prove several properties of nonmonotone functions analogous to the properties of non-self-dual and

*Condition (a) may be considered a special case of (c). The latter condition asserts that a regular DNF may not be simplified by means of the absorption law (2.20).

nonlinear functions obtained in the last two section.

5.14. Suppose that we wish to reduce the number of variables of a nonmonotone function by equating its variables. What is the least number of variables the resulting function may have and still be nonmonotone?

5.15. Suppose that we are given an arbitrary monotone function and any one of its constants. Consider the nonmonotone function which is the composition of this function and the constant; what is the least number of variables in such a composition?

5.16. Prove that a negation may be obtained from the composition of any nonmonotone function and constants.

Hints

5.2. (b) xy, $x \vee y$.

5.3. 0, 1, x.

5.5. 0, 1, x, y, xy, $x \vee y$. It is worth bearing in mind that a monotone function of two variables that is not a constant will be equal to zero on the null sequence and to the unit on the unit sequence, and may be specified arbitrarily on the other two sequences.

5.6. Whenever possible, the formulas must first be simplified. Note, too, that if two sequences are commensurate, there are fewer units in the preceding sequence (this is a necessary, though of course, not a sufficient condition). To prove monotonicity, it is often useful to consider a set of sequences on which the given function equals zero (or unit, depending on which sequences are longer) and to prove that the remaining sequences are either incommensurable with the given sequences or are longer (in the dual case, shorter). To prove nonmonotonicity, it is sufficient to consider a pair of sequences (cf. Problem 5.1).

Answer: (a), (d), (e), and (f).

5.7. Prove by induction on the rank of the composition.

5.9. Prove by induction on the number of variables. Expand $f(x_1, \dots, x_n)$ with respect to the last variable in the PDNF (2.16):

$$f(x_1, \dots, x_{n-1}, x_n) = \omega(x_1, \dots, x_{n-1})x_n \vee \psi(x_1, \dots, x_{n-1})\overline{x_n},$$

and prove the logical equivalence

$$f(x_1, \dots, x_{n-1}, x_n) = \omega(x_1, \dots, x_{n-1})x_n \vee \psi(x_1, \dots, x_{n-1}).$$

5.10. This assertion may be proved directly starting from the definitions of dual and monotone functions, though it is better to use Problem 5.9 and the law of duality.

5.11. Immediate corollary to Problem 5.9 and the algorithms for transforming formulas into DNF and CNF (cf. sec. 2.6 and 2.7).

5.14. A nonmonotone function of three variables may be obtained, but, in general, not one with fewer variables (give an example).

5.15. In general, two variables.

Solutions

5.1. The function $f(x_1, \dots, x_n)$ is nonmonotone if there exists a pair of sequences α and β such that $\alpha \prec \beta$ and

$$f(\alpha_1, \dots, \alpha_n) > f(\beta_1, \dots, \beta_n).$$

5.2. (a) Yes, since for any pair of sets $f(\alpha) = f(\beta)$.

(b) xy is monotone, since it is equal to 1 only on the set (1, 1) which all the other sets precede.

$x \vee y$ is also monotone, since it is equal to 0 only on the sequence (0, 0), which precedes all the other sequences;

$\bar{x}$ is nonmonotone since if $\alpha = (0)$ and $\beta = (1)$, we will have $\alpha \prec \beta$, but $\bar{\alpha} \succ \bar{\beta}$;

$x \to y$ is nonmonotone; suppose that $\alpha = (0, 0)$, $\beta = (1, 0)$, and $\alpha \prec \beta$; then $(0 \to 0) = 1 > (1 \to 0) = 0$;

$x \sim y$ is nonmonotone, as is clear from the sequences $\alpha = (0, 0)$ and $\beta = (1, 0)$.

5.3. We consider two cases.

(a) Suppose that the free term is equal to 1 and that the linear function contains at least one actual variable. Then by comparing the null set and the set containing precisely one unit at the spot corresponding to the actual variable, it follows that the function is nonmonotone. The case of the constant 1 remains.

(b) If the free term is equal to 0 and the linear function contains at least two actual arguments, nonmonotonicity follows if we consider the sequence that substitutes for the unit only these arguments and also the sequence that substitutes for the unit only one of these arguments. The functions 0 and x remain.

5.4. Since the null sequence precedes all other sequences, if a monotone function is equal to unit on this sequence, it is identically equal to unit (it may be less on all other sequences). The dual case may be considered analogously.

5.5. If we exclude the case of a constant, the monotone function $f(x, y)$ is equal to 0 on the null sequence and to 1 on the unit sequence. The two other sequences (0, 1) and (1, 0) follow the null sequence and precede the unit sequence, and are not commensurable. Therefore, the function may be assigned any values on these sequences without violating monotonicity. These values may be determined in any one of four ways, so that four monotone functions of two variables may be found that are not constants: x, y, xy, and $x \vee y$.

5.6. (a) We simplify the formula: $xy \vee xz \vee \bar{x}z = xy \vee z$. This function is equal to the zero on the sequences (0, 0, 0), (0, 1, 0), and (1, 0, 0). Except for (0, 0, 1), all the other sequences contain at least two units, that is, they can only be greater. The sequence $(0, 0, 1) \succ (0, 0, 0)$, and is not commensurable with the other sequences. That is, our function is monotone.

(b) The function is nonmonotone; compare the values of the function on the sequence $(0, 0) \prec (1, 0)$.

(c) By comparing the values on the sequences $(0, 0) \prec (1, 0)$, we find that the function is nonmonotone.

(d) $\overline{x \vee y} \sim \bar{x}\bar{y}$ is identically equal to unit, since $\overline{x \vee y} = \bar{x}\bar{y}$.

(e) $xy \vee x \vee \overline{xz} = x \vee z$ is a monotone function (Problem 5.2).

(f) The function is monotone, since it is equal to zero on sequences containing fewer than two units, and must be greater on all other sequences.

5.7. Suppose that we are given a system of monotone functions Φ. We must prove the monotonicity of functions that are compositions of functions from Φ.

By the condition, the assertion is true for compositions of rank 0, i.e., for functions from Φ. Suppose that it has been proven for compositions of rank k. Let us prove that it is true for compositions of rank $k+1$. Suppose that $\omega(x_1, \dots, x_n)$, $\psi(y_1, \dots, y_\ell) \in \Phi^{(k)}$. We must prove (cf. Definition 2.3) that the functions

$$\omega(x_1, \dots, x_{i-1}, y, x_{i+1}, \dots, x_k);$$

$$F(x_1, \dots, x_{i-1}, x_{i+1}, \dots, x_n; y_1, \dots, y_\ell)$$
$$= \omega(x_1, \dots, x_{i-1}, \psi(y_1, \dots, y_\ell), x_{i+1}, \dots, x_n)$$

are monotone. Recall that y and y_i may, in particular, coincide with some of the variables x_j. That the first of these functions is monotone follows by definition from the monotonicity of ω. Let us prove that F is monotone. We consider two commensurable sequences of values of its arguments:

$$\gamma' = (\alpha'_1, \dots, \alpha'_{i-1}, \alpha'_{i+1}, \dots, \alpha'_n; \beta'_1, \dots, \beta'_\ell);$$

$$\gamma'' = (\alpha''_1, \dots, \alpha''_{i-1}, \alpha''_{i+1}, \dots, \alpha''_n; \beta''_1, \dots, \beta''_\ell).$$

Suppose that $\gamma' \prec \gamma''$. Let us prove that $F(\gamma') \leqslant F(\gamma'')$. We have

$$F(y') = \omega(\delta'), \quad \text{where } \delta'_j = \alpha'_j \text{ if } j \neq i, \ \delta'_i = \psi(\beta');$$

$$F(y'') = \omega(\delta''), \quad \text{where } \delta''_j = \alpha''_j \text{ if } j \neq i, \ \delta''_i = \psi(\beta'').$$

Since ψ is a monotone function, and since $y' \preceq y''$ implies that $\beta' \preceq \beta''$, we have $\delta' \preceq \delta''$, i.e., $\omega(\delta'') = F(y'') \leqslant \omega(\delta'') = F(y'')$, since ω is a monotone function. Since any function from $\Phi^{(k+1)}$ may be represented in one of these two forms, the proof is complete.

5.8. The negation $\overline{x}$ is a monotonically decreasing function. Its composition with itself ($\overline{\overline{x}} = x$) is monotonically increasing.

5.9. Sufficiency has already been proved; necessity will be proved by induction on the number of variables. We begin with $n = 0$, in which case we are dealing only with constants; for $n = 1$, we further prove that $x = x \quad x$. Suppose that the assertion we wish to prove is valid for monotone functions of $(n\text{-}1)$ variables. Let us prove it for functions of n variables. Suppose that $f(x_1, \ldots, x_{n-1}, x_n)$ is a monotone function of n variables. We expand it in PDNF with respect to the last variable:

$$f(x_1, \ldots, x_{n-1}, x_n) = \omega(x_1, \ldots, x_{n-1})x_n$$
$$\vee \ \psi(x_1, \ldots, x_{n-1})\overline{x}_n ,$$

where

$$\omega(x_1, \ldots, x_{n-1}) = f(x_1, \ldots, x_{n-1}, 1),$$

$$\psi(x_1, \ldots, x_{n-1}) = f(x_1, \ldots, x_{n-1}, 0).$$

The functions ω and ψ are monotone, since they are compositions of a monotone function f and constants that are monotone functions (cf. Problem 5.7). Let us prove that the representation

$$f(x_1, \ldots, x_{n-1}, x_n) = \omega(x_1, \ldots, x_{n-1})x_n \qquad (5.1)$$
$$\vee \ \psi(x_1, \ldots, x_{n-1})$$

is valid. We must prove the logical equivalence

$$x_n\omega \vee \bar{x}_n\psi = x_n\omega \vee \psi.$$

Let us compare the set of sequences on which the left and right sides vanish. If the right side vanishes on some sequence $\alpha = (\alpha_1, \dots, \alpha_{n-1}, \alpha_n)$, then $\alpha_n\omega(\alpha_1, \dots, \alpha_{n-1}) = 0$ and $\psi(\alpha_1, \dots, \alpha_{n-1}) = 0$; that is, $\bar{\alpha}_n\psi(\alpha_1, \dots, \alpha_{n-1}) = 0$ as well, i.e. the left side is also equal to zero. Now suppose that the left side is equal to zero on the sequence $\alpha = (\alpha_1, \dots, \alpha_{n-1}, \alpha_n)$. Then $\alpha_n\omega(\alpha_1, \dots, \alpha_{n-1}) = 0$. Further, $f(\alpha_1, \dots, \alpha_{n-1}, \alpha_n) = 0$, so that $\psi(\alpha_1, \dots, \alpha_{n-1}) = f(\alpha_1, \dots, \alpha_{n-1}, 0)$, since $(\alpha_1, \dots, \alpha_{n-1}, \alpha_n) \geqslant (\alpha_1, \dots, \alpha_{n-1}, 0)$. As a result, the right side is equal to zero and the representation (5.1) for monotone functions is proved.

Let us now prove that f may be represented in the form of a composition of disjunctions and conjunctions, where f is not a constant. We first suppose that ω and ψ are also not constants. Then, since they are monotone and depend on $(n-1)$ variables, they may be considered (by the induction hypothesis) as compositions of disjunctions and conjunctions. From (5.1), it follows that f then possesses this property as well.

Now let us consider the case in which one or both functions ω and ψ are constants. Note that

$$\omega(x_1, \dots, x_{n-1}) \geqslant \psi(x_1, \dots, x_{n-1}).$$

Therefore, if ω is identically zero, ψ and f are identically zero; if* $\psi = 1$, then $f = 1$. If $\psi = 0$, whereas $\omega = 1$, then $f(x_1, \dots, x_n) = x_n$. But if $\psi = 0$ and $\omega(x_1, \dots, x_{n-1})$ is not a constant, or if $\psi = 1$ and $\omega(x_1, \dots, x_{n-1})$ is not a constant, by the induction hypothesis and (5.1) the desired result follows at once.

We advise the reader to carry out the proof using dual formulas. In particular, prove the following logical equivalence dual to (5.1):

$$f(x_1, \dots, x_{n-1}, x_n) = (\omega(x_1, \dots, x_{n-1}) \vee x_n)\psi(x_1, \dots, x_{n-1}).$$

*Recall that equivalence between functions is understood everywhere as an identity (cf. page 24).

where f is a monotone function, while $\omega(x_1, \dots, x_{n-1}) = f(x_1, \dots, x_{n-1}, 0)$ and $\psi(x_1, \dots, x_{n-1}) = f(x_1, \dots, x_{n-1}, 1)$. Equality (5.1) should be proved directly, and also derived from (5.1) by means of the law of duality.

5.10. Let us use the second suggestion mentioned in the hint. Since the constants are dual, we may limit ourselves to the case of a monotone function which is not a constant. By Problem 5.9, it may be represented in the form of a composition of conjunctions and disjunctions. To obtain a dual function using the law of duality, it is necessary to replace conjunctions by disjunctions and conversely, that is, by Problem 5.7, a monotone function is again obtained.

5.12. Consider two equivalent regular DNF. Let us prove that they coincide. Suppose that $x_{i_1} \ldots x_{i_k}$ is an arbitrary elementary conjunction occurring in the first DNF. It is sufficient to prove that it occurs in the second DNF. We substitute 0's for all the variables other than those occurring in this conjunction. Then all the conjunctions in the two DNF of degree at least that of the chosen conjunction vanish, since it must be possible to find variables in these conjunctions that do not occur in the chosen conjunction. In the first DNF, all the other conjunctions vanish, since otherwise they would be absorbed by the chosen conjunction (by Definition 5.3c). If there were no conjunction of the form $x_{i_1} \dots x_{i_k}$ in the second DNF, after substituting this term all that remains would be of lesser degree and would absorb it. Let us consider one of these conjunctions and substitute units for the variables occurring in it and zeros for the remaining variables. Then the first DNF vanishes and the second DNF will be equal to 1.

5.13. In a proper CNF there are distinct proper disjunctions that do not absorb each other.

5.14. Suppose that $f(x_1, \dots, x_n)$ is a nonmonotone function and let the sequences $\alpha = (\alpha_1, \dots, \alpha_n) \prec \beta = (\beta_1, \dots, \beta_n)$ be defined by

$$f(\alpha_1, \dots, \alpha_n) > f(\beta_1, \dots, \beta_n),$$

i.e.,

$$f(\alpha_1, \dots, \alpha_n) = 1; \quad f(\beta_1, \dots, \beta_n) = 0.$$

We divide the variables $x_1, \dots, x_n$ into three groups. In the first group are found variables x_i such that $\alpha_i = 0$ and $\beta_i = 1$; in the second group, variables x_i such that $\alpha_i = \beta_i = 0$; and in the third group, variables x_i such that $\alpha_i = \beta_i = 1$. Since $\alpha \prec \beta$, it is not possible for $\alpha_i = 1$ and $\beta_i = 0$. We equate variables within each such group (replacing them by the variables y_1, y_2, and y_3, respectively), obtaining a nonmonotone function of three variables $\omega(y_1, y_2, y_3)$: $\omega(0, 0, 1) = 1$; $\omega(1, 0, 1) = 0$.

It may be impossible to further decrease the number of variables, as can be seen from the example of a nonmonotone function of three variables: $x + y + z$. Arbitrarily equating variables leads to a function of a single variable that coincides with its argument (for example, x) and is monotone.

5.15. Suppose that we are given the constant 0 as an example. In $\omega(y_1, y_2, y_3)$, we make the substitution $y_2 = 0$, obtaining a nonmonotone function of two variables. The number of variables cannot, in general, be decreased, as is shown by the example of the function $x + y$.

Analogously, we may consider the case of the constant 1. Here, the number of variables in the function $x + y + 1$ cannot be decreased.

5.16. We substitute $y_2 = 0$ and $y_3 = 1$ for the arguments of the function $\omega(y_1, y_2, y_3)$ introduced in solving Problem 5.14, obtaining $\overline{y}_1$.

Chapter 6
FUNCTIONALLY CLOSED CLASSES AND POST'S THEOREM

The representability of the logical functions in one form or another (DNF, CNF, Zhegalkin polynomials) is among the most important topics we have discussed. We have in fact spoken everywhere of the possibility of representing the logical functions in the form of compositions of some fixed set of functions (Definition 2.3). The general problems may be stated as follows.

Suppose that we are given some system Φ of logical functions $\omega_1, \dots, \omega_k$.* Can we say that for every logical function, there exists a composition of functions from Φ that is logically equivalent to this function?**

Definition 6.1. A system of functions $\Phi = \{\omega_1, \dots, \omega_k\}$ is said to be *complete* if every logical function may be represented

*For the sake of simplicity, we are considering finite systems of functions, though as will be clear from what follows, this is not essential.

**By the results of sec. 2.4, this problem may be suitably interpreted for Boolean operations on any regular Boolean algebra. For example, in the algebra of sets the problem may be stated thus: In terms of which set-theoretic operations is it possible to express all such operations? We suggest the reader translate the examples of Problem 6.2 into the language of set theory.

directly by a composition of functions from Φ.

The completeness of the system of functions $\{xy, x \vee y, \overline{x}\}$ follows from the representability of functions in the form of PDNF. The case of functions that are identically zero does not lead to any restrictions, since $0 = x\overline{x}$.

6.1. The representability of functions by means of Zhegalkin polynomials is associated with the completeness of which system of functions?

6.2. Prove the completeness of the following systems of functions: (a) $xy, \overline{x}$; (b) $x \vee y, \overline{x}$; (c) $xy, x+y, 1$; (d) $\overline{x \vee y}$; (e) $\overline{xy}$; (f) $x + y, x \vee y$; (g) $x+y+z, xy, 0, 1$; (h) $x \to y, \overline{x}$; (i) $x \to y$, 0.

6.3. Prove that if some system of functions $\Phi = \{\omega_1, \ldots, \omega_k\}$ is complete, the system of dual functions $\Phi^+ = \{\omega_1^+, \ldots, \omega_k^+\}$ is also complete.

6.4. Prove that the following systems of functions are not complete: (a) $\overline{x}$; 1; (b) $xy, x \vee y$; (c) $x+y, \overline{x}$; (d) $xy \vee yz \vee xz, \overline{x}$; (e) $xy \vee yz \vee xz$, 0, 1.

From an analysis of the solution of Problem 6.4 we are led to the following approach for proving that a system of functions Φ is not complete. Some property must be found which is preserved under composition ("hereditary" property) and which all the functions of Φ, though not all the logical functions, possess. In fact, in that case a function that does not possess this hereditary property cannot be represented as a composition of functions from Φ. In our study of the hereditary properties of functions, it will be useful to apply the concept of a functionally closed class.

Definition 6.2. Every collection T of logical functions closed relative to composition (i.e., such that any composition of functions from T again belongs to T) is called a *functionally closed class.*

Obviously, the collection of functions that possesses some hereditary property is a functionally closed class and, conversely, a property that pertains to some functionally closed classes is hereditary. In other words, the two concepts reduce to each other.

6.5. Which of the following systems of functions are functionally closed classes:

(a) functions of one variable;
(b) functions of two variables;
(c) all the logical functions;
(d) linear functions;
(e) self-dual functions;
(f) monotone functions;
(g) monotonically decreasing functions;
(h) functions that preserve the additive identity (cf. Problem 2.3);
(i) functions that preserve the multiplicative identity;
(j) functions that preserve both the additive and multiplicative identities;
(k) functions that preserve the additive but not the multiplicative identity.

6.6. (1) Prove that an intersection of functionally closed classes is a functionally closed class.

(2) Prove that the set of functions dual to functions from a functionally closed class forms a functionally closed class (dual class).

6.7. Is a union of functionally closed classes a functionally closed class?

Definition 6.3. All the functionally closed classes, other than the empty class and the set of all the logical functions, are called *intrinsic functionally closed classes.*

6.8. Prove that the complement of an intrinsic functionally closed class (the collection of functions in which it does not occur) may not be a functionally closed class.

Definition 6.4. A minimal complete system of functions (i.e., a complete system of functions that becomes incomplete upon the removal of any one function) is called a *basis.*

6.9. Prove that the complete systems of Problem 6.2 are all bases.

Thus, a system of functions is complete if and only if for every intrinsic functionally closed class contained in it, a function can be found that does not occur in this class. We easily note that this condition is also sufficient.

6.10. Prove that a system of functions Φ is complete if and only if for every functionally closed class that does not coincide with the set of all functions, a function may be found in Φ that does not belong to this class.

It would be too much to expect that the assertion of Problem 6.10 could be used as a completeness criterion for a system of functions, since this would mean searching through all the functionally closed classes. Though the set of classes might be quite "accessible" (see the next section for a fuller discussion), it is not necessary to review all the classes to determine completeness. We only need consider the maximal functionally closed classes.

Definition 6.5. An intrinsic functionally closed class is said to be *pre-complete* if it is not contained in any functionally closed class other than itself and the class of all the logical functions.

6.11. Suppose that it is known that every intrinsic functionally closed class is contained in some pre-complete class.* Prove that a system of functions Φ is complete if and only if a function may be found for every pre-complete class in Φ that does not occur in Φ.

It turns out that all the pre-complete classes may be easily enumerated. They are $\mathbf{P}_0$, the class of functions that preserve the additive identity (zero) $\mathbf{P}_1$, the class of functions that preserve the multiplicative identity (unit) **L**, the class of linear functions; **M**, the class of monotone functions; and **S**, the class of self-dual functions.

It is more useful to us to prove the completeness criterion directly (the criterion may be obtained from the above result and Problem 6.11), rather than show that these classes are

*This assertion is proved below (Problem 6.18). In principle, there could exist an infinite sequence of ever-bigger functionally closed classes not entirely contained in any intrinsic functionally closed class.

pre-complete. By contrast, starting with the completeness criterion we may prove the pre-completeness of the classes $\mathbf{P}_0$, $\mathbf{P}_1$, **L**, **M**, and **S**, and the fact that every intrinsic functionally closed class is contained in one of these classes. We first prove an auxiliary assertion.

6.12. Prove that by equating the variables of every function that does not preserve the additive (multiplicative) identity, it is possible to obtain a function of one variable that possesses this property, i.e., $\overline{x}$ or 1 (correspondingly, $\overline{x}$ or 0).

We now formulate the completeness criterion.

6.13. Prove Post's theorem: A system of functions $\Phi = \{\omega_1, \ldots, \omega_n\}$ is complete if and only if for every one of the classes $\mathbf{P}_0$, $\mathbf{P}_1$, **L**, **M**, and **S** a function ω_i may be found in Φ that does not belong to Φ.

To check whether the premises of Post's theorem are satisfied by some system of functions $\Phi = \{\omega_1, \ldots, \omega_k\}$, we will compile tables, called Post tables. They have the following form.

Post Table

	$\mathbf{P}_0$	$\mathbf{P}_1$	S	L	M
ω_1					
ω_2					
...	...	...	...	...	...
ω_{k-1}					
ω_k					

In each cell of a Post table there is a plus or minus, depending on whether the function occurring in the given row occurs or does not occur in the given column. By Post's theorem, a system of functions is complete if and only if there is at least one minus in each column.

6.14. Solve Problem 6.2 using Post's theorem.

6.15. Using Post's theorem, formulate the incompleteness

criterion for a system of functions.

6.16. Prove that the following systems of functions are incomplete and decide whether the conditions of Post's theorem are essential: (a) 0, xy, $x+y+z$; (b) 1, xy, $x+y+z$; (c) $\overline{xy} \vee \overline{xz} \vee \overline{yz}$; (d) 0, 1, $x+y$; (e) 0, 1, xy.

6.17. Prove that none of the classes $\mathbf{P}_0$, $\mathbf{P}_1$, **S**, **L**, and **M** is contained in any other of these classes.

6.18. Prove that every intrinsic functionally closed class is contained in one of the classes $\mathbf{P}_0$, $\mathbf{P}_1$, **S**, **L**, or **M**.

6.19. Prove that the functionally closed classes $\mathbf{P}_0$, $\mathbf{P}_1$, **S**, **L**, and **M** are pre-complete and that there are no other pre-complete classes.

It is clear that by Post's theorem it is possible to determine whether a complete system is a basis. Again, this may be conveniently done using Post tables.

Now let us make some general remarks about bases, drawing on Post's theorem. We begin with a very simple idea.

6.20. Prove that a basis may not contain more than five functions.

In fact, we have the following more exact result:

6.21. Prove that a basis may not contain more than four functions.

By virtue of Problems 6.2 and 6.9, there exist bases with an arbitrary number of functions (but not more than four).

6.22. Starting with an arbitrary basis and equating the arguments of the functions occurring in this basis, prove that it is possible to obtain another basis in which all the functions depend on at most three variables; it is, in general, impossible to further decrease the number of variables.

Definition 6.6. If an incomplete system may be obtained by arbitrarily equating the variables of every function in some basis, the basis is said to be *minimal*.

6.23. Prove that there is a finite number of distinct minimal bases.

It turns out that there are a total of 48 minimal bases; a list of these bases may be found in [1]. Some of them are derived below.

Now let us investigate certain types of bases. We begin with bases consisting of a single function.

Definition 6.7. A logical function that constitutes a basis consisting of a single element is called a *generalized Sheffer function.*

6.24. How many generalized Sheffer functions of n variables are there?

6.25. Find all the minimal bases consisting of a single function.

Now, by contrast, let us consider bases containing the greatest possible number of functions.

6.26. Construct the Post table for a basis of four functions.

6.27. Find all the minimal bases consisting of four functions.

Hints

6.3. Use the law of duality.

6.4. Use Problems 3.6 (law of duality) and 5.7 and analogous considerations.

6.5. (a), (c), (d), (e), (f) (h), (i), and (j).

6.6. (2) follows from the law of duality.

6.7. No.

6.8. The Sheffer functions are contained in this complement (Problem 6.2, (d), (e)).

6.10. The completeness of a system of functions Φ is logically equivalent to the assertion that the functionally closed class generated by this system (the minimal basis containing it) coincides with the set of all functions.

6.13. By equating the variables, simplify the functions in Φ so that they again satisfy the conditions of Post's theorem. Prove that constants, negation, and conjunctions may be obtained from compositions of these functions. Use Problems 6.12, 3.11, 5.16, 4.11, and 4.12.

6.14. It is worth reviewing the preceding chapters (Chapters 3-5) to see how to determine whether a given function belongs to **S**, **L**, or **M**. The cases of $\mathbf{P}_0$ and $\mathbf{P}_1$ are trivial. Recall that, in particular, a function is linear or nonlinear depending on whether there are terms of degree greater than one in its Zhegalkin polynomial (cf. remark following Problem 4.6).

It is often useful to recall the relations between functionally closed classes. For example, of the linear functions only 0, 1, and x are monotone (Problem 5.3); functions with an odd number of variables are self-dual (Problem 4.6); functions whose free term is zero preserve the additive identity; functions with an even number of variables preserve the multiplicative identity if their free term is zero, and functions with an odd number of variables, if their free term is unit, preserve the multiplicative identity. All monotone functions other than the corresponding constants belong to $\mathbf{P}_0$ and $\mathbf{P}_1$ (Problem 5.4).

6.16. Recall that the premise of a theorem is said to be essential if the theorem turns into a false assertion once it is discarded. A premise may be nonessential for various reasons, for example, it may not relate to the case the theorem discusses, or it may be a consequence of other premises (the premises of the theorem are then not independent). Essential premises are sometimes confused with necessary premises.

6.18. Prove by contradiction using Post's theorem.

6.19. Follows from Problems 6.17 and 6.18.

6.21. A function that does not preserve the additive identity is either nonmonotone or is not self-dual.

6.24. $2^{2^n-2} - 2^{2^{n-1}-1}$. The only non-self-dual functions that do not preserve the additive and multiplicative identities are the generalized Sheffer functions.

6.25. Answer: $\overline{xy}$ and $\bar{x} \vee \bar{y}$. Use the preceding problem.

6.26.

	P_0	P_1	S	L	M
$\omega_1 = 1$	−	+	−	+	+
$\omega_2 = 0$	+	−	−	+	+
$\omega_3 = x_1 + \ldots + x_{2k+1}$, $k \geqslant 1$	+	+	+	+	−
ω_4	+	+	±	−	+

6.27. 0, 1, $x+y+z$, xy; 0, 1, $x+y+z$, $x \vee y$; 0, 1, $x+y+z$, $xy+yz+xz$.

Solutions

6.1. xy, $x+y$, 1.

6.2. (a), (b) cf. Problem 1.14; (d) cf. Problem 6.1; (d), (e) cf. Problem 1.16 (these functions are called Sheffer functions); (f) $x \vee y = xy+x+y$, therefore, $x \vee y + (x+y) = xy$ (cf. (c)); (g) $x+y+0 = x+y$, while $x+y$, xy, 1 is a complete system (cf. (c)); (h) $\bar{x} \to y = x \vee y$, while $x \vee y$, $\bar{x}$ is a complete system; (i) $x \to 0 = \bar{x}$ (cf. (h)).

6.3. If we wish to represent some function f in the form of a composition of functions from Φ^+, we first represent the dual function f^+ by a composition of functions from Φ, and then obtain the desired composition by introducing (by means of the law of duality) a composition of dual functions.

6.4. (a) A function of a large number of variables cannot be obtained by means of a composition of functions of a single variable.

(b) Both functions are monotone; that is, by Problem 5.7 any composition of the two functions is monotone nad no nonmonotone function may be obtained (e.g., $\overline{x}$).

(c) Both functions are linear, therefore, as can be readily seen, all their compositions are linear.

(d) Both functions are self-dual, therefore (by the corollary to the law of duality), their composition is self-dual.

(e) All the functions are monotone.

6.5. (b) $x \vee (y \vee z)$ is a function of three variables which is a composition of a disjunction; (e) follows from the law of duality; (f) cf. Problem 5.7; (g) cf. Problem 5.8; (h) use the same notation as in the solution of Problem 5.7. Substitute the null sequence in F. Since ψ preserves the additive identity, in this case the null sequence is substituted in ω, and since ω also preserves the additive identity, we find that F also preserves the additive identity; (i) same as for (h); (k) $x+y$ possesses this property, while $x+(y+z)$ preserves the multiplicative identity.

6.7. We take the classes of functions that preserve the additive identity and those that preserve the multiplicative identity. Their union will consist of functions that preserve the additive or multiplicative identities, and will include, in particular, the functions $x\overline{y}$ (which preserves the additive identity) and 1 (which preserves the multiplicative identity). Substituting 1 in $x\overline{y}$ for x, we obtain $\overline{y}$, which preserves neither the additive nor the multiplicative identity.

6.8. If a functionally closed class contains one of the Sheffer functions (Problem 6.2d, e), it coincides with the class of all the logical functions. Since the functionally closed class we are given is intrinsic, the Sheffer functions must be contained in its complement. But in that case this complement cannot be a functionally closed class, since it would otherwise coincide with the set of all classes, and the initial class would be empty, that is, non-intrinsic.

6.9. (a) xy is a monotone function, and $\overline{x}$ a function of a single variable (and is also linear and self-dual), i.e., these systems are incomplete. (b) same as in (a); (c) xy and 1 are monotone functions that preserve the multiplicative identity, $x+y$ and 1 are linear functions, and xy and $x+y$ preserve the additive identity; (f) same as (c); (g) xy, 0, and 1 are monotone functions, $x+y+z$, 0, and 1 are linear functions, $x+y+z$, xy, and 0 preserve the additive identity, and $x+y+z$, xy, and 1 preserve the multiplicative identity; (h) $x \to y$ preserves the multiplicative identity, $\bar{x}$ behaves as in (a); (i) $x \to y$ preserves the multiplicative identity, and 0 is a constant.

6.10. We have already spoken of necessity (all functions in Φ would otherwise belong to some intrinsic functionally closed class, and all their compositions would likewise belong to it).

Now let us prove sufficiency. Suppose the system Φ possesses the property stated in the premise of the problem. The set of functions that are compositions of functions in Φ is, obviously, a functionally closed class (note that this is the smallest functionally closed class containing Φ). This class cannot be intrinsic, as Φ is not in any intrinsic class. Since it is also non-empty, it contains all the logical functions.

6.11. Necessity follows from the necessity of the premise of Problem 6.10. Let us prove sufficiency. The functionally closed class generated by functions in Φ may not be contained in any pre-complete class since by our premise Φ is not contained in any pre-complete class. Since this class is, in addition, non-empty, it coincides with the class of all the logical functions.

6.12. Suppose that $\omega(x_1, \dots, x_n) \notin P_p$, and that $\omega(0, \dots, 0) = 1$. We equate all the variables. Then, if $\omega(1, \dots, 1) = 1$, we obtain 1; if $\omega(1, \dots, 1) = 0$, we obtain 0.

6.13. Necessity is self-evident. Let us prove sufficiency. We take from Φ a function that does not preserve zero and another function that does not preserve unit and equate the variables in Φ. By Problem 6.12, in the first case we obtain 1 or $\bar{x}$, and in the second case, 0 or $\bar{x}$. As a result, we obtain both constants or a negation (possibly all three). Suppose we have obtained constants. Let us prove that negation may then be represented in the form of a composition. Since the constants do not belong to the class S nor to either $\mathbf{P}_0$ or $\mathbf{P}_1$, whereas $\bar{x}$ is a linear function, for our purpose it is natural to use a nonmonotone function from Φ. In fact, by Problem 5.16 negation may be represented in the form of a composition of an arbitrary nonmonotone function and constants. Thus, in this case both constants and negation may be represented by means of a composition of functions from Φ.

Let us now consider a second case in which negation is obtained by equating the variables both of functions that do not belong to $\mathbf{P}_0$ and functions that do not belong to $\mathbf{P}_1$. We may show that constants may then be obtained. For this purpose, it is natural to use a non-self-dual function from which (by virtue of Problem 3.11) constants as well as negation may be obtained.

Thus, in both cases we have the functions 0, 1, and $\bar{x}$ and still have not used any nonlinear function. Applying Problem 4.11, we obtain a nonlinear function of two variables from the original nonlinear function and one of the constants. Then, using Problem 4.12, we obtain an arbitrary nonlinear function of two variables from this function and negation, for example, xy (or $x \vee y$ or the Sheffer function $\overline{xy}$). Since we have constructed a complete system of functions, the theorem is proved.

6.14. Compile the Post table:

		$\mathbf{P}_0$	$\mathbf{P}_1$	S	L	M
(a)	xy	+	+	−	−	+
	$\bar{x}$	−	−	+	+	−
(b)	$x \vee y$	+	+	−	−	+
	$\bar{x}$	−	−	+	+	−
(c)	xy	+	+	−	−	+
	$x + y$	+	−	−	+	−
	1	−	+	−	+	+
(d)	$\overline{x \vee y}$	−	−	−	−	−
(e)	$\overline{xy}$	−	−	−	−	−
(f)	$x + y$	+	−	−	+	−
(g)	$x \vee y$	+	+	−	−	+
	1	−	+	−	+	+

		P_0	P_1	S	L	M
(h)	$x + y + z$	+	+	+	+	−
	xy	+	+	−	−	+
	0	+	−	−	+	+
	1	−	+	−	+	+
(i)	$x \to y$	−	+	−	−	−
	$\overline{x}$	−	−	+	+	−
(j)	$x \to y$	−	+	−	−	−
	0	+	−	−	+	+

There is at least one minus sign for each of the nine systems of functions in each of the columns. That is, all the systems are complete.

6.15. A system of functions Φ is not complete if and only if it lies entirely in one of the classes $\mathbf{P}_0$, $\mathbf{P}_1$, **L**, **M**, and **S**.

6.16. Compile the Post table:

		P_0	P_1	S	L	M
(a)	0	+	−	−	+	+
	xy	+	+	−	−	+
	$x + y + z$	+	+	+	+	−
(b)	1	−	+	−	+	+
	xy	+	+	−	−	+
	$x + y + z$	+	+	+	+	−
(c)	$\overline{xy} \vee \overline{xz} \vee \overline{yz}$	−	−	+	−	−
(d)	0	+	−	−	+	+
	1	−	+	−	+	+
	$x + y$	+	−	−	+	−
(e)	0	+	−	−	+	+
	1	−	+	−	+	+
	xy	+	+	−	−	+

Let us now determine the properties of the function $\overline{xy} \vee \overline{xz} \vee \overline{yz}$. We easily verify that it belongs neither to $\mathbf{P}_0$ nor $\mathbf{P}_1$, in which case it does not belong to **M** (cf. Problem 5.4 and the hint to Problem 6.14). It is not a constant nor symmetric in all its variables. Therefore, all its variables are actual. It is easily verified that it coincides neither with $x+y+z$ nor $x+y+z+1$, i.e., it is nonlinear. Finally, by virtue of symmetricity, we may limit outselves to the sequences (0, 0, 0) and (1, 0, 0) in studying its self-duality.

Now let us see whether the premises of Post's theorem are essential. None of the systems considered in the problem are complete, since for each of them there is a column in the table consisting entirely of +'s. Note that there is precisely one such column for each system, and that these columns are distinct for distinct systems. Hence, it follows that none of the five classes may be discarded from the premises of Post's theorem. In fact, for each one of these classes a system of functions may be found in this class (which thus is not complete) that contains a function for each of the remaining four classes not belonging to the original system.

6.17. If any one class were contained in some other class, this class could be discarded from the statement of Post's theorem, which contradicts Problem 6.16.

6.18. If there were a functionally closed class T not in any of these five classes, a function could be found in T for each of them that did not belong to the particular class. But then (by Post's theorem), T would contain all the logical functions.

6.19. By Problem 6.18, the other classes may not be pre-complete. By Problem 6.17, each of the five classes is pre-complete.

6.20. Not more than five functions that satisfy Post's theorem may be selected from every complete system.

6.21. Consider a function that does not preserve zero. By Problem 5.4, it is either nonmonotone or is the additive identity, i.e., it is a non-self-dual function. Thus, it must be possible to find a function in the complete system that does not belong simultaneously to two pre-complete classes. In that case, at most three functions from the system we are considering may be adjoined to this function so as to satisfy the premises of Post's theorem. That is, there can be no more than four functions in the basis.

6.22. That the number of variables may be decreased follows from Problems 3.10, 4.10, 5.14, and 6.12. After equating the variables, functions are obtained that do not belong to $\mathbf{P}_0$ or $\mathbf{P}_1$, depend on at most a single variable, do not belong to **S**, depend on at most two variables and do not belong to **L** or **M**, and depend on at most three variables. There cannot be fewer than three variables, as is shown by the system of Problem 6.2(g).

6.23. There is a finite number of functions of three variables. The number of minimal functions is not greater than the number of subsets of this set which contain at most four elements.

6.24. The generalized Sheffer function $f(x_1, \dots, x_n)$ cannot preserve unit and zero: $f(0, \dots, 0) = 1$ and $f(1, \dots, 1) = 0$. It is thus automatically nonmonotone. There will be $2^{2^n - 1}$ such functions (values from $(2^n - 2)$ sequences may be specified arbitrarily; cf. Chapter 2, Problems 2.2 and 2.3). Moreover, f is a non-self-dual function. Let us find out how many functions f there are not belonging simultaneously to $\mathbf{P}_0$, $\mathbf{P}_1$, and **S**. There are $2^{2^{n-1}-1}$ functions not belonging to $\mathbf{P}_0$ and $\mathbf{P}_1$ which belong to **S** (they are determined by the values on $(2^{n-1} - 1)$ sequences). Therefore, the number to be determined is given as $2^{2^n-2} - 2^{2^{n-1}-1} = 2^{2^{n-1}-1}(2^{2^{n-1}} - 1)$. It remains for us to observe that if a function does not belong to $\mathbf{P}_0$, $\mathbf{P}_1$, and **S**, it belongs neither to **M** nor to **L** (cf. hint to Problem 6.14). In fact, suppose that f is a linear function, $f \notin \mathbf{P}_0$, $f \notin \mathbf{P}_1$. Then the free term in the decomposition of f is equal to 1 and the number of actual variables is odd. But such functions are self-dual. Thus, the number we have found is the number of generalized Sheffer functions.

6.25. Suppose that $f(x_1, \dots, x_n)$ is a generalized Sheffer function. In our proof of the preceding problem, we showed that this is logically equivalent to asserting that f is a non-self-dual function that preserves neither the additive nor the multiplicative identity. By Problem 3.10, the variables of f may be equated so as to obtain a non-self-dual function of two variables. It is easily verified that if the variables are equated arbitrarily, the property of not preserving the additive (or multiplicative) identity is retained. Therefore, once we have equated the variables, we again obtain a generalized Sheffer function. Thus, if a minimal basis consists of a single function, this function depends on two variables. It remains for us to find the generalized Sheffer functions of two variables. There are two such functions (as may be verified from the preceding problem): $\overline{xy}$ and $\overline{x} \vee \overline{y}$.

6.26. Compare the table in the hint to the problem. Suppose that $\Phi = \{\omega_1, \omega_2, \omega_3, \omega_4\}$ is a basis consisting of four functions, and let ω_1 be a function that does not preserve the additive identity ($\omega_1 \notin \mathbf{P}_0$). As we have already seen in the solution of Problem 6.21, ω_1 is necessarily either nonmonotone or non-self-dual. Clearly, if there are four functions in the basis, ω_1 will belong to at least three of them (and does not belong to either **M** or **S**). In particular, $\omega_1 \in \mathbf{P}_1$, i.e., $\omega_1(0, \ldots, 0) = 1$ and $\omega_1(1, \ldots, 1) = 1$, that is ω_1 is a non-self-dual function. It must then be monotone, and by Problem 5.4 there is only one way this could happen: $\omega_1 = 1$. It may be proved analogously that a function that does not preserve the multiplicative identity must be the additive identity. Thus, both constants will occur in the basis and we have completed two rows in the Post table. The other two functions in the basis necessarily preserve the additive identity and multiplicative identity; one of them must be monotone and nonlinear, and the other nonmonotone and linear. We begin with the latter function (suppose that it is ω_3). Since $\omega_3 \in \mathbf{P}_0$, the free term in the Zhegalkin polynomial is equal to 0, and since $\omega_3 \in \mathbf{P}_1$, it depends actually on an odd number of variables: $\omega_3 = x_1 + x_2 + \ldots + x_{2k+1}$, $k \geqslant 1$. Thus function is self-dual. We have thereby completed the third row in the Post table. The monotone, nonlinear function may be either non-self-dual (e.g., xy) or self-dual (e.g., $xy \vee xz \vee yz$). As a result, there are two possibilities for the fourth row in the Post table.

6.27. By Problem 6.26, in any, not necessarily minimal, basis consisting of four functions there are two functions which are constants. The third function has the form $x_1 + x_2 + \ldots + x_{2k+1}$; any function of this form may be used with $k \geqslant 1$. Let us equate all the variables beginning with x_3, thus obtaining the function $x_1 + x_2 + x_3$. The resulting function occurs in the same classes as does the initial function. No further equating of variables is possible. Thus, the third function in the minimal basis has the form $x_1 + x_2 + x_3$. Now let us consider the nonlinear monotone function ω_4. By Problem 6.22, if a basis is minimal, ω_4 depends on at most three variables. We have seen (Problem 5.5) that there are only two monotone nonlinear functions of two variables: xy and $x \vee y$. It remains for us to investigate the case of three variables. Nonlinear monotone functions of three variables ω_4 belonging to $\mathbf{P}_0$ and $\mathbf{P}_1$ must be found whose variables cannot be so equated that the function remains nonlinear. We represent ω_4 in the form of a Zhegalkin polynomial:

$$\omega_4(x, y, z) = axyz + b_1xy + b_2yz + b_3xz$$

$$+ c_1x + c_2y + c_3z.$$

The free term is equal to 0, since $\omega_4 \in \mathbf{P}_0$. Suppose that $a = 1$. Now equate any two arguments. Without any limitation on generality, we may assume that these arguments are y and z. Then the nonlinear part of the resulting polynomial will be $(1 + b_1 + b_3)xy$. The function represented by this polynomial will be nonlinear if and only if the coefficient $(1 + b_1 + b_3)$ is odd. Clearly, it is always possible to rename the variables of ω_4 so as to fulfill this condition. Thus, if $a = 1$, it is always possible to equate variables so as to obtain a nonlinear function. Therefore, $a = 0$.

Now suppose that one of the coefficients (say, b_3) is equal to 0. One of the other coefficients (say, b_1) must equal 1 (otherwise, ω_4 would be a linear function). Equating the variables x and z, we obtain as a result a function whose Zhegalkin polynomial is the term $xy(b_1 = 1)$, i.e., a nonlinear function, which is impossible. That is, $b_1 = b_2 = b_3 = 1$.

Now let us consider the coefficients c_i. Since $\omega_4 \in \mathbf{P}_1$, either all the coefficients are equal to 0 or only one of them is. Suppose that $c_1 = c_2 = 1$ and let $c_3 = 0$. Then $\omega_4\ (x, y, z) = xy + yz + xz + x + y$. The function ω_4 is not monotone, as is apparent from a comparison of its values on the sequences

(1, 0, 0) and (1, 0, 1). That is, this case is impossible, so $c_1 = c_2 = c_3 = 0$.

Thus, in the minimal basis ω_4 has the form

$$xy, \ x \vee y \quad \text{or} \quad xy + yz + xz = xy \vee yz \vee xz$$

(Problem 4.3c).

Chapter 7
THE GENERAL THEORY OF FUNCTIONALLY CLOSED CLASSES

1. *Functionally Closed Classes. The General Statement of the Completeness Problem.* In the preceding chapter, we saw that the completeness problem for a system of logical functions may be solved by means of the pre-complete functionally closed classes. In fact, a host of problems may be treated by studying other functionally closed classes (not necessarily pre-complete). We might expect that there would be an infinite number of distinct functionally closed classes, indeed that the set of such classes has the cardinality of the continuum, as these classes may be specified in a natural way by means of the systems of functions that generate them. Since the set of logical functions is countable, the set of distinct systems of functions has the cardinality of the continuum. Of course, distinct systems of functions often generate the same class. But there is no reason for assuming in advance that this identification is so rigorous that there is a finite or countable number of distinct classes. In fact, it turns out that the set of classes is countable. That this set is infinite is far from obvious (cf. conclusion of the section).

We now wish to describe the theoretical structure of the set of functionally closed classes, leaving its specific description (list of all classes) to the end of the chapter. The set of classes may be represented conveniently in the form of a tree whose vertices correspond to classes. If one class lies within another class, the vertex corresponding to the first class will be placed beneath the vertex corresponding to the second class, the two vertices being connected by line segments. The classes admit a natural arrangement by levels. At the base of the tree (level one) is found the class of all the logical functions (denoted **P**), in the second level the pre-complete classes, in the third level those not contained in any one class other than the pre-complete classes, and in general, in the $(i+1)$-th level, classes that do not belong to any of the preceding levels and not contained in any one class belonging to these preceding levels; there is an infinite number of levels. Basically, in each level there is a finite number of classes (this number is even bounded by some constant), and if one class is contained in some other class from the i-th level, either it belongs to the $(i+1)$-th level or is itself contained in some class of the $(i+1)$-th level (cf. footnote on page 88). In particular, this means that each class the i-th level has only a finite number (bounded by the same constant for all levels) of classes of the succeeding level, or that there is a finite basis in the i-th level; for all such classes the number of elements in the bases is bounded (cf. below). But there is no reason for assuming that by scanning the levels in this way, we will sooner or later have considered all the classes, i.e., it is not true that every class belongs to some level. However, it does turn out that all but a finite number of classes (these might be thought of in some sense as limiting classes, i.e., classes of level ∞) possess this property. Thus, the theoretical structure has the form shown in Figure 2, with the levels arranged from the top down.

In the figure, vertices are connected only for classes $Q_1 \subset Q_2$ that lack any intermediate class Q_3, i.e., $Q_1 \subset Q_3 \subset Q_2$, $Q_3 \neq Q_1$ and $Q_3 \neq Q_2$. The part of the figure beginning with the third level and moving down does not correspond to the actual structure (number of classes in a level, inclusions, etc.); here, we only wish to illustrate the general remarks given above. The last row in the figure corresponds to the limiting classes.

Now let us see how to use the table of classes to solve problems in algebraic logic. The general problem may be stated as follows. Suppose that $\mathfrak{U}$ is some (in general, infinite) collection of sets $R_1, \dots, R_n, \dots$ of logical functions. We wish to find an algorithm by means of which it would be possible to describe for each system of functions Φ whether all the functions from at least one set R_i occurring in $\mathfrak{U}$ may be obtained from the functions occurring in Φ by means of

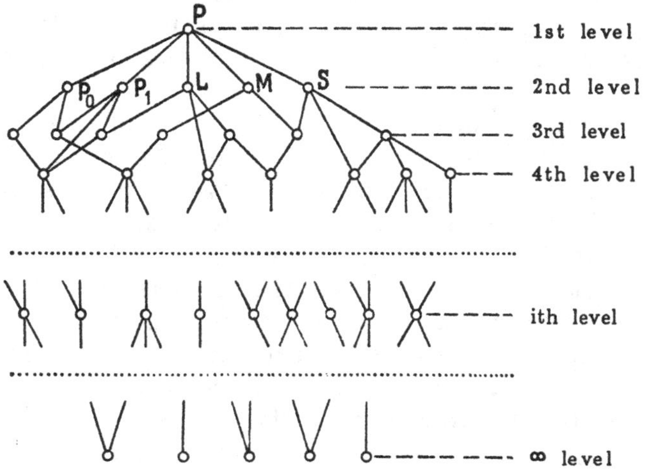

Figure 2

compositions. A system of functions possesing this property is said to be $\mathfrak{U}$-complete.

Whenever $\mathfrak{U}$ consists of a single set -- the set **P** of all the logical functions -- the problem is to determine whether Φ is complete (cf. Chapter 6). But there are many other variations on this problem. For example, we might wish to determine whether the system generates some functionally closed class Q (in which case $\mathfrak{U}$ contains the single element Q). If all the sets occurring in $\mathfrak{U}$ contain at least one function, may be considered a set of functions. In this case, the problem becomes that of deciding whether it is possible to represent at least one function from $\mathfrak{U}$ in the form of a composition of functions from Φ. By means of the structure of functionally closed classes it is in principle possible to solve these problems under the same conditions under which Post's theorem provides an answer to the completeness question.

A functionally closed class Q is said to be $\mathfrak{U}$-*pre-complete* if it does not contain any set $R_i \in \mathfrak{U}$ and if any class containing Q but not coinciding with Q contains at least one of the sets $R_i \in \mathfrak{U}$. Clearly, every class that does not contain any of the sets $R_i \in \mathfrak{U}$ is contained in at least one $\mathfrak{U}$-pre-complete class. The $\mathfrak{U}$-pre-complete classes may often be found in the following way. We first check to see which of the limiting classes contain the sets $R_i \in \mathfrak{U}$ and then, moving down from the class **P** and considering only classes not containing limiting classes that possess this property, we discard all the $\mathfrak{U}$-pre-complete classes. These classes are intermediate between classes containing some $R_i \in \mathfrak{U}$ and classes that do not contain any R_i.

This method cannot be applied if some of the limiting classes are $\mathfrak{U}$-pre-complete, since these classes cannot be reached in a finite number of steps by means of this procedure. Here we must make a special study. Note that even if all the functionally closed classes are known, it may be very difficult to determine whether at least one of the sets R_i belongs to some class. But this doesn't mean that knowing about the table turns the problem of finding the $\mathfrak{U}$-pre-complete classes into an entirely mechanical procedure. However, if the $\mathfrak{U}$-pre-complete classes have already been found, the $\mathfrak{U}$-completeness of some system Φ follows automatically (at least for finite sets Φ). Here it is important for us to have an algorithm for every class (as follows from the list of classes presented at the end of the chapter) that can be used to decide whether some function is or is not in this class. The algorithm may actually be quite cumbersome.

7.1. Prove that some system of functions Φ is $\mathfrak{U}$-complete if and only if for every $\mathfrak{U}$-pre-complete functionally closed class in Φ, there is a function that does not belong to Φ.

We still haven't said anything about how to find all the functionally closed classes. This rather difficult problem was solved by Post [1, 2]. Here we will present only the final result (at the end of the chapter) and consider a number of examples. Note that in the general result, it is most important to "guess" what might be all the classes, though it is an extremely difficult problem to prove that there are no other classes. Recall that in the problem of pre-complete classes (Chapter 6), once the classes have been "guessed at," the completeness theorem could be proved without any great difficulty.

2. *Extended Composition.* Now let us consider a single, simplified version of the problem of functionally closed classes. The problem will illustrate in many ways the situation that occurs in the general case, though it is significantly simpler.

Definition 7.1. We will say that a function f is obtained from the system Φ by means of *extended composition* if it may be obtained from Φ by means of composition (Definition 2.3) and substitution of constants.

In other words, the ordinary compositions of the system of functions $\Phi \cup \{0,1\}$ become extended compositions if Φ contains at least one function that is not a constant; otherwise (if only constants occur in Φ), Φ coincides with the set of extended compositions. We need only note that both constants may be obtained by means of substitutions from a function which is not a constant. (It is necessary to substitute for the argument the set of constants on which the function takes the corresponding value.) The only extended composition of a system of functions which reduccs to a constant is the system itself.

7.2. Find the pre-complete classes for extended composition.

7.3. How are the functionally closed classes related in terms of ordinary and extended composition?

Thus, we can complete the first two levels in the table of classes for extended composition (Figure 3).

Definition 7.2. If every function from some functionally closed class Q may be represented in the form of a composition of functions from some set Φ, we say that Φ is a *Q-complete class* (cf. the concept of $\mathfrak{U}$-completeness; here $\mathfrak{U}$ consists of the single set Q). A minimal Q-complete system Φ is called a *Q-basis*.

7.4. Prove that the system of functions $\{x+y, 1\}$ is an **L**-basis, and that the system consisting of the single function $x+y$ is an **L**-basis relative to extended composition.

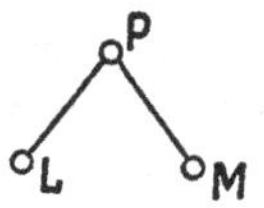

Figure 3

Remark. Unless otherwise stipulated, everywhere in this chapter in speaking of functionally closed classes we will be concerned with these classes from the standpoint of extended composition.

7.5. Find the **L**-pre-complete classes.*

7.6. Find all the functionally closed classes that occur in **L**, and draw the connection tree.

Thus, we have constructed the part of the table of classes comprising the class of linear functions **L**. Now we must carry out the analogous constructions for the class of monotone functions **M**. These constructions are somewhat more complicated than in the preceding case.

Definition 7.3.** Let **D** denote the class of functions consisting of disjunctions of an arbitrary number of variables $x_1 \vee x_2 \vee \ldots \vee x_k$ (including the single variable x), and let $\mathbf{D}^{01}$ be the class obtained from **D** by adjoining the constants.***

Analogously, we let **K** denote the class consisting of conjunctions of an arbitrary number of variables $x_2 x_2 \ldots x_k$ (including the single variable x as well), and $\mathbf{K}^{01}$, the class obtained by adjoining the constants.

7.7. Prove that $\mathbf{D}^{01}$ and $\mathbf{K}^{01}$ are the only **M**-pre-complete classes.

---

*That is, the maximal functionally closed classes in **L** that do not coincide with **L**. This is a special case of the $\mathfrak{U}$-pre-complete classes, which corresponds to a system $\mathfrak{U}$ consisting of the single set Q.

**In this definition we have introduced classes for ordinary composition. The classes $\mathbf{D}^{01}$ and $\mathbf{K}^{01}$ are also classes for extended composition.

***We will use the following notation for the functionally closed classes. If Q is a class, Q_0, Q_1, and Q_{01} will denote the intersections of Q with the classes $\mathbf{P}_0$ and $\mathbf{P}_1$, finally, the intersection of $\mathbf{P}_0$ and $\mathbf{P}_1$, i.e., $\mathbf{P}_{01} = \mathbf{P}_0\mathbf{P}_1$ (cf. hint to Problem 7.6); Q^0, Q^1, and Q^{01} will denote the classes obtained by extending Q by means of 0, 1, and both 0 and 1, respectively.

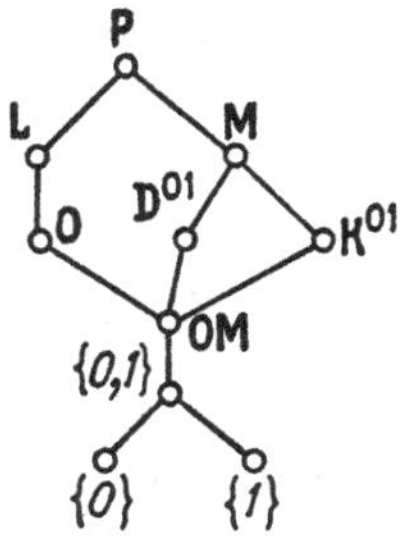

Figure 4

7.8. Find all the functionally closed classes contained in **M**. Draw the connection tree.

Thus, we have found all the functionally closed classes for extended composition, of which there are a finite number. We present the final table of classes in Figure 4. This table may be used to solve $\mathfrak{U}$-completeness problems for sets of functions relative to extended composition using the technique of sec. 7.1. Let us consider several examples.

7.9. In each of the cases that follow, find necessary and sufficient conditions on a system of logical functions Φ by means of the functions may be represented by extended compositions of functions from Φ:

(a) all the linear functions;
(b) all the monotone functions;
(c) the function $xy \vee yz \vee xz$;
(d) the function $x+y+z$;
(e) one or both of the functions $xy \vee yz \vee xz$ and $x+y+z$;
(f) both functions in (e);
(g) the function $\overline{x}$;
(h) every linear function or the function xy;
(i) xy or $\overline{x}$;
(j) xy and $\overline{x}$;
(k) xy or $x \vee y$ or $\overline{x}$.

In other words, it is necessary to find the $\mathfrak{U}$-completeness condition for the system Φ relative to extended composition whenever $\mathfrak{U}$ consists of (a) the single set **L**; (b) the single set **M**; and so on. We leave it to the reader to continue this list.

In the examples we have considered, all the sets occuring in $\mathfrak{U}$ may be described quite simply. Once this has been done, the $\mathfrak{U}$-pre-complete classes may be easily found at least in the case of extended composition. However, there may be elements in $\mathfrak{U}$ that can be described only with great difficulty and ineffectively, which makes it difficult to determine which classes contain elements of $\mathfrak{U}$ and which do not. To some extent, this circumstance is illustrated by the examples presented in the present and succeeding chapters (a more complicated problem of finding $\mathfrak{U}$-pre-complete classes may be found in ref. 3). Let us begin with a relatively simple example.

Definition 7.4. Suppose that Φ is some system of logical functions. We will say that the function f *may be represented in a self-dual fashion* (briefly, is *self-dual representable*) in terms of the system Φ if f and f^+ may be represented by a composition of functions from Φ.

Definition 7.5. A system of functions Φ is said to be *self-dual complete* if the set of functions that is self-dual representable in terms of Φ forms, together with the negation $\overline{x}$, a complete system.

In these definitions, we are keeping in mind both ordinary composition as well as extended composition.

7.10. Prove that every complete system of functions is self-dual complete; the converse is, in general, false (present an example).

7.11. Prove that (1) every self-dual complete system of functions forms, together with $\overline{x}$, a complete system; (2) every system of functions ω that forms, together with $\overline{x}$, a complete system, is *not* self-dual complete; (3) for a system Φ to be self-dual complete, it is sufficient (but not necessary!) that every function in Φ occur together with its dual.

7.12. Prove that the set of functions that is self-dual representable in terms of functions from some functionally closed class Q forms a functionally closed class that coincides with the intersection of Q and its dual class Q^+ (cf. Problem 6.6), i.e., QQ^+.

It is entirely understandable if the reader finds the concept of self-dual completeness somewhat artificial. In Chapter 8, we present as an example a problem that reduces to this concept in a more natural way.

In accordance with the conventions we have adopted, we now relate the concept of self-dual completeness to that of extended composition. The following problem deals with this case.

7.13. Find a necessary and sufficient condition for a system of functions to be self-dual complete.

With this problem, our discussion of extended composition concludes, though we will encounter further examples below (Chapters 8 and 9).

3. *Self-Dual Completeness.* As we have already mentioned, the general case is considerably more complicated than the case of extended composition. The remainder of this chapter will be devoted to this general case. It is often possible to solve a particular problem in terms of functionally closed classes without resorting to the table of all classes, by conjecturing in some way which of the classes are related to our problem. In this section we present such a problem, that is, a problem in which it is necessary to determine whether a system of functions is self-dual complete relative to ordinary composition. We begin by discussing how to solve the problem. However, we advise the reader to try to solve it on his own without making use of the hints that follow.

We will require certain new functionally closed classes.

Definition 7.6. Let $\mathbf{F}^{(2)}$ denote the set of functions $f(x_1, \dots, x_n)$ such that any two sequences $\alpha = (\alpha_1, \dots, \alpha_n)$ and $\beta = (\beta_1, \dots, \beta_n)$ on which $f(\alpha) = f(\beta) = 0$ have the same additive identity, i.e., $\alpha_i = \beta_i = 0$ for some $1 \leqslant i \leqslant n$. Similarly $G^{(2)}$ denotes the set of functions $f(x_1, \dots, x_n)$ such that any two sequences $\alpha = (\alpha_1, \dots, \alpha_n)$ and $\beta = (\beta_1, \dots, \beta_n)$ on which $f(\alpha) = f(\beta) = 1$ have the same multiplicative identity, i.e., $\alpha_i = \beta_i = 1$ for some i.

Note that:

(a) the functions $f \in \mathbf{F}^{(2)}$ preserve the multiplicative identity, i.e., $\mathbf{F}^{(2)} \subset \mathbf{P}_1$; similarly, $f \in \mathbf{G}^{(2)}$ preserve the additive identity, i.e., $\mathbf{G}^{(2)} \subset \mathbf{P}_0$;

(b) $1 \in \mathbf{F}^{(2)}$ and $0 \in \mathbf{G}^{(2)}$;

(c) $0 \notin \mathbf{F}^{(2)}$, since if we assume that $f = 0$ depends formally on certain variables, it turns out that F in that case does not satisfy Definition 7.6. Recall (Chapter 2, Definition 2.2) that we are not distinguishing between functions that differ only formally, and therefore consider properties of functions that do not vary when these variables are introduced. Analogously, $1 \notin \mathbf{G}^{(2)}$.

(d) The sets $\mathbf{F}^{(2)}$ and $\mathbf{G}^{(2)}$ are dual.

7.14. Prove that $\mathbf{F}^{(2)}$ and $\mathbf{G}^{(2)}$ are functionally closed classes.

We can now answer the question we have been concerned with.

7.15. Prove that the system of functions Φ is self-dual complete if and only if it contains the following functions: (a) $\omega_1 \notin \mathbf{L}$; (b) $\omega_2 \notin \mathbf{D}^{01}$; (c) $\omega_3 \notin \mathbf{K}^{01}$; (d) $\omega_4 \notin \mathbf{S}$; (e) $\omega_5 \notin \mathbf{F}^{(2)}$; (f) $\omega_6 \notin \mathbf{G}^{(2)}$.

If you were unable to guess the result stated in Problem 7.15, you should be able to prove it now without any further assistance. But if you still cannot do so, continue on and try to solve the following problems, which are designed to lead you to a solution of Problem 7.15 step-by-step.

7.16. Prove that every non-self-dual function belongs to at most one of the classes $\mathbf{F}^{(2)}$ or $\mathbf{G}^{(2)}$.

The next problem is a somewhat more rigorous statement of Problem 3.10.

7.17. List all the non-self-dual functions from which it is not possible to again obtain non-self-dual functions by equating the variables. Which of these functions do not occur in $\mathbf{F}^{(2)}$, but do occur in $\mathbf{G}^{(2)}$?

7.18. Suppose that $f \notin \mathbf{F}^{(2)}$. By equating the variables of f, is it possible to obtain a function which depends on some minimal number of variables and which does not occur in $\mathbf{F}^{(2)}$? Answer the analogous question for a function $f \notin \mathbf{G}^{(2)}$.

7.19. Suppose that a system of functions Φ contains the functions $\omega_1 \notin S$ and $\omega_2 \notin F^{(2)}$. Can a function of some minimal number of variables that does not occur in $F^{(2)}$ be represented in the form of a composition of functions from Φ? Similarly, suppose that the system of functions Φ contains the functions $\omega_1 \notin S$ and $\omega_2 \notin G^{(2)}$. Can a function of some minimal number of variables that does not occur in $G^{(2)}$ be represented in the form of a composition of functions from Φ?

You now have everything you need to know to solve Problem 7.15.

4. *Third Level in the Post System of Functionally Closed Classes.* In the present section, we discuss the problem of finding functionally closed classes, though do not solve it in its entirety. We already know about the pre-complete classes, which constitute the second level in the Post scheme. Now let us construct the next level and, in addition,* for every pre-complete class Q, find the Q-pre-complete classes. The results of the preceding two sections will be of great assistance here. For these new classes, we will adopt notation using the remarks in the third footnote on page 108 and in the hint to Problem 7.6. We begin with the classes of monotone and linear functions.

7.20. Find all the M-pre-complete classes.

7.21. Find all the L-pre-complete classes.

Before moving on to consider other Q-pre-complete classes, let us again recall the rationale underlying proofs of the Q-pre-completeness of the classes $R_1, \dots, R_m \subset Q$ for some functionally closed class Q (compare the solution of Problem 6.18 and the hint to Problem 7.7). It is necessary to prove that:

(1) none of the classes R_i coincides with Q;
(2) no one class R_i is contained in some other class R_j;

*As we will see below, this problem is actually more general (compare the solution of Problem 7.27).

(3) every system of functions Φ that contains the function $\omega_1 \notin R_i$ for every class R_i is Q-complete.

To determine whether a particular system Φ is Q-complete, we need only find certain standard Q-complete systems in Q (for example, $\{xy, x \vee y\}$ for the class **M**, and $\{x+y, 1\}$ for the class **L**), and then prove that all the functions in some standard system may be represented in terms of elements of Φ. It is also natural to begin our study of the problem of finding Q-pre-complete classes with a discussion of standard Q-complete systems.

7.22. Prove that the system of functions $\{xy, x+y+1\}$ is a $\mathbf{P}_1$-complete system (and even a $\mathbf{P}_1$-basis).

7.23. Prove that the system of functions $\{x \to y, xy\}$ is a $\mathbf{P}_1$-complete system and a $\mathbf{P}_1$-basis.

7.24. Find all the $\mathbf{P}_1$-pre-complete classes.

7.25. Formulate analogs to Problems 7.22, 7.23, and 7.24 for the class $\mathbf{P}_0$ of functions that preserve the additive identity. Solve the new problem by analogy with the solutions of the original problems, and also by deriving them from the corresponding problems for $\mathbf{P}_1$ by means of the law of duality (Chapter 3).

Let us proceed to a discussion of the class of self-dual functions **S**. We begin with the construction of a standard **S**-basis. As we shall see, this requires essentially new considerations. In all the preceding cases, we used normal forms or Zhegalkin polynomials in some way. These representations were related to bases consisting of non-self-dual functions and therefore could not be used directly in the case of self-dual functions.

7.26. Prove that the system of functions $\{xy \vee yz \vee xz, \overline{x}\}$ is an **S**-basis.

7.27. Find all the **S**-pre-complete classes.

We have now found all the Q-pre-complete classes for the pre-complete classes Q. Let us construct the connection tree

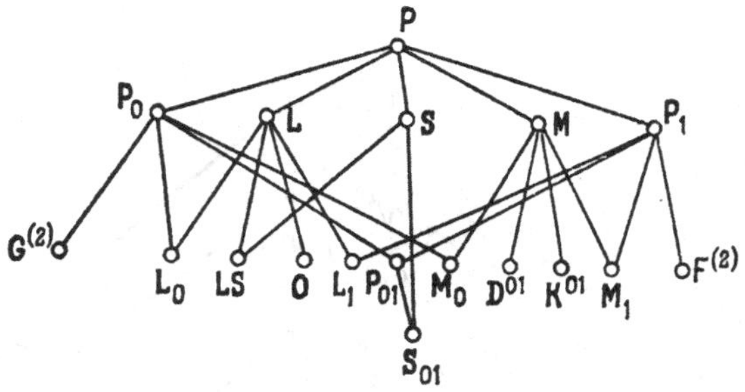

Figure 5

(Figure 5). Thus, we have found the third level in the Post system, which consists of all the Q-pre-complete classes, where the Q are themselves pre-complete classes, except for $\mathbf{S}_{01} \subset \mathbf{P}_{01}$.

5. *Post System of Functionally Closed Classes.* We present without proof the Post system of functionally closed classes (Figure 6).

Some explanation is in order. In the above system, the classes are not arranged by level (cf. Figure 2). However, the rule that places a class Q_1 in Q_2 below Q_2 in the system is retained. Therefore, it can be easily determined from the system which level a particular class Q belongs to. For this purpose, it is necessary to consider all possible chains of classes $\mathbf{P} = Q_1 \supset Q_2 \supset \ldots \supset Q_r = Q$ (the Q_i are all distinct). These chains have finite length, further there is a finite number of distinct chains. The maximal length of these chains r coincides with the number of the level in which Q is found. It is clear from the system that for every class Q, there is at most five Q-pre-complete classes (and only five such classes for $\mathbf{P}$).

Most of the classes occurring in the system have already been defined.* The classes $\mathbf{F}^{(k)}$ are determined by analogy with $\mathbf{F}^{(2)}$; that is, any k sequences on which $f \in \mathbf{F}^{(k)}$ is zero must share a common additive identity in some category. Since some of these sets may coincide, $\mathbf{F}^{(k)} \supset F^{(m)}$ if $k < m$. The classes $\mathbf{G}^{(k)}$ are dual to $\mathbf{F}^{(k)}$. The class $\mathbf{F}^{(\infty)}$ consists of functions in which all the sets on which these functions are zero share a common additive identity; $\mathbf{F}^{(\infty)} \subset \mathbf{F}^{(k)}$. The

*Our notation differs from that adopted in the books [1, 2].

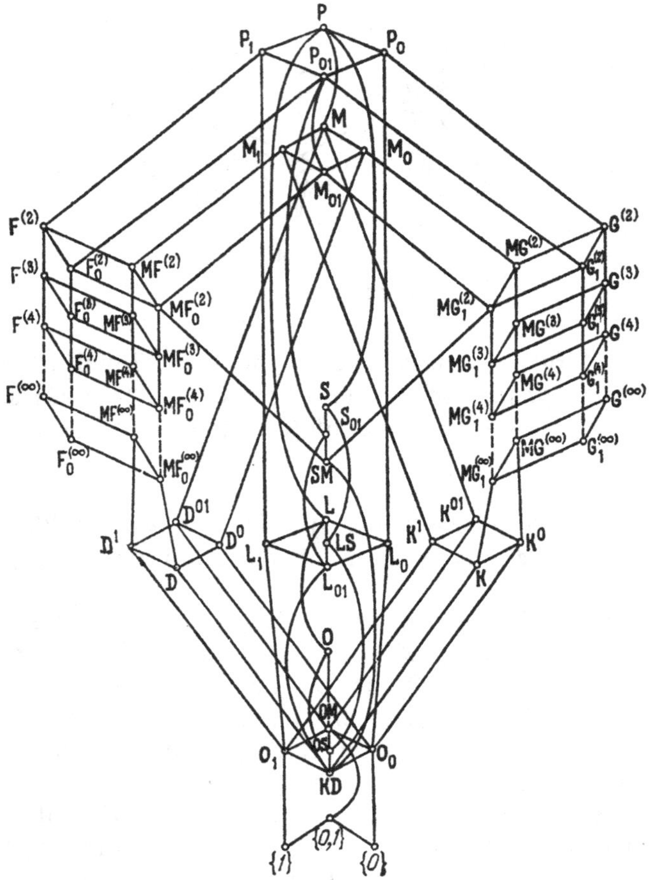

Figure 6

following limiting classes exist: $\mathbf{F}^{(\infty)}$, $\mathbf{F}_0^{(\infty)}$, $\mathbf{MF}^{(\infty)}$, $\mathbf{MF}_0^{(\infty)}$, $\mathbf{G}^{(\infty)}$, $\mathbf{G}_1^{(\infty)}$, $\mathbf{MG}^{(\infty)}$, $\mathbf{MG}_1^{(\infty)}$, $\mathbf{D}^1$, $\mathbf{D}$, $\mathbf{K}^0$, $\mathbf{K}$, $\mathbf{O}_1$, $\mathbf{O}_0$, **KD**, $\{1\}$, and $\{0\}$, i.e., there exist infinite chains of embedded classes connecting them to **P**. Let us describe the classes consisting of functions of a single variable:

$$\mathbf{O} = \{0, 1, x, \overline{x}\}, \quad \mathbf{OM} = \{0, 1, x\}, \quad \mathbf{OS} = \{x, \overline{x}\},$$

$$\mathbf{KD} = \{x\}, \quad \mathbf{O}_1 = \{1, x\}, \quad \mathbf{O}_0 = \{0, x\}.$$

7.28. Find self-dual completeness conditions by means of the system of Post classes (i.e., obtain the result of Problem 7.15).

7.29. Describe (using the Post system) bases in the class of self-dual monotone functions.

Hints

7.2. **L**, **M**

7.3. The functionally closed classes relative to extended composition are the functionally closed classes relative to ordinary composition (and no others) that contain both constants or consist of one and only one constant.

7.5. There is only one **L**-pre-complete class, the class **O** of functions of a single variable (all of which are linear). The class consists of the four functions x, $\overline{x} = x+1$, 0, and 1.

7.6. We have found the **L**-pre-complete class **O** (Problem 7.5); all the other classes are contained in it. These classes are **OM** = $\{x, 0, 1\}$, $\{0, 1\}$, $\{0\}$, and $\{1\}$. Note that **OM** is the class of monotone functions of a single variable (as well as the linear monotone functions). In what follows, we will routinely denote by QR the intersection of the classes Q and R (as in the example **OM** = **ML**, this notation is not unique). The connection tree is shown in Figure 7.

7.7. To prove the assertion, it is sufficient to show that if there are functions $\omega_1 \notin \mathbf{D}^{01}$ and $\omega_2 \notin \mathbf{K}^{01}$ in some system of functions $\Phi \subset \mathbf{M}$, Φ will be **M**-complete. We have already repeatedly used this technique of proving the $\mathfrak{U}$-pre-completeness of classes (cf. Problems 6.18 and 7.5). Here

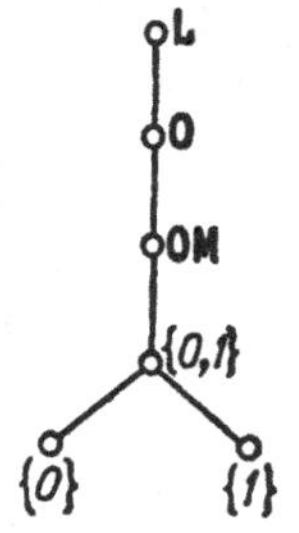

Figure 7

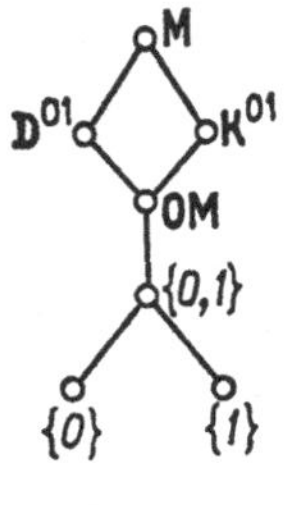

Figure 8

we need only note that neither $\mathbf{D}^{01}$ nor $\mathbf{K}^{01}$ is contained in the other.

To prove the **M**-completeness of $\{\omega_1,\omega_2\}$, it is sufficient to prove that disjunctions and conjunctions may be represented by means of their compositions, since the system is **M**-complete (Problem 5.9). In proving that xy and $x \vee y$ may be represented in terms of ω_1 and ω_2, it is convenient to use a representation of ω_1 and ω_2 in the form of proper DNF and CNF (Definition 5.3, Problems 5.12 and 5.13).

7.8. In both $\mathfrak{U}$-pre-complete classes, there is a unique pre-complete class $\mathbf{OM} = \{x, 1, 0\}$. Its subclasses are already known (Problem 7.5). The connection tree is depicted in Figure 8.

7.9. Note first that in (b) $\mathfrak{U}$ consists of the single set **M**; in (c), a single set consisting, in turn, of the single function $xy \vee yz \vee xz$; in (d), the single set $\{x+y+z\}$; in (e), the two sets $\{xy \vee yz \vee xz\}$ and $\{x+y+z\}$; in (f), the single set $\{xy \vee yz \vee xz, x+y+z\}$; in (g), the set $\{\overline{x}\}$; in (h), the set **L** and $\{xy\}$; in (i), the sets $\{xy\}$ and $\{\overline{x}\}$; in (j) the set $\{xy, \overline{x}\}$; and in (k), the sets $\{xy\}$, $\{x \vee y\}$, and $\{\overline{x}\}$.

By the reasoning in sec. 1.7, in $\mathfrak{U}$-completeness problems it is necessary to enumerate all the $\mathfrak{U}$-pre-complete classes. These classes are, in turn, the maximal classes that do not contain any sets from the system $\mathfrak{U}$. Maximality denotes that every larger class contains at least one of the sets of $\mathfrak{U}$.

7.10. The system $\{xy, x \vee y\}$ is self-dual complete, but not complete.

7.11. (2) The system $\{xy, \overline{x}\}$ is complete, though the system consisting of the single function xy is not self-dual complete.

(3) Sufficiency is self-evident, though, as is clear from the example of a complete (i.e., self-dual complete) system $\{\overline{xy}\}$, this condition is not necessary.

7.13. This problem may be reduced to a $\mathfrak{U}$-completeness problem as follows. A set R occurs in $\mathfrak{U}$ if the collection R of functions that occur in R together with its dual functions forms with $\bar{x}$ a complete system. There is an infinite number of such sets of functions, though it is possible to effectively determine whether some functionally closed class contains at least one of them.

Answer. Some system of functions Φ is self-dual complete (relative to extended composition) if and only if it contains the functions (a) $\omega_1 \notin \mathbf{L}$; (b) $\omega_2 \notin \mathbf{D}^{01}$; (c) $\omega_3 \notin \mathbf{K}^{01}$.

7.14. Proof by induction on the rank of the composition.

7.15. Compare the conditions of the next problems; also, see the hint to Problem 7.19.

7.16. Give a definition of a function that does not occur in the class $\mathbf{F}^{(2)}$ ($\mathbf{G}^{(2)}$), and use Problem 3.6.

7.17. The functions 0, xy, $\overline{xy}$, and $\overline{x \vee y}$ do not belong to the class $\mathbf{F}^{(2)}$; the functions 1, $x \vee y$, $\overline{xy}$, and $\overline{x \vee y}$ do not belong to $\mathbf{G}^{(2)}$. Use Problem 3.10.

7.18. It is always possible to obtain a function of at most three variables. It is in general impossible to decrease the number of variables any further (give examples).

7.19. It is always possible to obtain a function of two variables. A further decrease in the number of variables is in general impossible.

(*Hint For Problem 7.15.* To prove necessity, it may help to use Problem 7.13.)

Sufficiency. 1. It may be assumed that $\omega_4 \notin \mathbf{S}$ depends on two variables (Problem 3.10). For the sake of definiteness suppose that $\omega_4 \notin \mathbf{F}^{(2)}$ (Problem 7.16). Then, by Problem 7.17, we obtain one of the functions 0, xy, $\overline{xy}$, or $\overline{x \vee y}$. The last two functions are Sheffer functions. Cases must be considered in which we are given one of the functions $\psi_4 = 0$ or $\psi_4(x, y) = xy$.

2. Now considering $\omega_5 \notin G^{(2)}$ and using Problem 7.19, we may obtain one of the functions ψ_5: 1, $x \vee y$, or $x+y$ (or again the Sheffer function $\overline{x} \vee \overline{y}$).

3. Consider distinct combinations of ψ_4 and ψ_5. If we are dealing with the constants 0 and 1, our reasoning reduces to the case of extended composition (Problem 7.13). That the system $\{xy, x \vee y\}$ is self-dual complete has already been proved. The case $\{xy, x+y\}$ reduces in turn to this case.

4. If we are given $x+y$, then $x+x = 0$. From the nonlinear function ω_1 it is possible to obtain a nonlinear function of two variables (Problem 4.12); from the latter function, 1, xy, $x \vee y$, or $x\overline{y}$ may be derived, and in turn xy. These all reduce to the cases already analyzed.

5. If we are given 0 and $x \vee y$, then xy may be obtained from $\omega_2 \notin D^{01}$ if ω_2 is monotone. If ω_2 is not monotone, a nonmonotone function of two variables may be obtained from it, and either 1, $\overline{x}$, $x+y$ or $x\overline{y}$ may in turn be derived from it. In the second case, we obtain a complete system in the ordinary sense. The remaining cases have already been analyzed.

6. The case $\{1, xy\}$ is dual to case 5.

7.20. *Answer:* $\mathbf{D}^{01}$, $\mathbf{K}^{01}$, and $\mathbf{M}_0$ (the class of all monotone functions that preserve the additive identity, i.e., all the monotone functions, except for the multiplicative identity (Problem 5.4), and $\mathbf{M}_1$ (all the monotone functions, except for the additive identity). That $\mathbf{M}_0$ and $\mathbf{M}_1$ are closed follows from Problem 6.6.

7.21. As in the preceding problem, the **L**-pre-complete classes consist of the intersections of **L** with the other pre-complete classes, e.g., $\mathbf{L}_0 = \mathbf{L} \cap \mathbf{P}_0$, $\mathbf{L}_1 = \mathbf{L} \cap \mathbf{P}_1$, and $\mathbf{LS} = \mathbf{L} \cap \mathbf{S}$, and the class **O** of functions of a single variable that arises when considering the **L**-pre-complete classes relative to extended composition. Use Problem 4.6 (note, too, the hints to Problem 6.14). Recall, also, that the system of functions $\{x+y, 1\}$ is **L**-complete (Problem 7.4).

7.22. Use the representation of functions from $\mathbf{P}_1$ in the form of Zhegalkin polynomials. How can the Zhegalkin polynomials for functions from $\mathbf{P}_1$ be described?

7.23. Use the representation of functions from $\mathbf{P}_1$ in PCNF. How can they be described (cf. Problem 1.17)? In proving that the given system is a $\mathbf{P}_1$-basis, note especially the proof that the system $\{x \rightarrow y\}$ is not $\mathbf{P}_1$-complete. Use the class $\mathbf{F}^{(2)}$.

7.24. The intersections of $\mathbf{P}_1$ with the pre-complete classes $\mathbf{P}_0$, $\mathbf{L}$, and $\mathbf{M}$ are the $\mathbf{P}_1$-pre-complete classes, along with the class $\mathbf{F}^{(2)}$ already encountered in the preceding problem, i.e., $\mathbf{P}_{01}$, $\mathbf{L}_1$, $\mathbf{M}_1$, and $\mathbf{F}^{(2)}$.

Suppose that the system of functions $\Phi \subset \mathbf{P}_1$ contains a function for each of these classes that does not occur in the system. We present in outline a proof that Φ is $\mathbf{P}_1$-complete:

1. Obtain 1 from the function $\omega_1 \notin \mathbf{P}_{01}$.
2. Using Problem 5.15, derive $x \rightarrow y$ or $x+y+1$ from the function $\omega_2 \notin \mathbf{M}_1$ and 1.
3. Obtain xy or $x+y+1$ from the function $\omega_4 \notin \mathbf{F}^{(2)}$ and 1.
4. Given that $x \rightarrow y$ and xy, the system Φ is then $\mathbf{P}_1$-complete (Problem 7.23).
5. But if we are given $x+y+1$, obtain the conjunction xy from ω_2 using $x+y+1$. We again obtain a $\mathbf{P}_1$-complete system (Problem 7.22).

7.25. *Answer:* 1. The system of functions $\{x \quad y, x+y\}$ (as well as $\{xy, x+y\}$ and $\{\overline{x}y, x \vee y\}$) will be $\mathbf{P}_0$-complete.

2. The $\mathbf{P}_0$-complete classes are $\mathbf{P}_{01}$, $\mathbf{L}_0$, $\mathbf{M}_0$, and $\mathbf{G}^{(2)}$.

7.26. Suppose that $f(x_1, \dots, x_{n-1}, x_n) \in \mathbf{S}$. Then the functions $\omega(x_1, \dots, x_{n-1}) = f(x_1, \dots, x_{n-1}, 1)$ and $\psi(x_1, \dots, x_{n-1}) = f(x_1, \dots, x_{n-1}, 0)$ are dual: $\omega^+ = \psi$. We let $m(x, y, z)$ denote the function $xy+yz+xz = xy \vee yz \vee xz$. Note that $m(x, y, 1) = x \vee y$ and $m(x, y, 0) = xy$. Now suppose that the function ω may be represented somehow in the form of a composition of xy, $x \vee y$, and $\overline{x}$. (We may limit ourselves to one of the functions xy or $x \vee y$.) We now replace disjunctions everywhere in the composition representing ω by the functions $m(*, *, x_n)$, and replace conjunctions everywhere by the functions $m(*, *, \overline{x}_n)$. Let us describe this transformation in a somewhat simpler fashion. The function ω may be obtained (after several steps) from conjunctions, disjunctions, and negations. This substitution is performed successively on each step. If, for example, on some step a disjunction is replaced by $m(*, *, \overline{x}_n)$, the asterisks are replaced by functions obtained in the preceding step from functions that had been substituted for the arguments of the disjunctions. A rigorous definition should be given by induction, though this we will not do, instead limiting ourselves to an example. Suppose that

$$\omega(x_1,x_2,x_3,x_4) = ((x_1 \vee \overline{x}_2x_3) \vee \overline{x_1x_2})x_4.$$

Now let us perform the successive substitutions:

$$\overline{x}_2x_3 \Rightarrow m(\overline{x}_2,x_3,\overline{x}_5); \quad \overline{x_1x_2} \Rightarrow \overline{m(x_1,x_2,\overline{x}_5)};$$

$$x_1 \quad \overline{x}_2x_3 \Rightarrow m(x_1, m(\overline{x}_2,x_3,\overline{x}_5), x_5);$$

$$(x_1 \vee \overline{x}_2x_3) \vee \overline{x_1x_2}$$

$$\Rightarrow m(m(x_1,m(\overline{x}_2,x_3,\overline{x}_5), x_5), \overline{m(x_1,x_2,\overline{x}_5)}, x_5)$$

finally obtaining

$$\omega \Rightarrow m(m(m(x_1,m(\overline{x}_2,x_3,\overline{x}_5),x_5), \overline{m(x_1, x_2,\overline{x}_5)},x_5),x_4,\overline{x}_5).$$

Prove that $f(x_1, \dots, x_{n-1}, x_n)$ is obtained after performing this substitution in the composition representing the function $\omega(x_1, \dots, x_{n-1})$. You will have thereby proven that the system $\{m(x, y, z), \overline{x}\}$ is S-complete.

7.27. *Answer:* The classes **SL** = **S** ∩ **L** and $\mathbf{S}_{01} = \mathbf{S}_0 = \mathbf{S}_1$ are S-pre-complete.

Suppose that the functions $\omega_1 \notin \mathbf{SL}$ and $\omega_2 \notin \mathbf{S}_{01}$ are found in the system of functions Φ.

1. From ω_2 we may obtain $\overline{x}$.
2. By Problem 4.10, we obtain from the function ω_1 a self-dual nonlinear functions of three variables $\psi(x, y, z)$. Study the Zhegalkin polynomial for $\psi(x, y, z)$ using the self-duality condition. Its nonlinear part must have the form $xy+yz+xz$. Using $\overline{x}$, its linear part may be represented by zero.

7.28. The self-dual classes such that **P** is the unique class containing these classes and $\overline{x}$ may be taken as the collection of sets $\mathfrak{U}$ (cf. solution of Problem 7.13); these classes are **P**,

$\mathbf{P}_{01}$, **M**, and $\mathbf{M}_{01}$. Thereafter, follow the technique set forth in 1.

7.29. An **SM**-basis consists of a single element, i.e., any function from **SM**, except for x.

Solutions

7.1. Necessity follows from the fact that a $\mathfrak{U}$-pre-complete class does not contain any set $R_1 \in \mathfrak{U}$. To prove sufficiency, note that the functionally closed class generated by the system of functions Φ is not in any $\mathfrak{U}$-pre-complete class and that every class not in any of the sets $R_i \in \mathfrak{U}$ is contained in at least one $\mathfrak{U}$-pre-complete class.

7.2. Every class functionally closed relative to extended composition is functionally closed relative to ordinary composition. Similarly, a class which is pre-complete relative to extended composition is pre-complete relative to ordinary composition. It remains for us to note that the classes **L** and **M** are closed, while $\mathbf{P}_0$, $\mathbf{P}_1$, and **S** are not closed relative to extended composition. It might also be noted that a system of functions containing nonlinear and nonmonotone functions is complete relative to extended composition (by Post's theorem).

7.3. In solving the preceding problem, we have already noted that classes which are functionally closed relative to extended composition are closed in the ordinary sense. Thus, we must determine which of the functionally closed classes (in the sense of Chapter 6) are closed relative to extended composition. If a class contains at least one function that is not a constant, it contains both constants. (As we have already noted, they may be obtained from the function by means of extended composition.) There remain the classes consisting of the constants, of which there are three: {0, 1}, {0}, and {1}.

7.4. It is clear that a sum of any number of variables $x_1 + \ldots + x_{k-1} + x_k$ may be obtained by repeated application of the composition $x+y$. Substituting $x_k = 1$, we obtain an arbitrary linear function with free term equal to 1. It remains for us to note that $x+x = 0$. L-completeness is proved. Minimality is self-evident.

7.5. Let us prove that **O** is the only **L**-pre-complete class. For this purpose, it is sufficient to prove that every function $\omega \in \mathbf{L}$, $\omega \notin \mathbf{O}$ forms an **L**-complete system. Considering, if necessary, the composition of ω with itself, we may assume without loss of generality that ω contains at least three (actual) variables. Substituting 0 for all the variables other than any three of them (say, x_1, x_2, x_3), we obtain the function $x_1+x_2+x_3+a$. If $a = 1$, we substitute $x_3 = 1$; if $a = 0$, we substitute $x_3 = 0$. We obtain the function x_1+x_2, which is an **L**-basis (Problem 7.4).

7.6. We may add to **OM** only $\bar{x} \in \mathbf{O}$, after which we obtain **O**. Analogously, to {0, 1} we may add only x.

7.7. Suppose that $\omega_1 \notin \mathbf{D}^{01}$. Let us show that a conjunction may be obtained from it by means of extended composition. Represent ω_1 in the form of a proper DNF. Then, since $\omega_1 \notin \mathbf{D}^{01}$, we may find in ω_1 at least one elementary conjunction that contains more than one cofactor (say, $x_1 \ldots x_k$, $k \geqslant 2$). We substitute 0's in place of all the variables other than those that occur in the selected conjunction ($x_1 \ldots x_k$). All the elementary conjunctions other than $x_1 \ldots x_k$ are then equal to 0, since (by virtue of the fact that the DNF is proper) at least one variable other than $x_1 \ldots x_k$ may be found in each of them. As a result, we obtain the conjunction $x_1 \ldots x_k$. If we now substitute units for all the variables of this conjunction other than any two of them, we obtain a conjunction of two variables x_1x_2. All the variables other than x_1 may also be equated.

Analogously, using a proper CNF for $\omega_2 \notin \mathbf{K}^{01}$, it is possible to construct $x_1 \vee x_2$. Thus, by virtue of the **M**-completeness of the system $\{xy, x \vee y\}$, the system $\{\omega_1, \omega_2\}$ is also **M**-complete. It would also be possible to use the law of duality.

Remark 1. We have $\omega \in \mathbf{M}$, $\omega \notin \mathbf{D}^{01}$ if and only if there is at least one elementary conjunction of at least two variables in the proper DNF for ω. This follows from the uniqueness of the representation of a monotone function in the form of a proper DNF (cf. nonlinearity criterion for a function using

Zhegalkin polynomials in the remark on page 67). Analogously, $\omega \notin \mathbf{M}$, $\omega \notin \mathbf{K}^{01}$ if and only if there is at least one elementary disjunction containing at least two terms in the CNF for ω.

Remark 2. From the proof, it follows that by means of ordinary composition a conjunction may be obtained from any function $\omega \notin \mathbf{D}^{01}$, and that a disjunction may be obtained from any function $\omega \notin \mathbf{K}^{01}$ and 1.

7.9. The set Φ must contain the functions:

(a) $\omega_1 \notin \mathbf{O}$, $\omega_2 \notin \mathbf{M}$;
(b) $\omega_1 \notin \mathbf{L}$, $\omega_2 \notin \mathbf{D}^{01}$, $\omega_3 \notin \mathbf{K}^{01}$;
(c) $\omega_1 \notin \mathbf{L}$, $\omega_2 \notin \mathbf{D}^{01}$, $\omega_3 \notin \mathbf{K}^{01}$;
(d) $\omega_1 \notin \mathbf{O}$, $\omega_2 \notin \mathbf{M}$;
(e) $\omega_1 \notin \mathbf{O}$, $\omega_2 \notin \mathbf{D}^{01}$, $\omega_3 \notin \mathbf{K}^{01}$;
(f) $\omega_1 \notin \mathbf{L}$, $\omega_2 \notin \mathbf{M}$;
(g) $\omega \notin \mathbf{M}$;
(h) $\omega_1 \notin \mathbf{O}$, $\omega_2 \notin \mathbf{D}^{01}$;
(i) $\omega \notin \mathbf{D}^{01}$;
(j) $\omega_1 \notin \mathbf{L}$, $\omega_2 \notin \mathbf{M}$;
(k) $\omega \notin \mathbf{OM}$.

In each of these cases, we require that Φ contain functions that do not belong to the $\mathfrak{U}$-pre-complete classes.

7.10. Note the example in the hint. In this system, the function xy is self-dual representable, and forms (together with $\overline{x}$) a complete system.

7.11. (2) Note the example in the hint. The function y is the only function that is self-dual representable in terms of xy.

7.13. Note that if some set contains an arbitrary set in $\mathfrak{U}$, it itself occurs in $\mathfrak{U}$. Therefore, asking whether some system of functions occurring in $\mathfrak{U}$ is contained in some functionally closed class Q is the same as asking whether Q is itself in $\mathfrak{U}$, i.e., whether the class Q generated by functions occurring in Q together with their duals (self-dual representable in terms of Q; Problem 7.12) and the function $\overline{x}$ coincides with the class of all the logical functions $\mathbf{P}$. Since all the classes may be effectively described, this question is easily answered by means of the table. It is clear that **L**, $\mathbf{D}^{01}$, and $\mathbf{K}^{01}$ are the only maximal classes not in $\mathfrak{U}$ (since $\bar{x} \in \mathbf{L}$ and $\tilde{\mathbf{L}} = 1$ and since only the constants 0 and 1 and x are self-dual representable in $\mathbf{D}^{01}$; $\tilde{\mathbf{D}}^{01} = 0$, and similarly, $\tilde{\mathbf{K}}^{01} = 0$). The classes **P** and **M** containing them are obviously already self-dual complete. Thus, **L**, $\mathbf{D}^{01}$, and $\mathbf{K}^{01}$ are $\mathfrak{U}$-pre-complete classes, and we have obtained the result stated in the hint to this problem.

Remark. In solving the preceding problems, we saw that the system $\{xy, x \vee y\}$ is a typical example of a self-dual complete system of functions. From Problem 7.13, it follows that if we are given an arbitrary self-dual system Φ for extended composition, the functions xy and $x \vee y$ may be represented in terms of its elements. In fact, either the entire class **M** or, in general, all the functions (**P**) may be represented in terms of these elements.

7.14. The cases $\mathbf{F}^{(2)}$ and $\mathbf{G}^{(2)}$ may be considered analogously, and therefore we limit ourselves to $\mathbf{F}^{(2)}$ (we may also use the law of duality).

Suppose that Φ is some set of functions from $\mathbf{F}^{(2)}(\Phi \subset \mathbf{F}^{(2)})$ and that it is already known that compositions of rank k also belong to $\mathbf{F}^{(2)}(\Phi^{(k)} \subset \mathbf{F}^{(2)})$. Let us show that $\Phi^{(k+1)} \subset \mathbf{F}^{(2)}$. Suppose that $f \in \Phi^{(k+1)}$. If f is obtained from some function $\omega \in \Phi^{(k)}$ by renaming one of the variables (by Definition 2.3a), the result follows directly from the definition of $\mathbf{F}^{(2)}$. Now let $\omega(x_1, \ldots, x_n), \psi(y_1, \ldots, y_\ell) \in \Phi^{(k)}$ and

$$f(x_1, \ldots, x_{i-1}, x_{i+1}, \ldots, x_n; \; y_1, \ldots, y_\ell)$$

$$= \omega(x_1, \ldots, x_{i-1}, \psi(y_1, \ldots, y_\ell), x_{i+1}, \ldots, x_n).$$

We wish to prove that $f \in \mathbf{F}^{(2)}$. Suppose that we are given two sequences of values of its variables:

$$\Upsilon' = (\alpha_1', \ldots, \alpha_{i-1}', \alpha_{i+1}', \ldots, \alpha_n'; \beta_1', \ldots, \beta_\ell'),$$

$$\Upsilon'' = (\alpha_1'', \ldots, \alpha_{i-1}'', \alpha_{i+1}'', \ldots, \alpha_n''; \beta_1'', \ldots, \beta_\ell''),$$

on which $f(\Upsilon') = f(\Upsilon'') = 0$. We set $\alpha_i' = \psi(\beta_1', ..., \beta_m')$ and $\alpha_i'' = \psi(\beta_1'', ..., \beta_m'')$. Then the function ω is 0 on the sequences

$$\alpha' = (\alpha_1', ..., \alpha_{i-1}', \alpha_{i+1}', ..., \alpha_n')$$

and

$$\alpha'' = (\alpha_1'', ..., \alpha_{i-1}'', \alpha_i'', \alpha_{i+1}'', ..., \alpha_n'')$$

and since by the induction hypothesis $\omega \in F^{(2)}$, these sequences share a common additive identity in some place. If $\alpha_j' = \alpha_j'' = 0$ for $j \neq i$, then the sequences Υ' and Υ'' share a common additive identity. If $\alpha_i' = \alpha_i'' = 0$, i.e., if $\psi(\beta') = \psi(\beta'') = 0$, the sequences β' and β'' share a common additive identity, since $\psi \in F^{(2)}$ (by the induction hypothesis), that is, the sequences Υ' and Υ'' also share a common additive identity, i.e., $f \in F^{(2)}$. The proof is complete.

7.15. The solution to this problem follows the solution to Problem 7.19.

7.16. The function $f \notin F^{(2)}$ if there exist sequences α and β such that $f(\alpha) = f(\beta) = 0$ and if for every i either $\alpha_i \neq \beta_i$ or $\alpha_i = \beta_i = 1$. We may similarly give a definition of the function $f \notin G^{(2)}$.

Now suppose that $\omega \notin S$. Then there exists a sequence α such that $\omega(\alpha) = \omega(\overline{\alpha})$. The sequences α and $\overline{\alpha}$ do not share any common additive nor multiplicative identities. Therefore, if $\omega(\alpha) = \omega(\overline{\alpha}) = 0$, $\omega \notin F^{(2)}$; if $\omega(\alpha) = \omega(\overline{\alpha}) = 1$, $\Phi \notin G^{(2)}$.

Remark. The assertion we have proved may be given the following form. Suppose that Φ is some set of logical functions. We let $\overline{\Phi}$ denote the complement of Φ with respect to the set of all functions P. Then the assertion we have proved states that $\overline{S} \subset \overline{F^{(2)}} \cup \overline{G^{(2)}}$. By the ordinary relations of duality in set theory, this relation is equivalent to the relation $S \supset F^{(2)} \cap G^{(2)}$, i.e., functions that simultaneously belong to $F^{(2)}$ and $G^{(2)}$ are self-dual. (This also follows easily from the definitions.)

7.17. Suppose that $f \notin S$, and that the property does not hold if we arbitrarily equate the variables. Suppose, further, that f is not a constant. Then, by Problem 3.10, f depends on at most two variables. Thus, $f(x, y) \notin S$. Then for some sequence (α_1,α_2), $f(\alpha_1,\alpha_2) = f(\overline{\alpha}_1,\overline{\alpha}_2)$. If $\alpha_1 = \alpha_2$, by equating x and y, we obtain one of the constants. i.e., again a non-self-dual function. That is, $\alpha_1 \neq \alpha_2$. Suppose that $f(0, 1) = f(1,0) = 0$. By the reasoning just given, $f(0, 0) = f(1, 1)$, and there are two possibilities left: either $f(0, 0) = 0$, so that $f(x, y) = xy$, or $f(0, 0) = 1$, so that $f(x, y) = \overline{xy}$. We may consider the case $f(0, 1) = f(1, 0) = 1$ similarly. In this case, either $f(x, y) = x \vee y$, or $f(x, y) = \overline{x} \vee \overline{y}$. Obviously, it is impossible to equate the variables of these functions any further and still retain the non-self-dual property. Thus, there are six functions that possess the necessary properties: 0, 1, xy, $x \vee y$, $\overline{xy}$, and $\overline{x} \vee \overline{y}$. Whether these functions belong to $F^{(2)}$ and $G^{(2)}$ can be easily decided directly based on the definitions. The case of the constants has already been discussed. As an example, consider the function $f(x, y) = x \vee y$. Since $f(0, 1) = f(1, 0) = 1$, f does not occur in $G^{(2)}$. As for $\overline{xy}$ and $\overline{x} \vee \overline{y}$, these are Sheffer functions and therefore do not occur in any pre-complete class, that is, in general do not occur in any functionally closed class other than **P**.

7.18. As an example, consider the case $f(x_1, \dots, x_n) \notin G^{(2)}$. The case $f \notin F^{(2)}$ may be considered analogously, and in addition, may be obtained by means of the law of duality. In solving Problem 7.16, we gave a definition of the function $f \notin G^{(2)}$. There must exist sequences $\alpha = (\alpha_1, \dots, \alpha_n)$ and $\beta = (\beta_1, \dots, \beta_n)$ that do not share any common multiplicative identities and on which $f(\alpha) = f(\beta) = 1$. Let us divide the variables $x_1, \dots, x_n$ into three groups. In the first group will be those variables x_i for which $\alpha_i = 0$ and $\beta_i = 1$, in the second group those variables x_i for which $\alpha_i = 1$ and $\beta_i = 0$, and in the third group those variables x_i for which $\alpha_i = \beta_i = 0$. We equate the variables within each of these groups, obtaining the function $\omega(y_1, y_2, y_3)$ for which $\omega(0, 1, 0) = \omega(1, 0, 0) = 1$, i.e., $\omega \notin G^{(2)}$. Thus, we have proved the first part of the assertion: It is indeed possible to obtain a function of three variables.

Now let us present an example of a function of three variables in which it is impossible to further decrease the number of variables. We use the function $\omega(x, y, z)$ such that $\omega(0, 1, 0) = \omega(1, 0, 0) = 1$, with $\omega = 0$ on all other sets, i.e., the function $\omega \notin G^{(2)}$ "only because of the single pair of sequences" (0, 1, 0) and (1, 0, 0). Clearly, it is impossible to equate the variables in such a way that both these sequences correspond to given sequences of functions with already equated variables (since the sequences coincide only in one category). Let us make this reasoning more rigorous. We have $\omega(x,y,z) = \bar{x}y\bar{z} \vee x\bar{y}\bar{z} = \bar{z}(\bar{x}y \vee x\bar{y}) = \bar{z}(x+y)$. The variables x and y occur in ω symmetrically; therefore, we need only consider two possible "equatings": x with y, and x with z, and prove that in this we obtain functions that belong to $G^{(2)}$. (Thereby, equating all three variables is then ascribed to functions from $G^{(2)}$; this, incidentally, is also immediately clear.) We obtain $\omega(x, y, z) = 0 \in G^{(2)}$; $\omega(x, y, z) = \bar{x}(x+y) = \bar{x}y \in G^{(2)}$, since the only sequence (0, 1) on which $\bar{x}y = 1$ contains 1. We may analogously prove that the variables of the function $\overline{z(x+y)}$ cannot be equated in such a way as to again obtain functions that do not belong to $F^{(2)}$.

7.19. Let us show that it is possible to obtain a function in $F^{(2)}$ that depends on at most two variables. We equate the variables of ω_1 so as to obtain a non-self-dual function ψ_1 of two variables. By Problem 7.17, the resulting function either does not belong to $F^{(2)}$ or is the function 1 or the function $x \quad y$. In the first case, our goal has been reached. But if $\psi_1 \in F^{(2)}$, we consider $\omega_2 \notin F^{(2)}$. By the preceding problem, by equating the variables of ω_2 it is possible to obtain a function of three variables $\psi_2(x, y, z)$ such that $\psi_2(0, 1, 1) = \psi_2(1, 0, 1) = 0$. Depending on which of the functions 1 or $x \quad y$ we are dealing with, we consider the functions $\nu_1(x, y) = \psi_2(x, y, 1)$ or $\nu_2(x, y) = \psi_2(x, y, x \vee y)$. For both these functions, we have $\nu_i(0, 1) = \nu_i(1, 0) = 0$, i.e., $\nu_i \notin F^{(2)}$, that is, the desired function has been constructed. The example of the function $\omega(x, y) = xy$, $\omega \notin S$, $\omega \notin F^{(2)}$ shows that it is, in general, impossible to further decrease the number of variables.

The dual case ($\omega_2 \notin G^{(2)}$) is considered analogously or may be obtained from the above case by means of the law of duality.

7.15. 1. Necessity. The necessity of Conditions (a), (b), and (c) was basically checked in solving Problem 7.13. Let us verify the necessity of Condition (d). If all the functions in Φ were self-dual, all the functions representable in terms of these functions would also possess this property, and thereby would be self-dual representable. Since $\bar{x} \in S$, we do not obtain a complete system. Let us now consider the condition

$\omega_5 \notin \mathbf{F}^{(2)}$. Since $\mathbf{G}^{(2)}$ is the class dual to $\mathbf{F}^{(2)}$, the functions of $\mathbf{F}^{(2)} \cap \mathbf{G}^{(2)}$ are self-dual representable in terms of functions from $\mathbf{F}^{(2)}$ (cf. Problem 7.12). But by the remark following the solution of Problem 7.16, $\mathbf{F}^{(2)} \cap \mathbf{G}^{(2)} \subset \mathbf{S}$. Adjoining the functions $\overline{x} \in \mathbf{S}$ to $\mathbf{F}^{(2)} \cap \mathbf{G}^{(2)}$ does not take us outside $\mathbf{S}$. The necessity of Condition (f) may be verified analogously.

2. *Sufficiency.* Suppose that $\omega_4 \notin \mathbf{S}$ depends on two variables and, for the sake of definiteness, suppose that it does not belong to $\mathbf{F}^{(2)}$ (the case $\omega_4 \notin \mathbf{G}^{(2)}$ may be considered analogously). Then by Problem 7.17, ω_4 is one of the functions 0, xy, $\overline{xy}$, or $\overline{x \vee y}$. The last two functions are Sheffer functions, so that in these cases the system Φ is complete and, moreover, self-dual complete.

Thus, we may assume that we are dealing with either the function $\psi_4 = 0$ or the function $\psi_4(x, y) = xy$. By Problem 7.19, the function $\psi_5(x, y) \notin \mathbf{G}^{(2)}$ (which depends on two variables) may be obtained from $\omega_4 \notin \mathbf{S}$ and $\omega_5 \notin \mathbf{G}^{(2)}$. Let us look more carefully at what form the function ψ_5 may take. We require a pair of sequences that do not share any common multiplicative identities and on which $\psi_5 = 1$, i.e., either (0, 0) and (1, 1) or (0, 1) and (1, 0). In the first case, by equating the variables of ψ_5, we obtain a multiplicative identity. Suppose that this cannot be done. Then $\psi_5(0, 1) = \psi_5(1, 0) = 1$, and we cannot have $\psi_5(0, 0) = \psi_5(1, 1) = 1$. There remain three possibilities: $\psi_5(0, 0) = 0$, $\psi_5(1, 1) = 1$; $\psi_5(0, 0) = 1$, $\psi_5(1, 1) = 0$; and $\psi_5(0, 0) = 1$, $\psi_5(1, 1) = 0$. In these cases, the function ψ_5 has the form $x \quad y$, $x+y$, and $\overline{xy}$, respectively. In the latter case, we have a Sheffer function, and no further discussion is needed. Thus, on the one hand, we are dealing with one of the functions ψ_4; 0, or xy; on the other, one of the functions ψ_5: 1, $x \vee y$, or $x+y$.

Let us consider the different possible combinations. If we are given the constants 0 and 1, the result we have proved follows from the self-dual completeness criterion for extended composition (Problem 7.13). That the system $\{xy, x \vee y\}$ is self-dual complete has already been repeatedly noted. The disjunction $x \vee y = xy+x+y = \psi_5(xy, \psi_5(x, y))$ may be obtained from the functions $\psi_5(x, y) = x+y$ and $\psi_4(x, y) = xy$, i.e., this case reduces to the preceding case. There remain the dual cases $\{1, xy\}$ and $\{0, x \vee y\}$, and the case $x+y$. (Since $x+x = 0$, the presence of 0 does not add anything new.)

Let us begin with the last case. Then by Problem 4.11, the

nonlinear function $\psi_1(x, y) = xy+\alpha x+\beta y+\gamma$ (of two variables) may be obtained from the nonlinear function ω_1 and 0. If $\gamma = 1$, then $\omega_1(0, 0) = 1$, and we arrive at the case of two constants already analyzed. Thus, $\gamma = 0$. If $\alpha = \beta = 0$, $\psi_1(x, y) = xy$, and we again obtain the case $\{xy, x+y\}$. The case $\alpha = \beta = 1$ is analogous; thus $\psi_1(x, y) = x \vee y$. The remaining two cases are the same. Suppose that $\alpha = 1$ and $\beta = 0$; then $\psi_1(x,y) = x\bar{y}$. We have $\psi_1(x, \psi_1(x, y)) = x(\bar{x} \vee y) = xy$ (cf. (2.24)). In this case, we again arrive at a case already treated.

Suppose that we are given the functions 0 and $x \vee y$. Let us consider the function $\omega_2 \notin \mathbf{D}^{01}$. If ω_2 is monotone, by Remark 2 at the end of the solution to Problem 7.8, the conjunction xy may be obtained from ω_2 and 0, and in this case the proof is complete. But if $\omega_2 \notin \mathbf{M}$, by Problem 5.15 a nonmonotone function of two variables $(\psi_2(x, y))$ may be obtained from ω_2 and 0, and $\psi_2(1, 0) = 1$ and $\psi_2(1, 1) = 0$. If $\psi_2(0, 0) = 1$, by equating x and y, we obtain a negation, and as a result a complete system. Now suppose that $\psi_2(0, 0) = 0$. There are two possibilities to consider: either $\psi_2(x, y) = x+y$, or $\psi_2(x, y) = x\bar{y}$. The case $\{x+y, x \vee y\}$ has already been analyzed. As we saw in the preceding case, xy may be obtained from $x\bar{y}$, and so we have $x \vee y$ and xy. All the cases have been analyzed, and the proof is complete.

Remark. It is clear from the preceding proof that the remark at the end of the solution to Problem 7.13 may be carried over to the case of ordinary composition. (A disjunction and conjunction may always be obtained from a self-dual complete system.)

7.20. As we have repeatedly remarked (cf. hint to Problem 7.7), it is sufficient to prove that none of these classes is contained in any other (obvious), and that if there are monotone functions $\omega_1 \notin \mathbf{D}^{01}$, $\omega_2 \notin \mathbf{K}^{01}$, $\omega_3 \notin \mathbf{M}_0$, and $\omega_4 \notin \mathbf{M}_1$ in the system of functions Φ, the system is M-complete. The only function not occurring in $\mathbf{M}_0$ is $\omega_3 = 1$. Analogously, $\omega_4 = 0$. Thus, the M-completeness of Φ follows from the corresponding result for extended composition (Problem 7.7).

7.21. Suppose that the functions $\omega_1 \notin \mathbf{L}_0$, $\omega_2 \notin \mathbf{L}_1$, $\omega_3 \notin \mathbf{LS}$, and $\omega_4 \notin \mathbf{O}$ occur in the system of linear functions Φ. Let us show that the system Φ is then L-complete. In ω_4, we equate the greatest even number of variables so that at most one variable remains (only actual variables are considered). Since $x+x = 0$, we obtain a function ω_4 of two variables (x_1+x_2+a) or three variables $(x_1+x_2+x_3+a)$. We now equate the greatest even number of variables in the functions ω_1, ω_2, and ω_3,. From ω_1, we obtain the function $\psi_1 = 1$ or $\psi_1(x) = x+1$ (since the free term of ω_1 is 1); from ω_2, the function $\psi_2 = 0$ or $\psi_2(x) = x+1$ (since either the free term of ψ_2 is 0 and there is an even number of variables); from ω_3, the fuction $\psi_3 = 0$ or $\psi_3 = 1$ (since the number of variables of a self-dual linear function is even). Thus, either both functions ψ_1 and ψ_2 are constants or at least one of them is a negation. But we may then obtain from ψ_3 the other constant. Thus, we have both constants.

By means of composition, we may always obtain a function of the form $x_1+x_2+x_3 = x_1 + (x_2+x_3+b) + b$ from a function of the form x_1+x_2+b. Thus, we may assume that we have the function $x_1+x_2+x_3+a$. Substituting $x_3 = a$, we obtain x_1+x_2. It remains for us to note that $\{x_1+x_2, 1\}$ is an **L**-complete system.

7.22. The Zhegalkin polynomial for functions from $\mathbf{P}_1$ contain an odd number of monomials with nonzero coefficients (the free term is included among the monomials). This condition is necessary and sufficient. Since all the monomials of nonzero degree may be expressed in terms of the conjunction xy, the problem becomes a representability problem for linear functions from $\mathbf{P}_1$: $x_1+ \dots + x_k + a$, where $a = 1$ for even k and $a = 0$ for odd k. Clearly, all these functions may be represented in terms of $x+y+1$. (If $k = 0$, we have $x+x+1 = 1$; further, by sucessively substituting the function $x+y+1$ for one of the variables, we may obtained all the necessary functions; a rigorous proof may be conducted by means of induction on k.) Thus, the $\mathbf{P}_1$-completeness of the system $\{xy, x+y+1\}$ is proved. Obviously, the system is also a $\mathbf{P}_1$-basis, since only linear functions may be obtained from $x+y+1$, and only functions in **K** from xy.

7.23. Suppose that $f(x_1, \dots, x_n) \notin \mathbf{P}_1$. If $f = 1$, we have $1 = x \to x$. But if $f \neq 1$, f may be represented in the form of PDNF (formula (2.17)):

$$f(x_1, \dots, x_n) = \prod_{f(\bar{\sigma}_1, \dots, \bar{\sigma}_n)=0} x_1^{\sigma_1} \vee \dots \vee x_n^{\sigma_n}.$$

Clearly, PCNF that represents functions from $\mathbf{P}_1$ may be found from the fact that they lack the elementary disjunction $\bar{x}_1 \vee \bar{x}_2 \vee \ldots \vee \bar{x}_n$, in which all the variables occur as negations. Since conjunction occurs in the system $\{x \to y, xy\}$, to prove it is a $\mathbf{P}_1$-complete system, it is sufficient to prove that all elementary disjunctions $x_1^{\sigma_1} \vee \ldots \vee x_n^{\sigma_n}$, other than $\bar{x}_1 \vee \bar{x}_2 \vee \ldots \vee \bar{x}_n$, may be represented in terms of $x \to y$ and xy. We now show that all these disjunctions may be expressed in terms of the single function $x \to y$ alone.

As we already saw in solving Problem 1.17 (recall that $x \to y = \bar{x} \vee y$),

$$x \vee y = (x \to y) \to y = \overline{(\bar{x} \vee y)} \vee y = x\bar{y} \vee y.$$

Let us show that the elementary disjunction

$$\bar{x}_1 \vee \ldots \vee \bar{x}_m \vee y_1 \vee \ldots \vee y_\ell, \quad \ell \neq 0,$$

may be expressed in terms of $x \to y = \bar{x} \vee y$. The proof is conducted by induction on $n = m+\ell$. The basis of the induction (for example, for $n = 2$) has already been verified. Suppose that the representation has been proved for some number of variables less than n. It may be assumed that $n \geqslant 3$. Let us consider two cases. If $\ell \geqslant 2$, by the induction hypothesis we may represent $\bar{x}_1 \vee \ldots \vee \bar{x}_m \vee y_1 \vee \ldots \vee y_{\ell-1}$. Substituting the disjunction $y_{\ell-1} \vee y_\ell$ (which may also be represented) for $y_{\ell-1}$, we obtain the desired function. But if $\ell = 1$, $m \geqslant 2$, so $\bar{x}_1 \vee \ldots \vee \overline{x_{m-1}} \vee y_1$ may be represented. We obtain the desired function if we substitute $\overline{x_m} \vee y = x_m \to y$ in place of y_1. $\mathbf{P}_1$-completeness is thereby proved.

In solving the preceding problem, we had already noted that the single function xy does not form a $\mathbf{P}_1$-complete system. An analogous result holds for the function $x \to y$, though here it is not so obvious. In fact, $x \to y$ does not belong to any pre-complete class other than $\mathbf{P}_1$. Therefore, we must resort to one of the non-pre-complete functionally closed classes. A suitable class has already been encountered in our discussion: $\mathbf{F}^{(2)}$. We find that $x \to y \in \mathbf{F}^{(2)}$, since it is equal to 0 only on the sequence (1, 0) which contains 0. It is also clear that $\mathbf{F}^{(2)} \subset \mathbf{P}_1$, though $\mathbf{F}^{(2)}$ does not coincide with $\mathbf{P}_1$ ($xy \notin \mathbf{F}^{(2)}$, for example, but $xy \in \mathbf{P}_1$).

7.24. Let us show that the classes $\mathbf{P}_{01}$, $\mathbf{L}_1$, $\mathbf{M}_1$, and $\mathbf{F}^{(2)}$

form a complete system of $\mathbf{P}_1$-pre-complete classes. It is clear that $\mathbf{P}_{01}$, $\mathbf{L}_1$, and $\mathbf{M}_1$ do not coincide with $\mathbf{P}_1$, since this would contradict the pre-completeness of $\mathbf{P}_1$ (we would then have $\mathbf{P}_1 \subset \mathbf{P}_0$, $\mathbf{P}_1 \subset \mathbf{M}$, or $\mathbf{P}_1 \subset \mathbf{L}$); for $\mathbf{F}^{(2)}$, this was proved in solving the preceding problem. Without any difficulty, we may construct examples of functions that demonstrate that none of these classes is contained in any other class. (Construct these examples!)

Now let us show that if the functions $\omega_1 \notin \mathbf{P}_{01}$, $\omega_2 \notin \mathbf{L}_1$, $\omega_3 \notin \mathbf{M}_1$, and $\omega_4 \notin \mathbf{F}^{(2)}$ occur in the system of functions Φ, the system is $\mathbf{P}_1$-complete.

(1) By equating the variables of the function ω_1 which preserves the multiplicative identity but not the additive identity, we obtain the additive identity.

(2) We now consider the nonmonotone function ω_3. By equating its variables and substituting the constant 1, by Problem 5.15 we obtain a nonmonotone function of two variables $\psi_3(x, y)$, with $\psi_3(0, 0) = 1$ and $\psi_3(1, 0) = 0$. Further, $\psi_3(1, 1) = 1$, since ω_3 preserves the multiplicative identity, and this property is retained when the variables are equated and the multiplicative identity (unit) substituted. For the function ψ_3, there are two possibilities: either $\psi_3(0, 1) = 1$, so that $\psi_3(x, y) = x \to y$, or $\psi_3(0, 1) = 0$, so that $\psi_3(x, y) = x+y+1$.

(3) Now let us carry out the analogous transformations with $\omega_4 \notin \mathbf{F}^{(2)}$. Since $1 \notin \mathbf{S}$, by Problem 7.13 if we equate the variables and substitute 1, we obtain the function $\psi_4(x, y) \notin \mathbf{F}^{(2)}$: $\psi_4(0, 1) = \omega_4(1, 0) = 0$. Since $\omega_4 \in \mathbf{P}_1$, that is $\psi_4 \in \mathbf{P}_1$ as well, $\psi_4(1, 1) = 1$. If $\psi_4(0, 0) = 0$, $\psi_4(x, y) = xy$; but if $\psi_4(0, 0) = 1$, $\psi_4(x, y) = x+y+1$.

(4) Comparing (2) and (3), we find that we have either the function $x+y+1$ or the two functions xy and $x \to y$. In the latter case, we have a $\mathbf{P}_1$-complete system (by Problem 7.23).

(5) Suppose that we are given the function $x+y+1$. Consider the function $\omega_2 \notin \mathbf{L}_1$. Since we are given the constant 1, we obtain a nonlinear function of two variables $\psi_2(x, y) = xy + \alpha x + \beta y + \gamma$ from ω_2 (Problem 4.11). If $\alpha \neq 0$, substituting the function $\psi_2(x, y)$ for y in $x+y+1$, we obtain a nonlinear function of two variables with $\alpha = 0$. Analogously, the coefficient of y becomes 0; we obtain the function $xy + \delta$. This function must preserve the multiplicative identity, since we obtained it by means of a composition of functions from

$\mathbf{P}_1$. Therefore, $\delta = 0$, and we have obtained the conjunction xy (incidentally, $xy+1$ is a Sheffer function). By Problem 7.22, we then have a $\mathbf{P}_1$-complete system. The proof is complete.

Remark. In a number of the problems considered above, particularly Problem 7.23, we saw that intersections of pre-complete classes Q with other pre-complete classes play an important role in the study of Q-pre-complete classes for classes Q. However, not every Q-pre-complete class has such a form (for example, the class $\mathbf{F}^{(2)}$ in $\mathbf{P}_1$). On the other hand, not every intersection of Q with pre-complete classes is Q-pre-complete. For example, the class $\mathbf{S}_1 = \mathbf{S} \cap \mathbf{P}_1$ is not $\mathbf{P}_1$-pre-complete; $\mathbf{S}_1$ is contained, in particular, in $\mathbf{P}_{01}$ without coinciding with it. Other examples may be presented.

7.26. If $x_n = 1$, the functions m turn into disjunctions whenever they replace the disjunctions ($m(*, *, x_n)$), and into conjunctions whenever they replace conjunctions, i.e., by substituting $x_n = 1$ we obtain a representation of $\omega(x_1, \dots, x_{n-1})$. If we substitute $x_n = 0$, the dual function $\omega^+(x_1, \dots, x_{n-1}) = \psi(x_1, \dots, x_{n-1})$ is obtained (by the law of duality). Since f coincides with ω if $x_n = 1$ and with ψ if $x_n = 0$, we have in fact obtained f.

It is clear that neither $m(x, y, z)$ nor $\bar{x}$ forms an $\mathbf{S}$-complete system by itself ($m(x, y, z) = xy \vee yz \vee xz$ is a monotone function and $\bar{x}$ is a function of a single variable).

7.27. Suppose that the functions $\omega_1 \notin \mathbf{SL}$ and $\omega_2 \notin \mathbf{S}_{01}$ occur in the system Φ.

1. By equating the variables of ω_2, we obtain the function $\bar{x}$.

2. By Problem 4.10, the variables of ω_1 may be equated so as to obtain a nonlinear function of three variables $\psi(x, y, z)$. The function ψ is also self-dual. Note that ψ depends actually on all three variables (i.e., they are actual variables), since there is no nonlinear self-dual function of fewer variables (the only self-dual functions of two variables are x, y, $\bar{x}$, and $\bar{y}$). Let

$$\psi(x, y, z) = axyz + b_1xy + b_2yz + b_3xz$$
$$+ c_1x + c_2y + c_3z + d.$$

Consider the dual function $\psi^+(x, y, z) = \overline{\psi(\overline{x}, \overline{y}, \overline{z})} = \psi(x+1, y+1, z+1) + 1$. Suppose that

$$\psi^+(x, y, z) = a'xyz + b'_1xy + b'_2yz + b'_3xz$$
$$+ c'_1x + c'_2y + c'_3z + d'.$$

It is clear that $a = a'$ and $b'_i = b+a$.* Since $\psi = \psi^+$, and since a representation in the form of a Zhegalkin polynomial is unique, $b'_i = b_i$, that is, $a = 0$. Further, $c'_i = c_i+b_j+b_k$, where i, j, and k are all distinct. Hence, it follows that all the sums $b_j+b_k = 0$, so that either all the $b_i = 1$ or all the $b_i = 0$. The latter case must be excluded, since ψ is a nonlinear function. Finally, $d' = d+c_1+c_2+c_3$, i.e., $c_1+c_2+c_3 = 0$. Thus, either all the $c_i = 0$ or only one of them is 0 (for the sake of definiteness, say $c_3 = 0$). Thus,

$$\psi(x, y, z) = xy + yz + xz + c(x+y) + d.$$

If $c = 1$, we consider $\nu(x, y, z) = \psi(x, y, \overline{z})$. We obtain

$$\nu(x, y, z) = xy + yz + xz + d.$$

If now $d = 1$, by considering $\overline{\nu(x, y, z)}$, we obtain the function $xy+yz+xz$. That is, by the preceding problem the system Φ is S-complete. It is easily verified that neither **SL** nor $\mathbf{S}_{01}$ is contained in the other and that neither coincides with **S**.

Remark. As we already mentioned in the remark appended to the solution of Problem 7.25, the class $\mathbf{S}_{01} = \mathbf{S}_1 = \mathbf{S}$ is not $\mathbf{P}_1$-complete. Now we see that it is **S**-pre-complete. Thus, the intersection of the two pre-complete classes **S** and $\mathbf{P}_1$ is pre-complete in **S**, but not in $\mathbf{P}_1$. Using the terminology introduced in sec. 7.1, we may say that $\mathbf{S}_{01}$ does not belong to the third level of the Post tables, though it is **S**-pre-complete, since it is contained in a class $\mathbf{P}_{01}$ from the third level.

7.28. The set of functions self-dual representable in terms of some system of functions is, by Problem 7.12, a self-dual functionally closed class, i.e., it coincides with its dual class.

* We are everywhere considering sums modulo 2.

(Do not confuse this with the subclass of the class of self-dual functions.) There is a finite number of such classes, which form the middle column of the Post table: **P**, $\mathbf{P}_{01}$, **M**, $\mathbf{M}_{01}$, **S**, $\mathbf{S}_{01}$, **SM**, **L**, **SL**, and so on. The set of classes Q such that the only class containing Q and $\bar{x}$ is the class **P** of all functions may be understood as a set $\mathfrak{U}$ for which $\mathfrak{U}$-completeness coincides with self-dual completeness. Since $\bar{x} \in \mathbf{S}$ and $\bar{x} \in \mathbf{L}$, it is then clear that **L** and **S** and the classes contained in them do not possess this property. On the other hand, the remaining classes (**P**, $\mathbf{P}_{01}$, **M**, and $\mathbf{M}_{01}$) do possess this property. Thus, consists of these four classes. It is clear that the limiting classes do not contain any of them. Therefore, by moving down from **P**, we find all the classes containing them: **P**, $\mathbf{P}_0$, $\mathbf{P}_1$, $\mathbf{P}_{01}$, **M**, $\mathbf{M}_0$, $\mathbf{M}_1$, and $\mathbf{M}_{01}$. It remains for us to take all the Q-pre-complete classes for these classes Q, discarding those that are contained in any other classes. We obtain the classes **L**, **S**, $\mathbf{F}^{(2)}$, $\mathbf{G}^{(2)}$, $\mathbf{D}^{01}$, and $\mathbf{K}^{01}$, i.e., the classes indicated in the condition to Problem 7.15.

7.29. There is a single SM-pre-complete class in **SM**, namely the class **KD**, which consists of the single function x. Therefore, any other function from **SM** (for example, $xy \vee yz \vee xz$) forms an **SM**-basis. Try to prove this without using the Post table.

Chapter 8
NETWORKS OF FUNCTIONAL ELEMENTS

1. *Functional Elements Without Delay. Networks of Functional Elements Without Delay.* Suppose we are given a device (Figure 9) whose real internal structure is of no interest to us. All that we know is that it has n ordered "inputs" (indexed, for example, by numbers from 1 to n) and a single "output". Two signals (denoting, for example, the absence of an electrical current or its presence), which we provisionally denote by means of the symbols 0 and 1, may be fed to each input, and one of the signals 0 or 1 may appear at the output in response to each set of signals fed to the inputs; in addition, the set of signals at the inputs uniquely determines the output signal. Such a device is called a *functional element.** It is clear that a logical function $f(x_1, \dots, x_n)$

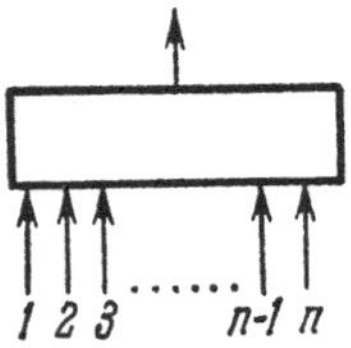

Figure 9

*In this section we consider functional elements without time delay, i.e., we assume that the output signal appears at the same moment as the input signals. In the next section we will drop the assumptions that the functional elements produce an instantaneous response.

corresponds to each such functional element (we say that the functional element *realizes* the function f), which is constructed by associating with the input numbered i ($1 \leqslant i \leqslant n$) the variable i and with each (binary) sequence of values of these variables the quantity $f(x_1, \dots, x_n)$, which is equal to 0 or 1, depending on which signal appears at the output when this set is fed to the inputs of the functional element.

If we are given several functional elements, new compound functional elements may be obtained from them by connecting an input of one functional element to the output of another (Figure 10). The resulting device may be treated as a new functional element whose output is the output of the first element (f) and whose inputs are all the other free inputs of the two elements, i.e., all the inputs of the second element (ω) along with the inputs of the first element that were not connected to the output of the second element. Signals then fed to all the free inputs of f enter these inputs directly, while the signal appearing at the output of ω enter the other inputs.

Besides this operation, the inputs of a functional element may also be equated (Figure 10). A new functional element with the same output and same inputs, other than those that have been equated (now considered the same input) is thereby created. Feeding a set of signals to the inputs of this new element is equivalent to feeding a set of signals to the inputs of f such that the same signal is fed to all the equated inputs. Further, certain (distinct) inputs of some element f may be connected to the outputs of other elements $\omega_1, \dots, \omega_k$, and certain groups of inputs of f (not connected to the outputs ω_j) such that no input is simultaneously in more than one group, may be equated (an example is given in Figure 12). Clearly,

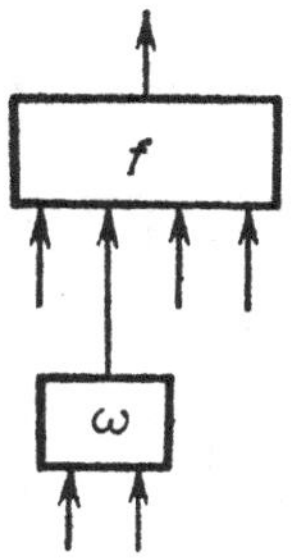

Figure 10

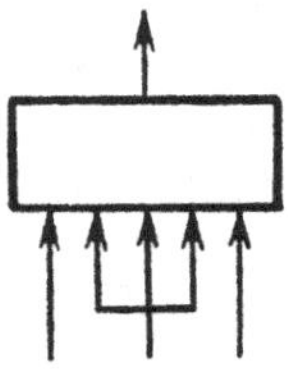

Figure 11

the function realized by this compound functional element is a composition of functions realized by the same functional elements from which it is constructed. That is, if the output of the element that realizes the function ω_i is connected to some of the inputs of the element that realizes f, ω_i must be substituted for the corresponding argument of f. The same argument is made to correspond to all equated inputs in the same group.

Connections obtained by multiple repetitions of the constructions shown in Figures 10 and 11 are called *feasible* connections of functional elements, or *networks*. Networks obtained at some step may be subsequently used as functional elements, in turn connected together according to the same rules. Let us now give a rigorous description of the class of feasible connections. A suitable definition may be given by induction, by analogy with Definition 2.3 (compositions of logical functions). A number of differences will be discussed below.

Definition 8.1. (a) Every functional element is a *network*. The inputs of this element are the inputs of the network, and its output is the output of the network.

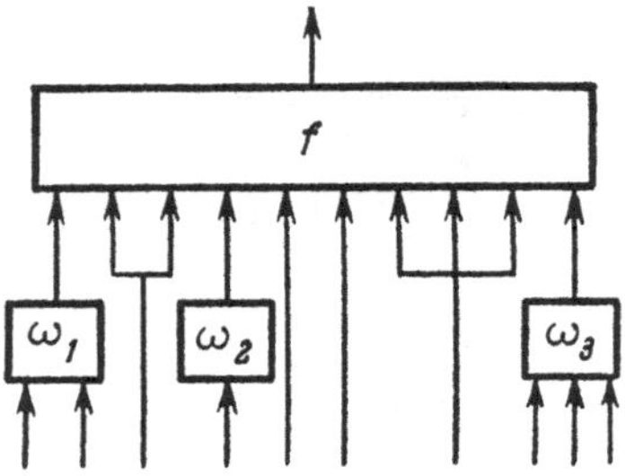

Figure 12

(b) If S_0 is a network and if two of its inputs are connected together, the construction S obtained as a result is also a network. In this case, S has the same output as S_0, while its inputs are those inputs of S_0 that have not been connected together, along with one additional input corresponding to the pair of connected inputs of S_0.

(c) If S_0 and S_1 are networks, the construction S obtained by connecting some input of S_0 to the output of S_1 in such a way that the output of S_0 becomes the output of S and all the inputs of S_1 together with all the inputs of S_0 -- except for the input connected to the output of S_1 -- become the inputs of S, will also be a network.

(d) Every network may be constructed from functional elements in a finite number of steps by means of the constructions described in (b) and (c).

Formally speaking, this definition differs to some extent from Definition 2.3, mainly because the concept of the rank of a network (number of steps in which a network is obtained from functional elements) has been discarded. Step (d) is equivalent to the last sentence in Definition 2.3; the operation described in (c) corresponds to the operation described in Definition 2.3(b). The operation described in (b) corresponds to the process of equating a set of variables. There is no need for us to introduce an analog of the general operation of renaming a variable, however. In associating a logical function with some functional element that realizes it, we are thereby associating arbitrary variables with its inputs. Thus, functions that differ only in the notation used for their variables (according to the definitions given in Chapter 2, they are not, in general, logically equivalent) are realized by the *same* functional element. In this sense, the correspondence between functions and elements is not one-to-one.

Let us show how to carry out induction arguments for networks. To prove some assertion by means of induction, it is first necessary to verify its validity for functional elements (basis of induction). Then, assuming that the assertion is valid for a network S_0, we prove it for the network S obtained from S_0 by connecting inputs (Definition 8.1b), and from the fact that it is valid for S_0 and S_1 we conclude that it is valid for the network S constructed in Definition 8.1c (inductive transition). Various network characteristics may be introduced in an analogous fashion.

8.1. Define a logical function realized by some network. Prove that it is uniquely defined to within a renaming of the variables that does not reduce to actually equating the variables.

It is convenient to assume that logically equivalent functions are realized by the same element. For this purpose, we introduce the concept of a dummy input.

Definition 8.2. We will say that an input x of a functional element f is a *dummy input* if, for any set of signals to the other inputs, the output signal is independent of the input signal (i.e., the corresponding variable of the function realized by this element is a dummy variable).

Two functional elements that differ perhaps in the enumeration of the inputs only or in terms of dummy inputs are said to be *materially equivalent*. Thus, any number of dummy inputs may be formally introduced (or discarded) without changing the functional element (relative to material equivalence of functional elements).

Below, in proving that some connection of elements is feasible (i.e., is a network), we will not always follow Definition 8.1 (dividing its construction into elementary steps; cf. (b) and (c) of the definition), and instead may use larger steps (such as that depicted in Figure 12) if they may be broken down into elementary operations in an obvious way.

Now let us see how the completeness problem for a system of functions may be rephrased in the language of networks. Suppose that we are given some system Φ of functional elements $\omega_1, \dots, \omega_n$ (more precisely, types of functional elements, i.e., we are given an unbounded number of elements that realize each of the functions ω_j). We wish to know under what conditions imposed on this system is it possible to realize an arbitrary logical function by means of a network consisting of elements of Φ. According to Problem 8.1, connections of elements that correspond to compositions of the functions realized by them (and only these types of connections) are feasible. Therefore, an answer is provided by Post's theorem.

Thus, we see that the language of networks of functional elements is materially equivalent to the language of compositions of logical functions. However, in the case of networks there are new problems that cannot be expressed in terms of compositions in a natural way. We now wish to focus our attention on one of these problems.

It would be wrong to assume that any connection of functional elements whatsoever is feasible. There are several obvious constraints, for example, it may not be possible to connect the outputs; there may be an isolated output in a network, and so on. But there are also less obvious constraints. As an example, suppose that we are given one and only one element all of whose inputs, except for one, are dummy inputs, and suppose that the element realizes the negation of the argument corresponding to this function. Let us connect this actual input to the output of the element (Figure 13).

It is clear that a functional element is not obtained in this case, since we are not able to assign any value to the output. This is because, on the one hand, the output must coincide with the value at the actual input (they are connected), and on the other, they must be distinct (the initial element realizes $\bar{x}$). The treatment of composite propositions in Chapter 1 (page 14) presented a similar situation. It might be objected that there is no free output in the device depicted in Figure 13. But this is a minor point, since we could consider a connection (Figure 14) which, though it has a free output, nevertheless possesses the above property. From Problem 8.1, it is at once evident that the connections of Figures 13 and 14 are not networks. That is, a signal to one of the inputs of the element $\bar{x}$ depends on the signal to its output, which cannot happen in the case of feasible connections of elements.

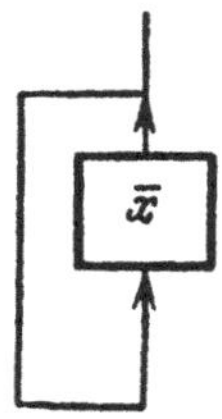

Figure 13

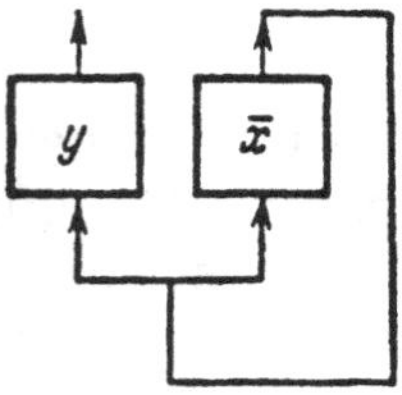

Figure 14

would have to be proved by induction; we will discuss this question somewhat later.)

8.2. Which of the connections* in Figure 15 are networks?

By means of Definition 8.1, it is possible to prove that an arbitrary connection is a network. However, to prove that some connection is not a network, we have to find certain properties of networks the given connection does not satisfy (cf. solution of preceding problem). In solving such problems, it is a good idea to apply a criterion by means of which we may decide whether a connection is a network. We now derive this criterion (Problem 8.3).

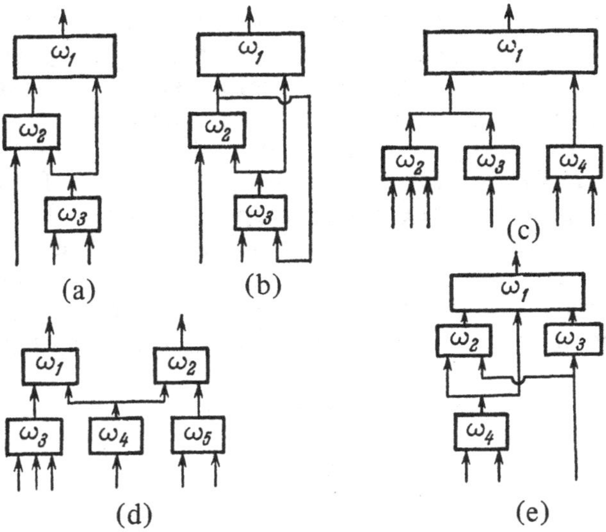

Figure 15

*The notation  indicates that the two paths are not connected.

Definition 8.3. Suppose that we are given functional elements connected in some way.* A set of elements $\omega_1, \ldots, \omega_k$ is said to be a *loop* if the output of ω_1 is connected to some input of ω_2, the output of ω_2 to some input of ω_3, and so on, the output of ω_{k-1} being connected to some input of ω_k, and the output of ω_k to some input of ω_1. In this case, we also say that the connection of elements exhibits *feedback*. Very simple examples of loops are shown in Figures 13 and 14.

8.3. Prove that a connection S of a finite number of functional elements $\omega_1, \ldots, \omega_m$ is a network if and only if:

(1) there is one and only one element in the set ω_i with a free output (i.e., with output not connected to any of the inputs of the elements ω_j);**

(2) the input of each element ω_i may be connected to at most one output of the elements ω_j;

(3) there is no feedback (loops) in S.

8.4. Which of the connections shown in Figure 16 are networks?

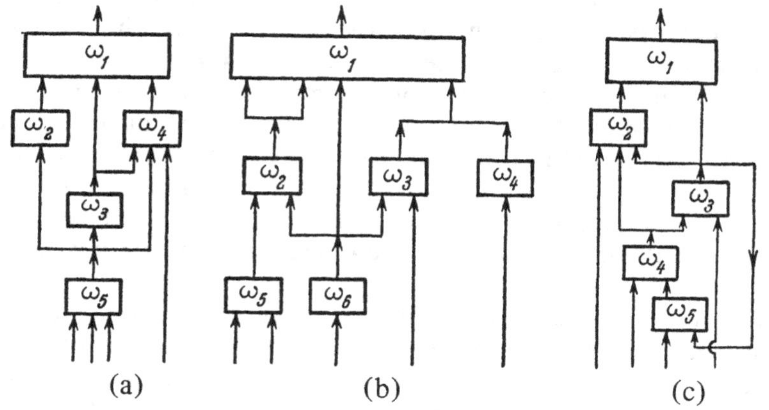

Figure 16

*In connecting functional elements, the input of one element is connected to the output of another. There are no other constraints.

**It is often useful to discard this constraint and consider networks with several outputs. In that case, each of the outputs realizes its own logical function.

The reader should have no trouble constructing other examples of connections of functional elements which are networks as well as connections which are not networks.

2. *Multi-Cycle Networks of Functional Elements.* There are two ways of thinking about networks of functional elements. We might assume (cf. footnote on page 138) that the elements respond instantaneously. In this case, at the same time that certain signals are fed to the inputs of a network, a resultant signal is generated at its output. We might also assume that some time is needed to obtain the output signal of the functional element. Then, signals fed to the inputs of the network may not all arrive simultaneously, since in this case signals that have arrived at different inputs of the interior elements may have traveled along different paths, not all of which have the same number of elements. The time required by the different elements to process the input signals may also differ. However, it may be assumed that signals are fed to the inputs of the network over an arbitrarily lengthy period of time, so that signals that have arrived at certain inputs of the interior elements earlier than at other inputs of these elements could continue to arrive until the required signals have reached the other inputs. As a result, after some period of time a signal corresponding to signals that had been fed to the inputs of the network appears at the output of the network. Afterwards (and sometimes even earlier), it may be decided to stop feeding signals to the inputs, and if so desired, the network may be used to compute the value of the function that realizes it on a new sequence of values of the arguments.

The first of these assumptions (instantaneous operation of the elements) is not natural from the standpoint of a realization of functional elements. The second assumption is fully in accord with the actual situation, though it has one serious drawback: the need to feed input signals within a designated period of time, as well as the fact that over this period of time signals appear at the output of the network that do not correspond to the signals being fed to its inputs. This circumstance may also be expressed in the following way. Under these assumptions on the operation of the functional elements, networks consisting of functional elements may not be considered some kind of new functional element. We should be more precise and say exactly what we mean by a functional element under the new conditions (taking into account the factor of operating time). A *functional element* is a device (schematically depicted in Figure 9) that produces an entirely determined signal at its output after some period of time (which is fixed for the given element) when a sequence of signals is fed to its inputs; the output signal is the value of the function realized on this sequence. If a new sequence of signals is fed to the inputs at the next moment of time, ν units of time after the sequence has been fed to the inputs a signal corresponding to it appears at the output, i.e., sequences of signals fed successively to the inputs of the element are processed independently. We assume here that time varies discretely, i.e., time takes natural numbers as values ($t = 1, 2, \ldots, k, \ldots$). A unit of time is referred to as a *cycle*. In other words, we shall study the state of the inputs and outputs of the elements only at moments of time that are multiples of a single cycle. The length of time ν between the moment the input signals are fed and the moment an output signal appears is called the *delay time* of the functional element. It is now clear that a network consisting of functional elements whose operation is comprehended by the second of the two ways described above is not, in general, a functional element.

In treating compound networks, it may be difficult to compute the time (number of cycles) during which signals must be fed to the inputs; when two or more networks are connected together, this length of time for the inputs of the compound network will, as a rule, not be the same as for the inputs of the individual networks, and so on.

We will therefore limit ourselves to networks which are functional elements in the new sense. In constructing such networks, an important role is played by special elements, called *delay elements*, by means of which signals may be fed to all the inputs of each element simultaneously. The delay elements are functional elements which realize the function x, i.e., the same signal appears at the output as at the input, though after some delay (after a certain number of cycles).

Let us introduce a constraint that greatly simplifies the discussion. We assume that *all the functional elements from which networks may be constructed are single-cycle elements*, i.e., there is a single cycle between the time when signals are fed to the input and the time when the resulting signal reaches the output. The signal is fed to the input of the element instantaneously; a new signal is fed one cycle later.

Definition 8.4. A network of functional elements S realizes a designated logical function f with delay ν if its inputs may be equated to the arguments of f in the following way. If some sequence of signals is fed to the inputs of S at any moment of time, a signal appears at the output of S ν cycles later. This signal corresponds to the value of f as a function of the values of the arguments corresponding to the given signals. Such a network S may be considered a functional element with time delay ν which realizes the function f.

A network that realizes some logical function will be said to be a *proper network.*

A network of functional elements that operates instantaneously will henceforth be called a *zero-cycle network*; networks consisting of single-cycle functional elements (we agree to consider only these types of functional elements) will be called *multi-cycle networks*, or simply networks.

8.5. Give an example of an improper network.

Remark 1. If all the elements of a proper network (consisting of single-cycle elements) were zero-cycle networks, the resulting zero-cycle network would realize the same logical function as the initial multi-cycle network.

Remark 2. It would be wrong to assume that the delay ν after which a proper network S realizes some function f is equal to the maximal number of successively connected elements of the network. For example, a network consisting of more than one functional element may have a one-cycle delay. Figure 17 shows an example of such a network. Here, if ω_1 realizes $xy(z \vee t \vee s)$, ω_2 realizes uv, ω_3 realizes $\overline{p}$, and ω_4 realizes $\overline{r}$, and the entire network realizes xy after a

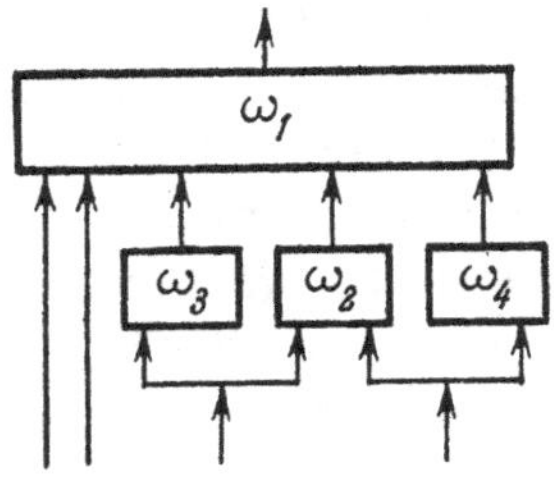

Figure 17

one-cycle delay. It is typical for the last two inputs of a network to be dummy inputs of the entire network, though actual inputs of its interior elements. The signals to the inputs of element ω_1 at time t depend both on the signals to the first two inputs of S at the same moment t and on the signals to the two other inputs of S at the preceding moment t-1, further, all the signals are actual signals; however, the output signal of ω_1 depends only formally on the signals to the last two inputs. In general, if the delay is equal to ν, the output signal of the network at time t depends only on the signals to its inputs at time t-ν, whereas the corresponding signals to the inputs and outputs of the interior elements may depend on signals fed previously to its inputs (as in the example considered above) and on signals fed later on (give an example), though the output signal of the network itself depends on these signals only formally.

Remark 3. All the inputs of elements or networks which realize constants are dummy inputs. All such networks may be asumed to have delay 0 (like, incidentally, any other network). In fact, independent of the time between the arrival of signals to the inputs and the time when of the processed signal is output, the output signal already is present at the output (at any moment of time). Henceforth, we will assume that if some constant can, in general, be realized by a network created from a designated set of elements, we are dealing with the zero-cycle functional element that realizes it.

When solving problems, it is sometimes useful to bear in mind the following fact.

8.6. Prove that every proper network S may be transformed (but remain proper) in such a way that it realizes the same logical function with the same delay as before and will then consist of functional elements of the same types, even though only inputs of elements that are the inputs of the network have been equated in it.

In the problems considered below, we may limit ourselves to networks that possess the properties listed in Problem 8.6 (though this is not necessary). We routinely make use of the sequence of networks S_i constructed in solving Problem 8.3.

Note that once the constraints stated in Problem 8.6 have been imposed, the set M_i will consist of all the elements whose outputs are connected to the inputs of S_i (they may not be connected to the inputs of any other elements). Inputs of S_i that are not conected to the outputs of elements of M_i are the inputs of the entire network S. If $S_i = S$ but $S_{i-1} \neq S$, we will say that i is the *depth* of the network S. As we have already noted (Remark 2), depth does not, in general, coincide with delay.

Let us now consider the basic problem of this section: the completeness of the system of single-cycle functional elements. The statement of the problem is similar to the preceding problem.

Definition 8.5. A system Φ of (single-cycle) functional elements* that realize the functions $\omega_1, \ldots, \omega_n$ is said to be *complete* if every one of the logical functions may be realized (each with some delay) by a network created from these elements.

Let us begin by considering some examples.

8.7. (1) Suppose that some system of functional elements Φ contains a delay element ($\omega_1 = x$) and suppose that the functions realized by its elements form a complete system (in the sense of the logical functions). Then the system Φ is complete. Prove that the first of these conditions is not necessary.

(2) Suppose that Φ consists of a delay element ω_1 (which realizes x) and a Sheffer element ω_2 which realizes $\overline{xy}$. Construct networks which realize: (a) $\overline{x}$; (b) xy; (c) $x \vee y$; (d) 1; (e) 0; (f) $x+y$. Find the length of the delay.

8.8. Prove that the system of elements $\{x\overline{y}, 1\}$ is complete.**

Thus, we see that the presence of delay elements does not

---

*More precisely, as we have already noted (page 142), types of functional elements.

**We will often write "the functional element $f(x_1, \ldots, x_n)$" instead of "the functional element that realizes the function $f(x_1, \ldots, x_n)$."

guarantee the completeness of a system of functional elements. We might then ask whether the delay elements may be discarded, i.e., whether the completeness conditions for single-cycle functional elements are the same as for the corresponding logical functions. We advise the reader to attempt to analyze this question before proceeding (it is answered below).

8.9. Is there a complete system that consists of a functional element ω realizing the Sheffer function $\overline{xy}$?

In solving this problem, the reader should check and see whether the completeness conditions for single-cycle functional elements are the same as the completeness conditions for the functions they realize. Let us consider one more example.

8.10. Is it possible to obtain a complete system if we adjoin to an element ω which realizes the Sheffer functions $\overline{xy}$ the elements which realize the constants?

You should be able to solve the completeness problem through careful consideration of the examples in Problems 8.9 and 8.10.

8.11. Find necessary and sufficient completeness conditions for a system of single-cycle functional elements.

Note that the completeness problem becomes more complicated if we discard the single-cycle property for the elementary functional elements, and instead deal with an arbitrary delay time [11]. In this case, the completeness condition contains an infinite number of premises.

We have associated with proper networks of single-cycle elements certain compositions of the corresponding functions. Therefore, we have actually considered the completeness problem for a certain contraction of ordinary composition (cf. extended composition considered in sec 7.2). A description of this class of compositions was given in terms of networks (definition of a proper network), though it could have been presented without resorting to the language of networks. In concluding this section, we suggest that the reader consider certain other contractions of composition (not related to networks), and determine what might be the completeness conditions for these contractions. It should be clear at the outset that whereas the introduction of extended composition leads to a decrease in the number of classes (and, in particular, pre-complete classes), a contraction of the class of compositions leads, in general, to an increase.

Definition 8.6. A composition of a system of functions Φ is said to be *global* if it may be obtained from elements of Φ by successive application of the operations of renaming the variables (and, in particular, by equating them) and substitution to certain functions $\omega_1, ..., \omega_n \in \Phi$ in place of all the arguments of some function $f(x_1, ..., x_n) \in \Phi$.

The constraint states that it is not permissible to replace only some arguments of a function. However, if there is a function x in Φ, any composition may be considered global (since the argument y may then be considered invariant by virtue of the substitution $\omega(y) = y$).

8.12. Give an example of a system of functions complete in the ordinary sense, but not complete relative to global composition.

8.13. Find completeness conditions for a system of functions relative to global composition.

8.14. Let us say that a substitution of functions $\omega_1, ..., \omega_k$ for certain arguments of a function f is a *restricted substitution* if the resulting function depends formally on all the arguments of the functions $\omega_1, ..., \omega_k$. Extend the concept of global composition, admitting not only substitution for all the arguments, but also restricted substitution. What are the conditions for this class of compositions?

3. *Automata Without Feedback.* In the preceding section, we showed that networks consisting of single-cycle functional elements realize logical functions only in special cases (if the networks are proper). In that event, how might we describe the operation of a network consisting of single-cycle elements in the general case? We will assume that each moment of time, certain signals are fed to the inputs of the network. It

is clear that the network's output signal depends on the signals which have been fed to its inputs in the course of the several preceding moments of time.

We will use the concept of an upper index, introduced in the remark appended to the solution of Problem 8.7. By analogy with this concept, we now introduce the concept of a lower index by means of induction: The *lower index* of an element ω of a network is one greater than the smallest lower index of the elements whose outputs are connected to the inputs of ω. In other words, the lower index of ω is equal to the minimal number of elements which some signal fed to the input of a network requires before reaching the output of ω. We let $\eta = \eta(S)$ and $\mu = \mu(S)$ denote the lower and upper indices of the output element of a network S, respectively. Then it is immediately clear from the definition (and can be easily proved by induction) that the output signal of S at some moment of time t may depend only on the input signals of S at moments from t-μ to t-η, inclusively. Thus, at time t the output signal is a function of $\rho = \mu$-η+1 successive sequences of input signals:

$$F(x_1^0, \dots, x_n^0; x_1^1, \dots, x_n^1; \dots; x_1^{\rho-1}, \dots, x_n^{\rho-1}),$$

where F is a logical function of ρn variables, and $(x_1^i, \dots, x_n^i)$ is the sequence of signals at time t-μ+i. If S is a proper network, F may depend actually only on one of the sequences of signals $(x_1^i, \dots, x_n^i)$ corresponding to time t-ν (i.e., $i = \eta$-ν, and $\eta \geqslant \nu$ necessarily), and S realizes this function $F(x_1^i, \dots, x_n^i)$ $(i = \eta$-$\nu)$ after delay ν.

Definition 8.7. Suppose that we are given a device with n inputs and a single output (Figure 9) such that at each moment the signal 0 or 1 is fed to its inputs and the signal 0 or 1 appears at its output; which output appears is a function of the ρ successive sequences of signals fed to the inputs at moments from t-μ to t-η, where $\rho = \mu$-η+1:

$$F(x_1^0, \dots, x_n^0; x_1^1, \dots, x_n^1; \dots; x_1^{\rho-1}, \dots, x_n^{\rho-1}),$$

where $(x_1^i, \dots, x_n^i)$ is the sequence fed at time t-μ+1. Then this device constitutes an *automaton without feedback*. The logical function F is called its *characteristic function*, ρ is its *index*, and η its *delay time*.

Automata without feedback are said to be *equivalent* if their characteristic functions F_1 and F_2 differ only by dummy variables; the corresponding variables of F_1 and F_2 are associated with the same moments of time relative to the times of the output signals (corresponding to the values of F_1 and F_2). That is, an automaton does not change if it is formally assumed that F depends on the input signals even at arbitrary moments of time.

A functional element is an automaton without feedback. Its characteristic function coincides with the function it realizes, and its index and delay are equal to one.

From this line of reasoning, it is clear that every network of single-cycle functional elements is an automaton without feedback. In fact, it is not essential here to assume that the elements are single-cycle. It is only necessary to compute the upper and lower indices μ and η differently. That is, in the inductive definition it is necessary to add to the maximum (correspondingly, the minimum) not 1, but the length of the delay of the particular element; in particular, for single-cycle elements, it is necessary to add 1. We might therefore ask whether the opposite assertion holds: Can every automaton without feedback be represented (to within equivalence) by some network formed of functional elements (still without any constraints on the structure of the elements).

8.15. Prove that every automaton without feedback may be represented by some network formed of functional elements.

Now let us consider networks constructed from a limited collection of functional elements. Again, we will limit ourselves to the case of single-cycle elements.

Definition 8.8. A network S of functional elements *realizes* some automaton without feedback A if the automaton represented by this network differs from A possibly only by the delay time η.

If A is a functional element, we obtain the definition of a realization of a logical function by means of a network formed of functional elements (cf. Definition 8.5).

Definition 8.9. A system of functional elements is said to be *weakly automaton-complete* if every automaton without feedback may be realized by a network formed from these elements.

8.16. Prove that the weak automaton-completeness condition for a system of functional elements and the completeness condition for the logical functions coincide. In particular, necessary and sufficient conditions for the weak automaton-completeness of a system of elements are given simultaneously in Problem 8.11.

We have considered the realization of automata by means of networks consisting of single-cycle functional elements. We will not deal with the more general statement of the problem, in which the basic elements have arbitrary delay times. On the other hand, not only functional elements, but entirely arbitrary automata as well, may be connected into networks. In this case, problems arise similar to those considered above, in particular, completeness problems.

4. *Networks of Functional Elements with Feedback. General Concept of a Finite Automaton.* Everywhere in this chapter we have considered networks of functional elements without feedback. This constraint first appeared in our treatment of zero-cycle networks, since otherwise it would have been impossible to provide a consistent description of their operation. However, this constraint is no longer justified once we pass on to multi-cycle networks. Consider, for example, the network shown in Figure 13; we assume that $\overline{x}$ is a single-cycle element. Then, if the signal 1 appears at the output at some moment t, at the same moment 1 will appear at the input of $\overline{x}$ and, at the $(i+1)$-th moment 0 will appear at the output, etc. As a result, the sequence 1, 0, 1, 0, 1, 0, ... appears at the output, and the contradiction present if we assume that the element is a zero-cycle element vanishes. This network may be called quite naturally a bell circuit, since a bell utilizes a sequence of signals which at each cycle are replaced by their opposite.

It can be easily proved that, in general, it is still possible to describe the operation of networks with nonzero delay that do not satisfy the third condition of Problem 8.3 (networks with feedback) in an entirely consistent fashion. In this section, we will understand by the term "network" (sometimes speaking of "networks with feedback") connections that

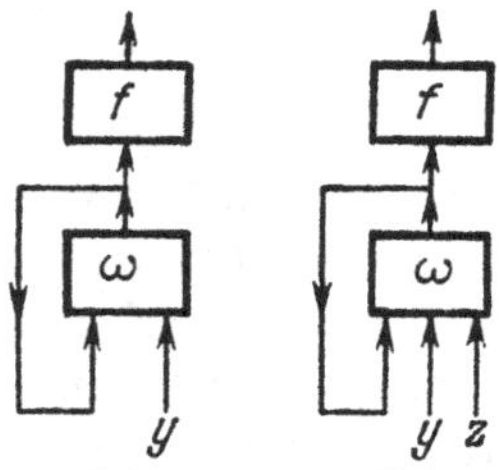

(a) (b)

Figure 18

satisfy only the first two conditions of Problem 8.3. As before, we limit ourselves to the case of single-cycle functional elements. In this case, we can no longer claim as we did in the preceding section that the signal at the network output depends only on the input signals that have been fed in over some fixed period of time. Because of feedback, the output signal may depend on the signals at the outputs of the network elements, which either cannot be functions of the input signals or may be functions of the input signals only at arbitrarily remote moments of time. For example, the elements of a loop (Definition 8.3) may not, in general, have any inputs that are the inputs of the network (free inputs). In Figure 18a, the element ω realizes the function $x \vee y$, and in Figure 18b, the element ψ realizes the function $(x \vee y)z$; both are single-cycle elements, and f is a delay element (we have included it so as to provide the network with a free output). Then, if an isolated signal is fed to the input y of the network (a) at an arbitrarily remote point in time, the same signal will be permanently present at its output (the signal y is "remembered"). In network (b), the signal y is remembered if $z = 1$. By feeding $z = 0$, we "clear the memory" and thereafter the network can "remember" a new signal (fed when $z = 1$). In actual networks, the "memory [storage] devices" are realized at once by means of loops.

The state of the internal elements of a network may be taken into account in a natural way if we wish to describe the operational characteristics of a network with feedback. We do so in the following fashion.

Suppose that S is a network (in general, with feedback), and let $\omega_0, \omega_1, \dots, \omega_k$ be its elements, with ω_0 being an output element.* We let $\omega_i(t)$ denote the output signal of ω_i at time t ($\omega_i(t) = 0$ or 1). The value of $\omega_0(t)$ indicates the output signal of the network at time t. The set of values $\omega(t) = \{\omega_0(t), \omega_1(t), \dots, \omega_k(t)\}$ is called the *state* of S at time t. By $s_1(t), s_2(t), \dots, s_n(t)$ we denote the signals fed to the inputs of S at time t, and by $s(t) = \{s_1(t), \dots, s_n(t)\}$, the set of these signals. Then the state $\omega(t+1)$ of the network at time $t+1$ is determined uniquely by the state of the network $\omega(t)$ at the preceding moment and the input sequence $s(t)$ at that time, i.e.,

$$\omega(t+1) = \Phi(\omega(t), s(t)).$$

In other words, the output signals of the elements at time $t+1$ depend on the input signals to the elements at time t, i.e., on the output signals of the elements and the input signals of the network at this moment. More precisely, Φ constitutes a set of $k+1$ logical functions $\Phi_0, \Phi_1, \dots, \Phi_k$ of $k+n+1$ variables. In fact, not all of these variables are dummy variables; only those arguments that correspond to functional elements whose outputs are connected to the inputs of ω_i and the input signals of the network that are fed directly to the inputs of ω_i are actual to Φ_i. If we are left with only these arguments, Φ_i coincides with the function realized by the element ω_i, and the formula given above simply shows which signals are fed to the outputs of the elements ω_i at time $t+1$, depending on which signals have been fed to its inputs at time t. Note, too, that a set of binary sequences of length $k+1$ that may serve as a state of S cannot be the same as the set of all binary sequences of length $k+1$. Thus, the values of the functions Φ_i on designated sequences may be undefined (the Φ_i are said to be *partially defined*). Note that the present description of the operation of a network without feedback is not the same as that given in the preceding section.

Let us now define what we mean by a finite automaton.

Definition 8.10. Suppose that Ω is some set of binary sequences of length $k+1 > 0$. We will say that a *finite automaton* (Ω,n) with *n inputs* is defined on the set of

*We may assume that the output of ω_0 is connected to the inputs of the other elements.

feasible states of Ω if there is a sequence Φ consisting of $k+1$ partially defined logical functions Φ_i of $n+k+1$ variables. Further, these functions are defined for all binary sequences of length $k+n+1$ in which the first $k+1$ elements form a sequence occurring in Ω, moreover the sequence of values of Φ on these sequences belongs to Ω. The number of elements in Ω is called the *memory* of the automaton.

We will say that the operation of the automaton (Ω,n) is defined if some initial state $\omega^0 \in \Omega$ and some natural number ν, which we call the *delay time*, are defined, and if a sequence of input signals $s(t) = \{s_1(t), \dots, s_n(t)\}$ of length n is fed in at each moment.

If the operation of an automaton has been determined, its sequence of states for $t \geq \nu$ may be determined by the formula

$$\omega(t+\nu) = \Phi(\omega(t), s(t)), \quad \omega(0) = \omega^0.$$

This formula is called the *equation of state* of the automaton (Ω,n). It is clear that at any moment of time the state of the automaton $\omega(t) \in \Omega$. The sequence $\omega_0(t)$ $(t \geq \nu)$ is called the *output* of the automaton (*result of its operation*). If $k = 0$ and if Φ depends only on $s(t)$, then becomes a logical function.

Using consistent notation, a network (with feedback) S consisting of single-cycle elements is an automaton with $\nu = 1$; ω^0 are the signals at the outputs of the elements of S when $t = 0$; as Ω, it is best to take all possible sequences of (simultaneous) signals to the outputs of the elements.

8.17. Prove that the operation of every finite automaton with delay time $\nu = 1$ may be represented by a network with feedback consisting of single-cycle functional elements.

If we limit the set of functional elements from which the networks are to be constructed (by establishing a system of basic elements), it can then be easily verified that there are automata that cannot be represented by a network formed from these elements (even when $\nu = 1$). However, we will associate several automata with any one network rather than just one. That is, suppose that $\Xi = \{\omega_0, \dots, \omega_k\}$ is some set of elements occurring in S (perhaps not all the elements), and that ω_0 is the output element of S, and let the signal at the output of each of these elements at time $t+\nu$ be determined by the signals at time t to the inputs of the network ($s(t) = \{s_1(t), \dots, s_n(t)\}$) and the signals to the outputs of the elements $\omega_0, \dots, \omega_k$ taken from our set ($\omega(t) = \{\omega_0(t), \dots, \omega_k(t)\}$). In this case, the set Ξ of elements $\omega_0, \omega_1, \dots, \omega_k$ is said to be *ν-closed*. For a proper network, the output element ω_0 forms a ν-closed set (ν is the delay time). We meet with a ν-closed set whenever we are dealing with networks $S_0, \dots, S_k$ that realize some logical functions with delay ν, and if a new network S may be constructed from them as if from functional elements. In this network, the output elements $\omega_0, \dots, \omega_k$ of the sub-networks $S_0, \dots, S_k$ then form a ν-closed set.

Every ν-closed set of elements Ξ generates in a natural way a finite automaton that operates with delay ν. We will say that this automaton is *realized* by a network S with delay ν relative to the set of elements Ξ.*

Definition 8.11. A system of functional elements is said to be *automaton-complete* if every finite automaton may be realized (to within delay time) by a network S created from the elements of this system with some delay ν relative to some ν-closed subset Ξ of elements of S.

8.18. Prove that the automaton-completeness of a system of functional elements is the same as the completeness condition relative to the logical functions (that is, by Problem 8.16, the condition of weak automaton completeness).

The hint to Problem 8.11 thus provides necessary and sufficient conditions for the automaton completeness of a system of elements.

As in the case of automata without feedback, note that we could consider networks of functional elements with arbitrary delay times, as well as networks consisting of automata, though this complicates the completeness problem. For systems of automata, there is in general no algorithm by means of which we could decide whether a given system is complete [2].

---

*If Ξ consists only of the output element ω_0 of S, and if the output of ω_0 is free (i.e., not connected to the input of any element), then relative to Ξ, S realizes an automaton which is a logical function.

There is a considerable number of studies devoted to automata theory [3, 4, 5]. There are also several different definitions of an automaton which are in some natural sense equivalent. Very complicated devices (for example, high-speed computers) may be considered as automata (possibly with a very large number of states), and this point of view is useful in many problems. Here we have only dealt with the design of automata using certain standard parts. Automata theory considers other problems as well, for example, the structure of an automaton expressed in terms of output sequences of signals corresponding to various sequences of sets of input signals.

The learning problem for automata is of considerable interest. One possible statement of the problem is as follows [5]. An automaton is in some environment in which "reward" or "penalty" signals are fed (in the form of binary sequences) to its inputs; which type of signal depends on its output signal with probability that depends in turn on this signal. Further, the principle according to which these signals are fed and the corresponding probabilities are not precisely known (there is some set of possibilities). It is necessary to construct an automaton which will, if placed in some environment (precisely what kind of environment is not known in advance), begin to produce (with probability nearly equal to 1) a signal that to a large degree is "rewarding" after a certain number of cycles. Note that the environment may change from time to time. The reward signals may be fed by adjoining the outputs of other automata to the inputs of the given automaton; such a linkage may be reciprocal, thereby giving rise to an "automaton game". For an exact statement of this problem, see ref. 5.

5. *Networks of Functional Elements for Double Lines.* In this section, we will present one more way of realizing logical functions by means of functional elements; the present technique was developed by John von Neumann [6]. For the sake of simplicity, we will assume that the functional elements are zero-cycle elements. With each variable x we associate (Figure 19) an ordered pair of wires (*double lines*) through one of which x is fed, and through the other $\overline{x}$.

Suppose that we are given two networks of functional elements, each with n inputs, and suppose that a one-to-one correspondence may be established between the inputs of the

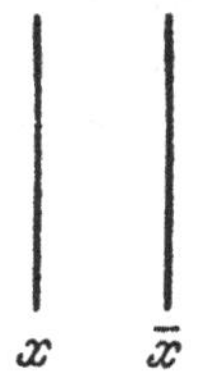

Figure 19

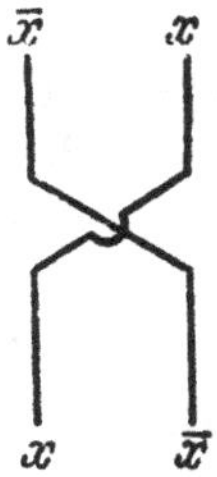

Figure 20

two networks such that if the values of the arguments $x_1, \ldots,$ x_n are fed to the inputs of the first network and their negations $\overline{x}_1, \ldots, \overline{x}_n$ to the corresponding inputs of the second network, the value of $f(x_1, \ldots, x_n)$ is obtained at the output of the first network, and the value of $\overline{f(x_1, \ldots, x_n)}$ at the output of the second network. In this case, we will say that the network realizes the function $f(x_1, \ldots, x_n)$ in double lines, and that a pair of corresponding inputs of the networks may be considered a double line that represents the corresponding argument. Moreover, if some variable is already represented in double lines, its negation may be obtained without resorting to functional elements (Figure 20).

Therefore, in creating a realization of functions by means of networks of double lines it is also possible to feed the negations of the arguments to the arguments of the first network, and the arguments themselves to the corresponding inputs of the second network. It would also be possible to construct in this way the negation of a function, though this could be done by interchanging the first and second networks and introducing the negations of the arguments. The general form of a network of double lines may be obtained by repeatedly combining these constructions. (We will not bother making this concept more rigorous.) However, a system of functional elements should not be assumed to be "double-line complete" if its elements realize functions that form (together with the negation) a complete system.

8.19. Prove that a functional element that realizes the conjunction operation does not form a system which is double-line complete.

The completeness condition for a system of double-line

elements may be obtained from the following assertion.

8.20. Prove that a system of functional elements is double-line complete if and only if the system of functions realized by these elements is self-dual complete (Definition 7.5).

The condition of Problem 7.15 therefore includes necessary and sufficient completeness conditions for a system of double-line functional elements.

Networks of double lines may be considered under the assumption that there are elements that realize the constants (in von Neumann's terminology, "active" and "grounded" elements that realize 1 and 0, respectively). In this case, the completeness problem for the elements also reduces to the self-dual completeness of the functions, but now relative to extended composition (so that in this case the completeness criterion is given in the solution of Problem 7.13).

Hints

8.1. Determine the function to be realized by induction.

8.2. To prove that some connection of elements is a network, we need only construct it in an arbitrary number of steps from initial elements by means of the operations given in Definition 8.1.

To prove the contrapositive assertion, we need only note that the connection possesses some property that the network does not (there are several free outputs; some of the outputs of the elements are connected to each other; the signal to the input of some element depends on the signal at its output). Strictly speaking, the necessity of these conditions should be proved by means of induction. These assertions will be justified below (Problem 8.3).

Answer: Connections (a) and (e) are networks, while the others are not.

8.3. Prove necessity by induction (cf. Definition 8.1 and the accompanying explanation).

To prove sufficiency, it is best to start with an element whose output is free, and to gradually enlarge the network by adjoining to it on each step elements that are connected with only the outputs of the preceding network. In the proof, note especially that in such a construction we will have used up all the elements at some step.

8.4. To determine whether some connection of elements is a network, as in the preceding problem it is best to successively construct the networks $S_0, S_1, \dots, S_k$, and so on. If at some step it is found that Condition (1) or (2) of Problem 8.3 does not hold or that S_i coincides with S_{i+1} without coinciding with S, then S is not a network. But if S_i coincides with S for some i, then S is a network.

Answer: In example (a), S is a network.

8.6. Use the representation of S in the form of a sequence of networks S_i constructed by solving Problem 8.3 (rearrange the induction with respect to i).

8.7. (1) Using part 1, for any function ω it is possible to construct a zero-cycle network S that realizes ω. If the elements in this network are single-cycle elements, S in general is not a proper network. However, S may be transformed by inserting delay elements in such a way as to obtain a new network that realizes ω. This transformation may be carried out by induction on the construction of S (Definition 8.1).

8.9. No, there is no such complete system. For example, it is impossible to obtain a constant. Prove that a network formed of the elements of ω with a single input (all the inputs are equated) realizes either x or $\overline{x}$ depending on whether the delay is even or odd.

8.10. *Answer.* No, no such complete system may be obtained. For example, it is impossible to realize the function $x\overline{y}$. Suppose that we have a network consisting of the elements given in this type of condition which realizes a function of two variables $\omega(x, y)$ with delay ν. Then if one of the variables is replaced by a constant, we obtain either a constant or x or y if ν is even, and any one of the functions 0, 1, $\overline{x}$, or $\overline{y}$ if ν is odd. The solution is analogous to the solution of Problem 8.9.

8.11. We introduce some notation. Suppose that Q is the

set of logical functions $f(x_1, \ldots, x_n)$ which do not preserve the additive or multiplicative identities ($f \notin \mathbf{P}_0$, $f \notin \mathbf{P}_1$):

$$f(0, \ldots, 0) = 1, \quad f(1, \ldots, 1) = 0,$$

i.e., a negation is obtained if we equate all the variables of f; **R** is the set of functions such that if any of the variables are "defined" (i.e., if constants are substituted for them) and if the remaining variables are equated, any one of the functions 0, 1, or $\bar{x}$ may be obtained (i.e., x is not obtained).

A system of single-cycle functional elements is complete if and only if the set of functions realized by these elements includes:

(a) the complete system of logical functions (i.e., the functions that satisfy the conditions of Post's theorem, Chapter 6, page 89);

(b) a function that does not belong to the set **Q**;

(c) a function that does not belong to the set **R**.

Use the solutions of Problems 8.9 and 8.10. Prove that if these conditions are satisfied, it is always possible to obtain a single-cycle delay.

8.12. Consider the system $\{x\bar{y}, 1\}$, which is complete in the ordinary sense. It is impossible to represent global composition, for example, x or $\bar{x}$, by means of this system. Any global composition of designated functions take identical values on the null and unit sets. Prove that the collection **T** of such functions ($f(0, \ldots, 0) = f(1, \ldots, 1)$) is closed relative to global composition.

8.13. The pre-complete classes for ordinary composition and the class **T** introduced in the hint to the preceding problem are the pre-complete classes for global composition.

8.14. The completeness conditions for compositions that satisfy the conditions of the problem coincide with those for ordinary compositions. Bear in mind that a substitution of constants is always a restricted composition.

8.15. Consider a single-cycle functional element that realizes the function F. Construct the desired automaton from this element using single-cycle delay elements.

8.16. Every weakly automaton-complete system of functional elements is complete relative to the logical functions, since the functional elements are a special case of automata without feedback. The converse assertion may be obtained using the construction of the preceding problem.

8.17. Use functional elements that realize Φ_i, and connect them in accordance with the equation of state of the automaton.

8.18. On the one hand, the validity of the assertion follows from the fact that the logical functions are a special case of automata.

To realize the automaton, it is sufficient to apply the network of Problem 8.17 and replace all the elements in it by networks that realize the same function with identical delay and by elements created from those available to us (use Remark 1 appended to the solution of Problem 8.11).

8.19. It is impossible to realize $x \vee y$ or even xy, since in the latter case we would simultaneously realize $\overline{xy}$.

8.20. If S realizes some function f in double lines, then some function ω not the same as f, possibly only negations of its arguments, is self-dual representable in terms of functions realized by elements of S.

Solutions

8.1. We construct the function by induction. It is immediately clear that the function is uniquely defined to within a renaming of variables that does not lead to their being equated.

1. *Basis of Induction.* Functions realized by the functional elements have already been defined.

2. *Inductive transition* (a) If the network S_0 realizes the function $\omega(x_1, \dots, x_n)$, then the network S_1 constructed in

Definition 8.1(b) realizes the function obtained by equating the variables x_i and x_j corresponding to the combined inputs.

(b) Suppose that S_0 realizes the function $\psi(x_1, \dots, x_n)$ and S_1, the function $\psi(y_1, \dots, y_\ell)$. We select notation so that the variables $x_1, \dots, x_k, y_1, \dots, y_\ell$ are all distinct. Then the network S constructed in Definition 8.1(b) realizes $\omega(x_1, \dots, x_{i-1}, \psi(y_1, \dots, y_\ell), x_{i+1}, \dots, x_n)$, i.e., the function ψ replaces the argument x_i corresponding to the input of S_0 connected to the output of S_i. It is extremely important that x_j and y_m be distinct, since the variables would otherwise be equated in an unacceptable way.

This construction shows that a network may be thought of as a functional element that realizes an already constructed function.

8.2. (a) We first consider the network depicted in Figure 21. In this network we then equate the last two inputs and connect the equated input to the output of the element that realizes ω_3.

(b) The signal to an input of element ω_2 depends on the output of element ω_3, while the signal to an input of ω_3 depends on the signal to ω_2. As a result, the signal to an input of ω_2 depends on the signal to its output (similarly for ω_3). This cannot happen in a network.

(c) The outputs of ω_2 and ω_3 are connected.

(d) There are two free outputs (that of ω_1 and ω_2).

(e) Consider the network of Figure 22. In this network, we connect in pairs the first input to the third, and the second to the fourth. The first of the equated inputs we now connect to the output of ω_4.

8.3. *Necessity.* The proof is conducted by induction on the construction of a network S.

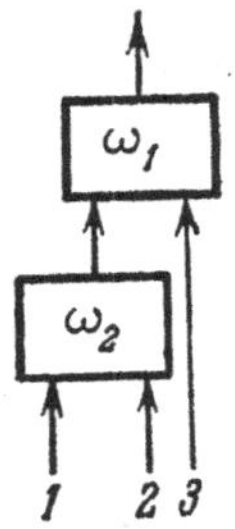

Figure 21

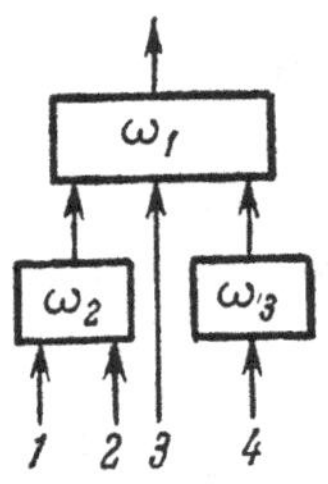

Figure 22

1. That Conditions (1) - (3) are valid for a functional element is self-evident.

2. (a) Suppose that Conditions (1) - (3) are satisfied by some network S_0. We verify them for the network S obtained from S_0 by equating inputs (Definition 8.1(b)). It is clear that in this case the free outputs remain free and that no new free outputs are created, and also that no two outputs may end up at one input. Moreover, no loop $\{\omega_1, \ldots, \omega_k\}$ could result, since these elements would then form a loop in S_0 as well. In fact, we connected only the inputs of S_0, i.e., the free inputs of the functional elements, and no new connections of inputs and outputs of elements could have arisen.

(b) Now suppose that S has been constructed from networks S_0 and S_1 as in Definition 8.1(b), and that it is known that S_0 and S_1 satisfy Conditions (1) - (3). Then the only output of elements of S, namely the output of S_0, will be free. Further, since in connecting S_0 and S_1 we connect only some of the inputs of the elements in S_0 to the output of an element in S_1, there can be no connection between the outputs of elements in S_0 and S_1. In addition, if there were a loop in S, it would necessarily comprise both elements of S_0 and elements of S_1 (since neither of these networks has any loops). But in that case, as can be easily seen, there has to be an element $\omega_i \in S_0$ whose output is connected to the input of some element $\omega_j \in S_1$. But there can be no such connections. Thus, Conditions (1) - (3) are satisfied by S. Necessity is proved.

Sufficiency. Suppose that the functional elements are connected in such a way that Conditions (1) - (3) hold. Let us prove that we then have a network S. We show that S can be constructed by induction by means of feasible connections. In S, there is first a unique functional element (call it ω_0) with a free output (Condition 1). This element we make the initial network of S_0. Now consider those functional elements whose outputs are connected only to the inputs of S (i.e., ω_0). We let S_1 denote the network obtained from S_0 by connecting these elements, and if any element ω_i is connected to several inputs of S_0, these inputs must first be equated, which may be carried out in a consistent maner by virtue of Condition 2. Again by Condition 2, the remaining elements in S may be connected only to those inputs of elements in S_1 that are inputs of S_1 itself. Now connecting to S_1 those elements of S that are connected only to inputs of S_1 and, where necessary, first equating the inputs of S_1, we obtain a network S_2, and so on. If S_i has already been constructed, we consider the set M_i of elements of S whose outputs are connected only to inputs of S_i. Note that the outputs of elements of S not occurring in S_i cannot be connected to inputs of elements of S_i that are not inputs of the entire network S_i (by Condition 2). We now equate those inputs of S_i that are conected to the same element of M_i. We connect the outputs of elements in M_i to the corresponding inputs of the resulting network. The network constructed as a result is denoted S_{i+1}.

If by some step we have used up all the elements of S_1 then, since all the S_i are networks, and since the inputs of elements of S_i are connected to the inputs of the same elements as in S, we will have completed the proof. Thus, it is necessary to prove that if S_i does not coincide with S, then S_{i+1} does not coincide with S_i, i.e., each succeeding network contains more elements than the preceding network, so that at some step S_i coincides with S. Let us suppose the contrary, i.e., that S_i does not coincide with S, though S_i and S_{i+1} do, that is, M_i is empty. This will mean that the output of every element $\omega \notin S_i$ is connected to the input of at least one element that also does not belong to S_i. Let us consider some sequence of elements $\psi_1, \psi_2, \ldots, \psi_j, \ldots$, where none of the $\psi_j \in S_i$, and let some input of ψ_{j+1} be connected to an output of ψ_j. By the preceding remark it is possible to construct such a sequence. But since there is only a finite number of elements in this sequence, some of the elements are encountered twice. The segment of the sequence between two identical elements (beginning with this element) is then a loop. We have arrived at a contradiction with Condition 3. **Q.E.D.**

8.4. (a) cf. Figure 23. The networks S_3 and S coincide.

(b) S is not a network, since the last input of ω_1 is connected to the outputs of ω_3 and ω_4 (which violates Condition 2).

(c) cf. Figure 24. The set M_1 is empty and S_1 coincides with S_2; that is, S is not a network. The elements ω_3, ω_4, and ω_5 form a loop.

8.5. We could, for example, consider the network depicted in Figure 25, where ω_1 realizes xy and ω_2 realizes zt. If the

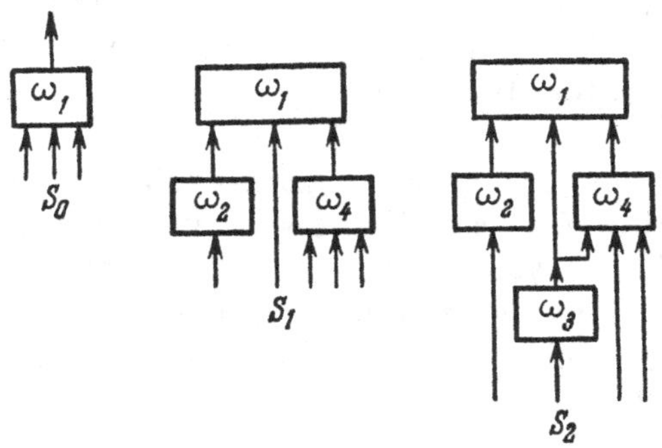

Figure 23

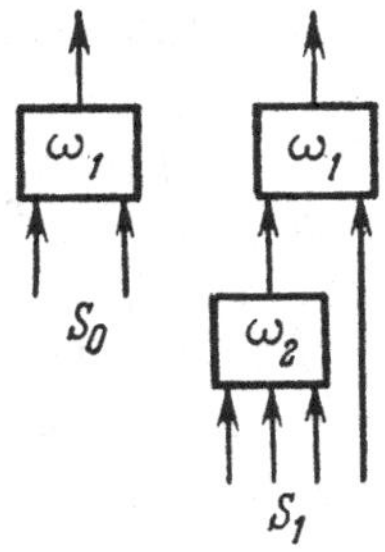

Figure 24

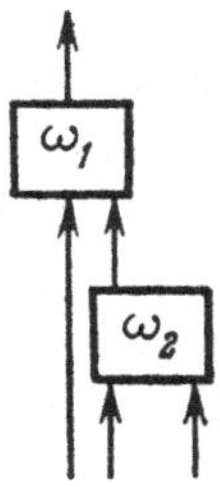

Figure 25

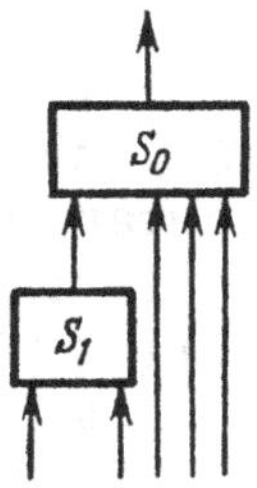

Figure 26

elements ω_1 and ω_2 are single-cycle elements, the signal to the output of this network depends actually on signals to inputs of the network arriving at different moments of time.

8.6. We use the notation introduced in the solution of Problem 8.3. The transformation will be performed by induction on i. The network S_0 does not have to be transformed. Suppose that S_1 has been transformed into a network $\tilde{S}_i$ that possesses the desired properties. If in the transition to S_{i+1}, some of the inputs of S_i are equated and if in this case the output of an element $\omega \in M_i$ is fed to it, the corresponding inputs of $\tilde{S}_i$ will not be equated, and we connect to these inputs the outputs of the corresponding number of elements of ω in which we equate inputs of the same type, obtaining as a result the network $\tilde{S}_{i+1}$. Here the signal to the output of S_{i+1} at time t depends on the signals to the inputs of $\tilde{S}_{i+1}$ at the same times, likewise the signal to the output of $\tilde{S}_{i+1}$ depends on the signals to the corresponding inputs of S_{i+1} (from the construction, it is clear that there is a one-to-one correspondence between the inputs of S_{i+1} and

$\tilde{S}_{i+1}$). Hence it follows that the network $\tilde{S}$ constructed on the last step will be proper (it is necessary to consider $\tilde{S}_i$ for a value of i such that S_i coincides with S).

8.7. Suppose that we are given a zero-cycle network S that realizes the function f. We will show how to construct a network for f from single-cycle elements of the same type as those that occur in S along with delay elements. The latter will be realized based on the inductive definition of a zero-cycle network (Definition 8.1).

1. *Basis of the Induction.* If S consists of a single functional element, it does not have to be transformed: $\tilde{S} = S$.

2. *Inductive Transition.* (a) If S is obtained from S_0 by equating the inputs (Definition 8.1(a)) and if we are already able to transform S_0 ($\tilde{S}_0$), to reconstruct S we need only equate the inputs of $\tilde{S}_0$ that correspond to those inputs of S_0 that were equated in obtaining S. Clearly, the resulting network $\tilde{S}$ may be treated as a functional element that realizes the same function as the zero-cycle network S.

(b) Suppose that the network S of Figure 26 is obtained by connecting S_0 and S_1 (Definition 8.1(c)). Further, suppose that S_0 and S_1 ($\tilde{S}_0$ and $\tilde{S}_1$) have already been transformed and that there is a ν-cycle delay in $\tilde{S}_0$, and a μ-cycle delay in $\tilde{S}_1$. To obtain $\tilde{S}$, in S we then replace S_0 by $\tilde{S}_0$ and S_1 by $\tilde{S}_1$, and connect in sequence μ delay elements to each input of $\tilde{S}_0$, except for the input connected to an output of $\tilde{S}_1$ (Figure 27).

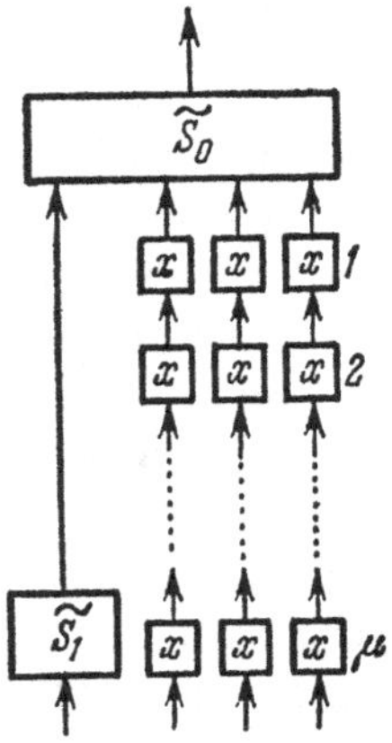

Figure 27

As a result, signals from the output of $\tilde{S}_1$ and signals fed directly to the inputs of the entire network $\tilde{S}$ (other than the inputs of $\tilde{S}_1$) are fed simultaneously to the inputs of S_0 with delay μ. It can be easily verified that the resulting network possesses the desired property, and that its delay will be $\nu + \mu$.

Remark. Let us show once again how a possible transformation of a zero-cycle network S may be performed. We define by induction the concept of the *upper index* of an element occurring in a network. The upper index of an element all of whose inputs are the inputs of the given network is equal to one. Suppose that the upper indices of all the elements whose inputs are connected to the inputs of ω have already been defined. Then the upper index of ω is equal to the greatest of these indices plus one. As in the solution of Problem 8.3, the absence of feedback demonstrates that the definition is correct. In general, the upper index of a functional element ω is equal to the maximal number of elements which signals fed to an input of S must pass through before reaching an output of ω (i.e., the element ω is also counted). We will not bother restating this definition more formally. It may be easily seen that if the upper indices of all the elements whose outputs are connected to the inputs of some element are equal, then the network is a proper network (this is the sufficient condition). Hence follows a method of transforming S into $\tilde{S}$: It is necessary to insert between the inputs of the elements and the outputs connected to them a number of successively connected delay elements so as to satisfy the above condition (equality of upper indices).

In conclusion, let us note that in certain cases Φ need not contain any delay elements. Thus, a system of elements that realizes xy and $\bar{x}$ is complete, since a (one-cycle) delay may be obtained from the element that realizes the conjunction by equating the inputs. Note that a connection of two elements $\bar{x}$ yields a two-cycle delay $\bar{\bar{x}} = x$, but as will be clear from what follows, this we cannot, in general, establish.

(2) cf. Figure 28.

Note that a transformation of the network by means of delay elements is required in examples (d), (e), and (f).

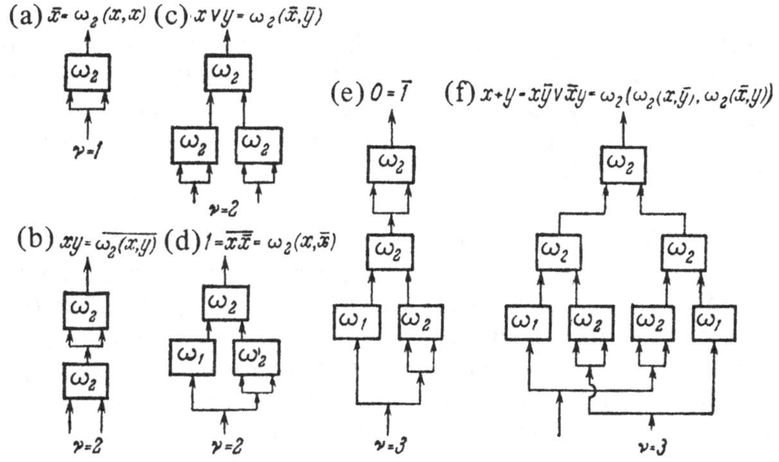

Figure 28

8.8. An element that realizes the multiplicative identity may be considered a zero-cycle element (cf. Remark 3 following Definition 8.4), likewise the network that realizes the additive identity obtained by equating the inputs of ω that realize $x\bar{y}$. Connecting the output of this network to the second input of ω (corresponding to y), we obtain a network that realizes x with delay 1; this network may be used as a single-cycle delay element. Completeness now follows from Problem 8.7 and the completeness of the system of logical functions $\{x\bar{y}, 1\}$.

8.9. Suppose that we are given a proper network S that realizes some function f with delay ν. We equate all its inputs. The signal y to the output of the resulting network at time t may depend actually only on the signal to the input of the network at time $t - \nu$ (we denote this signal by x), and cannot depend on the signals to the input of the network at any other times. Without changing y, therefore, any signals may be fed to the input of the network at any time other than $t - \nu$. At time $t - \nu + k$ (k any integer), signal x is fed if k is even, and signal $\bar{x}$ if k is odd. Since ω realizes negation once the inputs have been equated, we then find that at every moment, the signal that appears at the outputs of all the elements in the network is the same as the input signal of the network at this time (i.e., the same signal is fed to the network at every moment of time). A rigorous proof may be conducted by induction on the upper index of the element (cf. remark following the solution of Problem 8.6). If this has been proved for elements with upper index at most k, then at every moment the same signals will arrive at all the inputs of elements with upper index $k + 1$. (These signals are either the input signals of the network of the output signals of elements with upper index less than $k + 1$.) At the next moment, the opposite signal will appear at the output of the network, i.e., the signal which at this moment is fed to the input of the network. As a result, at time t the same signal will appear at the output of the network as at its input, i.e., $y = x$ if ν is even, and $y = \overline{x}$ if ν is odd. Thus, a network with equated inputs realizes x if the delay is even, and realizes $\overline{x}$ if the delay is odd.

Remark. We may once again construct (cf. remark following the solution of Problem 8.6) a two-cycle delay, though, as is clear from the solution of this problem, unlike the case of a one-cycle delay (cf. Problem 8.7), the existence of a two-cycle delay does not yield a complete system of *elements* which realize a complete system of *functions*.

8.10. Suppose that we are given some proper network S formed from these elements which realizes a function ω with delay ν. We connect to some of the inputs of the network the outputs of elements that realize the constants (which may be thought of as lacking inputs), and equate the remaining inputs. We obtain a network with delay ν that realizes a function of one variable obtained form ω by substituting the constants for the corresponding variables and equating the remaining variables. Suppose that the signal x is fed to the input of the resulting network S at time $t - \nu$. Then the signal at the output at time t depends only on x and is independent of the signals to the input of the network at any other time. As in the solution of the preceding problem, at time $t - \nu + k$ the signal x is fed if k is even, and the signal $\overline{x}$ if k is odd. Then either the same signal, or a signal that coincides with the signal fed to the input at this time, will always be present at the outputs of all the elements of the network. The proof may be conducted in the same way as in the preceding problem. We need only note that substituting a constant for any argument of a Sheffer function results either in a constant or a negation. As a result, $\tilde{S}$ realizes either a constant (if the same signal is always fed to its output), or x if ν is even, or $\overline{x}$ if ν is odd.

If we are given a function (for example, $x\overline{y}$) in which it is possible to substitute constants for some of the variables in two different ways, and to equate the rest of the variables so that x and $\overline{y}$ are obtained (for example, $x\overline{0} = x$, $1\overline{y} = \overline{y}$), then a network consisting of the elements cited in the premise cannot be realized. (By virtue of what we have just proved, such a network can have neither an odd nor an even delay.)

8.11. *Necessity.* The necessity of Condition (a) is self-evident (cf. Remark 1 following Definition 8.4).

The necessity of Condition (b) was basically proved in solving Problem 8.9. There we used only the fact that $\overline{xy} \in$ **Q**. That is, we proved that if all the elements of a network realize functions in **Q**, once the variables have been equated any one of these functions coincides with x if the delay is even, and coincides with $\overline{x}$ if it is odd, i.e., the constants cannot be realized.

Similarly, in solving Problem 8.10 we used only the fact that $x\overline{y} \in$ **R**. We proved that if all the elements of a network realize functions from R, once an arbitrary number of the variables have been found and the rest equated, the function realized by the network either becomes a constant, or the function x or the function $\overline{x}$, depending on whether the delay is even or odd.

Sufficiency. By equating the inputs of elements that realize the functions $\omega_1 \notin \mathbf{P}_0$ and $\omega_2 \notin \mathbf{P}_1$ (which exist by (a)), we obtain elements that realize the constants or an (single-cycle) element that realizes negation. By equating the inputs of the element that realizes the function $\omega_3 \notin \mathbf{Q}$, we obtain either a one-cycle delay element (in which case the system is complete by Problem 8.7) or an element that realizes some constant. If the case of a one-cycle delay element is discarded, we would then be able to realize both constants, since if we did not obtain them on the first step, then by means of negation (realized in this case on the first step) and one of the constants (obtained on the second step) the other constant may be obtained. Recall that networks that realize the constants have zero delay.

Further, some of the arguments of the function $\omega_4 \notin \mathbf{R}$ may be replaced by constants and the remaining arguments equated so as to obtain x. Performing the corresponding operation with the element that realizes ω_4 yields a one-cycle delay. We can now apply Problem 8.7.

Remark 1. Since we have proved that a one-cycle delay may be obtained from every complete system of functional elements, in this case every logical function may be realized with arbitrary delay not less than some fixed number (by successively connecting the delay elements to the output of the circuit).

Remark 2. The systems of functions Q and R not only are not closed relative to composition, but in addition by means of elements which realize the functions of any one of these systems it is possible to realize functions not occurring in them (for example, the function x, of course only with even delay; cf. remark following the solutions of Problems 8.9 and 8.10). This we may attribute to the fact that though we may be able to realize some function, this does not mean it can be simply supposed that we are given the element which realizes it, since in our discussions the elementary functional elements are assumed to be one-cycle elements.

8.12. That the class T is closed relative to equating of variables is self-evident. Let us now consider the function $f(\omega_1(\ldots), \omega_2(\ldots), \ldots, \omega_k(\ldots))$, where the ω_i have been substituted for all the arguments of f. Then, assuming that $\omega_i \in$ T, if null or unit sequences are substituted for them the same sequences are substituted for the arguments of f, and the composition belongs to T. Here we are not using the fact that $f \in$ T. It remains for us to note that $x\bar{y} \in$ T and $1 \in$ T.

The class T is not closed relative to ordinary composition ($x\bar{y} \in$ T and $1 \in$ T, but $1\bar{y} = \bar{y} \notin$ T).

8.13. Suppose that we are given a system of functions complete in the ordinary sense, and containing a function ψ that does not belong to a class T. By equating the variables of ψ, we obtain x or $\bar{x}$. Since $\bar{\bar{x}}$ is a global composition, in both cases we have x, so that, as we have already noted, a system of functions complete in the ordinary sense is sufficient.

8.14. Suppose that we are given a system Φ complete in the ordinary sense. By equating the variables of the functions $\omega_1 \notin \mathbf{P}_0$ and $\omega_2 \notin \mathbf{P}_1$, we obtain x, and the ordinary composition reduces to global composition (cf. preceding proof). But if we obtain the constants, we may take any function $\omega \in \Phi$ that is not a constant. Suppose that x is its actual variable. In place of the remaining variables, we substitute a sequence such that the value of ψ depends on the value of x (Problem 1.9). We obtain either x or $\overline{x}$, but these cases have already been considered.

8.15. We consider a one-cycle functional element that realizes the function F, and successively connect $\mu - i - 1$ one-cycle delay elements to the input corresponding to the argument x_j^i. The desired automaton is obtained.

8.16. From one standpoint, the assertion is self-evident, since the functional elements are a special case of automata without feedback. Now suppose that we are given a complete system of functional elements Φ and some automaton without feedback $\mathfrak{U}$. We construct a network S that realizes its characteristic function F. Suppose that S realizes F with delay ρ. From the solution of Problem 8.11, we may construct one-cycle delay elements using elements from Φ. To the inputs of S we successively adjoin the same number of connected delay elements as in the preceding problem. The resulting network is an automaton that differs from only in delay time (its delay will be $\rho - 1$ cycles longer). We have thereby realized the automaton $\mathfrak{U}$.

8.17. Suppose that $\mathfrak{U}(\Omega,n)$ is a finite automaton and let Ω consist of certain sequences of length $k + 1$. We create a network of functional elements $\omega_0, \omega_1, \ldots, \omega_k$ which realize functions that coincide with the corresponding functions $\Phi_0, \Phi_1, \ldots, \Phi_k$ in their domain of definition. The output of the element ω_0 is understood as the output of S. In accordance with the equation of state of $\mathfrak{U}$, the inputs of each element ω_i are connected to the outputs of all the elements ω_j; we equate the corresponding inputs from the remaining inputs of distinct elements ω_i so that the outputs of S are obtained (Figure 29).

Thus, the external form of the constructed network depends only on the numbers k and n. If we now assume that when $t = 0$, the signals at the outputs of ω_i form a designated

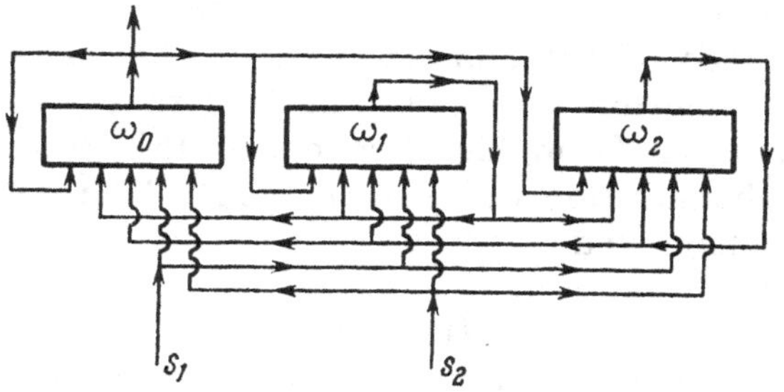

Figure 29

initial sequence ω^0 and that at every moment corresponding signals $s(t)$ are fed to the inputs of S, S will then represent the operation of $\mathfrak{A}$ under the assigned conditions (ω^0, $\nu = 1$, $s(t)$).

Remark. S may be simplified as follows. Suppose that $f(x_1, \ldots, x_n)$ is a partially defined logical function with domain of definition Ω. We will say that $x_{i_1}, \ldots, x_{i_\ell}$ is a system of actual variables if the functions f take the same values on all sequences in Ω on which the values of these variables coincide. (In the case of everywhere defined functions, this definition is in accord with the concept of an actual variable given in Chapters 1 and 2.) It can be easily seen that there is a minimal system of actual variables. In constructing S in Problem 8.17, elements ω_i may be taken in which the inputs correspond to a minimal system of actual variables Φ_i. As before, the elements are connected and the inputs equated in accordance with the equation of state of the automaton.

8.18. Since some of the automata are logical functions, and since the definition of a realization of functions considered as automata does not differ from that given above, an automaton-complete system of functional elements is complete also relative to the logical functions.

Now suppose that we are given a system of elements complete relative to the logical functions, and let $\mathfrak{A}$ be some automaton. We consider the network S which represents it (it was constructed in the solution of Problem 8.17). The functional elements occurring in it are replaced by networks that realize the same functions with the same delay time ν and created from elements of the given system. This may be done by virtue of the completeness of this system relative to the logical functions, and also the fact that a one-cycle delay may be obtained from the elements of this system (Remark 1 following the solution of Problem 8.11). The desired network has been constructed.

8.19. We prove that, for example, the disjunction $x \vee y$ cannot be realized. Suppose that we are given a network which realizes $x \vee y$ in double lines. We take the output which itself realizes $x \vee y$ along with the corresponding output element. This element ω_1 must realize the conjunction (since there are no other elements). Let us consider elements ω_1 and ω_2 whose outputs are connected to the inputs of ω_1. The function $x \vee y$ is again realized at the output of one of them. (It can be easily verified that functions realized on the outputs of ω_2 and ω_3 must equal 1 on the sequences (1, 1), (1, 0), and (0, 1), and at least one of them will equal 0 on (0, 0).) Continuing this process further, we end up with a network consisting of a single element that realizes $x \vee y$, and, thereby, arrive at a contradiction. A rigorous proof may be conducted by induction (for example, on the number of elements in the network).

Remark. Without the need for any additional elements, we may construct a negation of a function already represented in double lines. However, this is not the same thing as having a functional element that realizes negation.

8.20. 1. Note that all the functions that may be self-dual represented by the system of functions Φ may be realized in double lines using elements that realize Φ. In fact, it is sufficient to use networks consisting of functional elements which realize f and f^+ and to feed the arguments themselves to the inputs of the network for f and their negations to the inputs for f^+. If we are given a system of elements that realize a self-dual complete system of functions, we could realize in double lines a system of functions that would become complete upon the addition of negation. But as we have seen, negation may be realized in double lines without the use of functional elements.

2. If some function f can be represented in double lines by a network, it can then be easily seen that the parts of the network connected to each of the outputs realize the dual functions ω and ω^+ (each of them has n inputs, and if the opposite signals are fed to the corresponding inputs, the opposite signals will also appear at the outputs; this may be proved by induction.)

In this case, the function ω may differ from f only by negations of arguments or the value of the function (the latter is not essential). Hence it follows that if an arbitrary logical function f can be realized in double lines, the system of corresponding functions ω is self-dual complete, that is, the initial system of elements realizes a self-dual complete system of functions.

Chapter 9

RELAY-CONTACT NETWORKS. ESTIMATING THE COMPLEXITY OF A CONTACT NETWORK

1. *Relay-Contact Networks.* In this chapter, we will learn about another method of realizing the logical functions, through the use of what are known as relay-contact networks. This method cannot be described within the framework of the theory of networks formed from functional elements (cf. below, end of present section). We will discuss an actual set-up of relay-contact networks, though bear in mind that we are not interested in any particular realization of the functional elements.

As a start, note that if we associate with a logical variable x a conductor along which current is or is not flowing depending on whether $x = 1$ or $x = 0$, we may then associate with a series connection of conductors a conjunction of variables, and with a parallel connection, a disjunction (Figure 30). By repeatedly applying series-parallel connections, we may construct networks. It is, however, clear that in this case we realize only monotone functions. To realize arbitrary functions, it would be sufficient if we could realize negation. This may be done using a device called a *break (negative) contact relay.* It is shown in Figure 31. If there is no current through the coil winding A ($x = 0$), the spring pulls the contact B upwards and the network is closed. As a result, there is a current at the output C. But if $x = 1$ and there is a current through A, the contact B is pulled downward and there is no current at C. Thus, the function $\overline{x}$

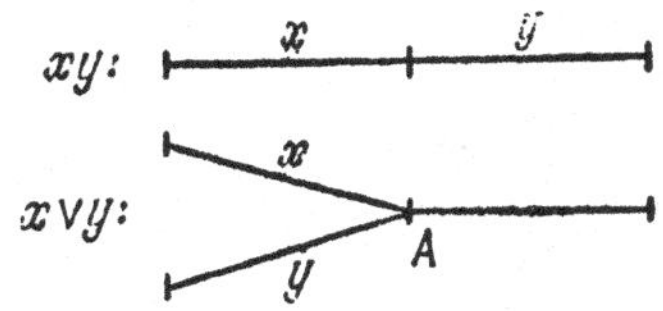

Figure 30

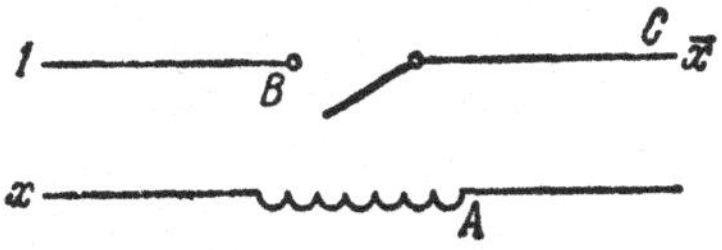

Figure 31

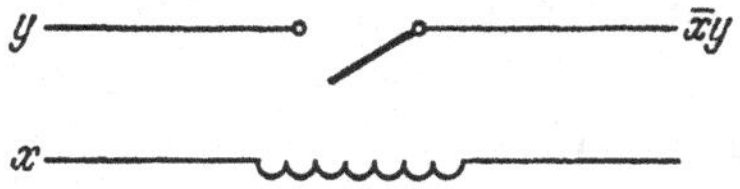

Figure 32

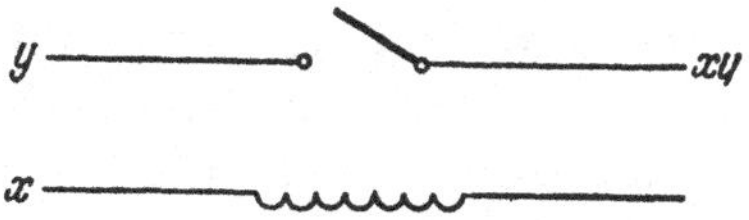

Figure 33

is realized. Note that if some variable y is fed to B instead of 1, we realize the function $\bar{x}y$ (Figure 32).

We may also consider a *make* (*positive*) *contact relay* (Figure 33). In this case, the contact is closed if there is a current through the coil winding. At the output of the network, the function xy is realized.*

When connecting conductors, it should be kept in mind that current propagates in all directions. Therefore, if $x \vee y$ is realized (Figure 30), when $x = 1$ and $y = 0$ current from A propagates in all directions, including through the conductor corresponding to the variable y (even though $y = 0$). Clearly, in this case the operation of a complicated network may be distorted. This can be avoided in any number of ways, for example, by means of special devices that transmit current in only one direction. We could use a relay for this purpose; with a make contact relay disjunction is realized by means of the network shown in Figure 34. In this case, the result at the output of $x \vee y$ does not alter the values of x and y, since coil windings not connected directly to the network output serve to represent their values. We will limit ourselves to this example without showing how to correct the situation in the general case.

*In our figures we are not showing the springs that pull the contacts this way or that. That there is a break (correspondingly, make) contact is indicated by drawing the conductor in which it is found nearer to (correspondingly, further from) the coil.

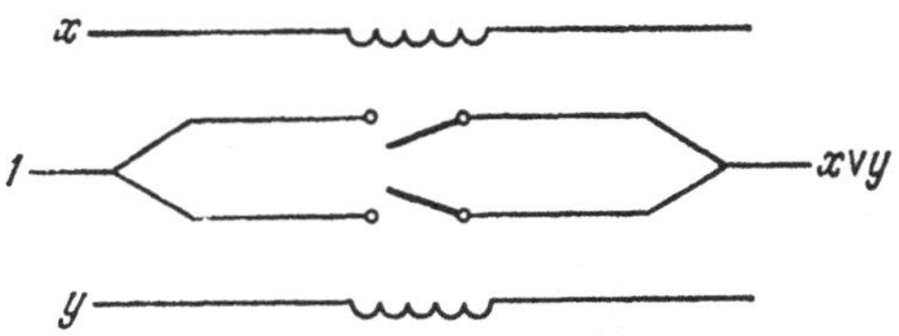

Figure 34

Until now we have assumed that signals remain in the conductor for an arbitrarily lengthy period of time. Now let us discard this assumption, and discuss the operational life of a network constructed from a relay (called a relay-contact network). We assume that current propagates instantaneously, and that a single cycle elapses in response to the relay (closure of the contact). This means that in Figures 32 and 33, signal y will arrive one cycle after signal x; the output signal ($\bar{x}y$ or xy) appears at the same time as y. Thus the time the network requires to process signals must be taken into account, and this time may have to be changed without changing the functions the network realizes. As in the case of a multi-cycle network of functional elements, this may be done by means of delay elements. Here the delay element is a make-contact relay (Figure 33) to whose contact 1 is fed ($y = 1$). There is a delay of one cycle.

If the signal $f(x_1, \ldots, x_n)$ appears at the output of a relay-contact network ν cycles after the signals $x_1, \ldots, x_n$ have been fed to its inputs regardless of which signals have been fed to the inputs at other moments of time, we say that the *network realizes the function* $f(x_1, \ldots, x_n)$ *with delay* ν. Here it is assumed that if two sequences of signals are fed at two successive moments of time, ν cycles later an output signal appears that corresponds to the value of f on the corresponding sequence. Briefly, successive sequences of signals are processed independently.

We will not be giving a rigorous definition of such a relay-contact network here; the above explanation is sufficient for our purposes.

9.1. (1) Prove that every logical function may be realized by a relay-contact network with some delay.

(2) Prove that in this case we cannot limit ourselves to break-contact relays.

Digression on Feedback. Since relay-contact networks are multi-cycle circuits, feedback-type effects may be observed (cf. Chapter 8, page 145). For example, the network in Figure 35 is similar to a bell circuit. Here there are two relays, one with a make contact and the other with a break contact, and each contact is connected to the coil of the other. Then, if at the initial moment of time t there is a current in both coils, at time $t + 1$ contact A is closed and contact B is open, i.e., there is no current in the coil C, but there will be current in D; at time $t + 2$, both contacts are open and there is no current in the coils; afterwards, at time $t + 3$, A is open, B closed, and correspondingly, there is current in C but no current in D. As a result, at time $t + 4$ both contacts are closed and there is a current in both coils, i.e., we have returned to the initial state. Thus, no matter what the initial state is, the network returns to it after four cycles (its operating period is four cycles); every two cycles, the state of both contacts changes to its opposite.

If a relay is connected in a similar way to contacts of the same type, for example, make contact relays (Figure 36), everything depends on the initial state. If at the initial moment of time, the contacts are closed, they remain closed forever; but if one of them is closed and the other open, at every moment the state of the contact changes to its opposite. In the case of relays with negative contacts connected in this way, the state in which both contacts are open is the stable state.

9.2. Realize the following functions by means of relay-contact networks (a) $xy \vee \overline{z}$; (b) $\overline{xy} \vee zt$; (c) $xy \vee yz \vee xz$.

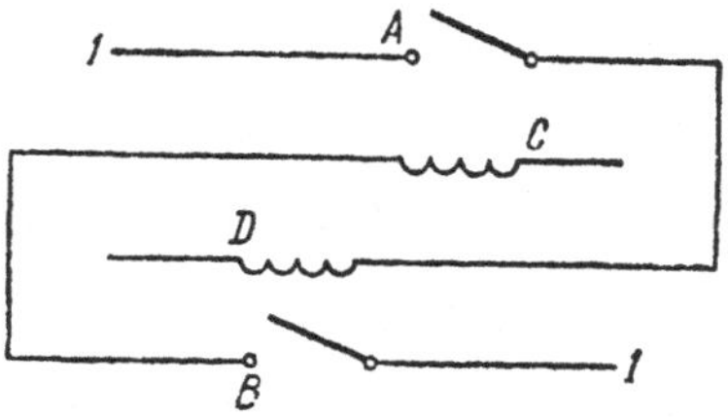

Figure 35

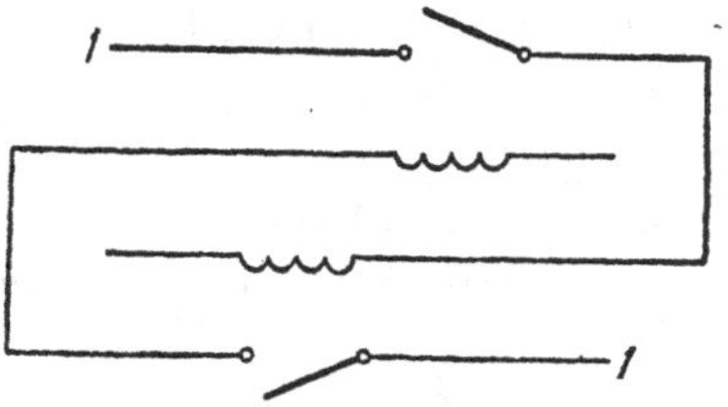

Figure 36

Relay-contact networks should not be thought of simply as networks of functional elements that realize xy, $\overline{xy}$, and 1. One slight difference is that y must be fed to the inputs of the relay one cycle later than x (and not at the same time, as in the case of a functional element). This convention may be dispensed with if y is connected to the output of a relay that causes a one-cycle delay, in which case it is important that signal $y - 1$ not be delayed. A more critical difference is that in networks consisting of functional elements, every available signal may be "multiplied" without the need for additional devices (the output of the element may be connected to an arbitrary number of inputs). By contrast, to multiply a signal in a relay-contact network, it is necessary to feed it to a coil winding equipped with the corresponding number of positive contacts.* On the other hand, in relay-contact networks it is possible to connect conductors that realize disjunction; a special element is required to do this if we use networks of functional elements.

2. *Contact Networks.* From now on, we will limit our discussion to networks in which only the contacts are connected (there are no connections between the relay coils and the contacts). Such networks are called contact networks.** We introduce some rules for representing the contact networks. A contact is represented by segments whose end-points are called *poles*, while the segment itself is called a *dipole*. A dipole is supplied with the variable symbol x if the contact is a make contact, and with the variable symbol $\overline{x}$ if it is a break contact; x is the variable that is realized by the corresponding coil. Dipoles are connected by poles. It may be assumed that with every variable we may associate a coil which is itself connected to an arbitrary number of contacts (cf. end of last section).

As a result, a contact network may be visualized as a *graph*, or set of points (the *vertices* of the graph) connected by

---

*Note that an arbitrary number of contacts, both positive and negative, may be linked to one and the same coil.

**The coils of the contact networks are not connected, and so the values of the variables cannot change (cf. remark for Figure 34).

*segments** (the *edges* of the graph). To the edges of a graph, we assign the symbols of the logical variables or their negations; the edges correspond to the contacts. We associate with the vertices connections of contacts corresponding to the segments (dipoles) that converge at these vertices. If there is current at one of the contacts leading to a vertex, current propagates through all the contacts closed at that moment for which the given vertex is a pole. Finally, there are two special vertices in a graph: the *input vertex* and the *output vertex*. Current is always being fed to the input, i.e., a current is fed to the corresponding poles of the contacts leading to this vertex. It may be assumed that current is always being fed to several vertices of a network, though in that case these vertices could be equated without changing the network. Current is never fed from the outside to the other poles. If a current is fed to the windings of some of the coils, after one cycle the make contacts corresponding to them close, and the break contacts open; the opposite pattern is observed at the contacts of the other coils. If there is a current at the network output, we say that the network is conductive for these values of the variables (states of the coil windings); otherwise, it is not conductive. Thus a contact network operates in one cycle. Figure 37 presents examples of contact networks. Here the empty circle denotes the input of the network, and the shaded circle, its output.

9.3. Which of the networks obtained in solving Problem 9.2 are contact networks? Draw the networks corresponding to them using the notation introduced in this section.

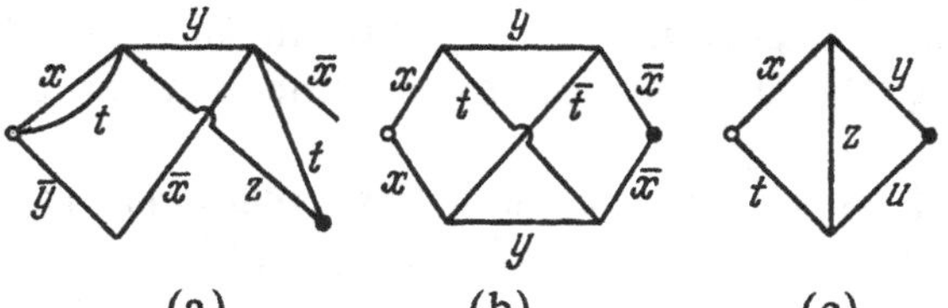

(a) (b) (c)

Figure 37

*Two vertices may be connected by several different segments (and the segments need not be straight lines). In addition, the segments may "coincide" at a point which is not a vertex (cf. Figure 37 above).

Let us now discuss the logical function that realizes a contact network. This function is equal to unit for those values of the arguments for which the network is conductive, and to zero for those values for which it is not conductive. The *conduction function* of a network is the function realized by the network. Note that the conduction function does not change if the input and output of the network are interchanged.

The DNF for the conduction function may be constructed from the network in a natural way. We understand a *circuit* to refer to any sequence of contacts in which it is possible to order the poles (origin and exit) of each contact such that the input of the network is the origin of the first contact, the origin of each succeeding contact coincides with the exit of the preceding contact, and the exit of the last contact coincides with the exit of the network. The same contact may occur several times in a network, and for each occurrence in the network, the poles are ordered differently. Geometrically speaking, a circuit is a connected sequence of edges of a graph issuing from an input to an output. There will be current at the output of a network for some set of values of the variables if and only if all the contacts in at least one circuit are closed. Hence it follows that if we associate with each circuit Γ a conjunction $\mathfrak{U}_\Gamma$ of variables or, correspondingly, negations of variables assigned to the contacts occurring in Γ, the disjunction of these conjunctions over all circuits in the network will coincide with the conduction function of the network. This line of reasoning requires some refinement, since, in fact, there is an infinite number of possible circuits. To overcome this difficulty, note that the conduction function can in fact be obtained by simply taking the disjunction over some, rather than all the circuits.

A circuit Γ is said to be *actual* [*sushchestvennyi*] if it does not pass through any vertex of a graph twice; more precisely, the input and output of the circuit each have one contact with this vertex as pole, while there are two contacts with this vertex as pole for every other vertex of the circuit.

9.4. Prove that there is only a finite number of actual circuits in a network.*

*We are only considering networks with a finite number of contacts.

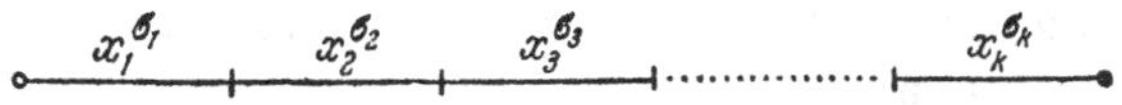

Figure 38

9.5. Prove that a disjunction of conjunctions corresponding to actual circuits is equivalent to the conduction function of a network.

The conduction function of a network may be constructed using the result of Problem 9.5.

9.6. Find the conduction functions for the networks in Figure 37.

Let us now consider how to handle the inverse problem, i.e., constructing a network that realizes some function from the function itself. Based on the above discussion, we may proceed by representing the function in DNF. With each elementary conjunction $x_1^{\sigma_1} x_2^{\sigma_2} \ldots x_k^{\sigma_k}$ occurring in DNF, we associate a network consisting of successively connected contacts $x_1^{\sigma_1}, x_2^{\sigma_2}, \ldots, x_k^{\sigma_k}$ (Figure 38). Now we equate, first, the inputs of all these networks with each other (in the case of all elementary conjunctions occurring in the DNF), and then the outputs of all these networks with each other. The resulting network realizes the designated function; the only actual circuits are those which correspond to the initial elementary conjunctions.

9.7. Using the above algorithm, realize the following functions by means of contact networks: (a) $(y \vee z) \rightarrow x\bar{y}$; (b) $z\bar{y} \sim yx$; (c) $x + y + z$.

In solving Problem 9.7, we tried to simplify the DNF by means of (2.20) - (2.25). These formulas sometimes make it easier to construct the conduction function when using the algorithm.

In realizing a function by means of contact networks, we have proceeded using a representation of the function in DNF. It is natural to expect that the function could also be realized using CNF.

9.9. Find a method of realizing functions by means of CNF analogous to the method presented above for realizing functions by means of DNF.

9.10. Using the method found in solving Problem 9.9, realize the functions of Problem 9.7.

In realizing the logical functions by means of CNF and DNF, we have applied two operations to the contact networks, namely parallel and series connections of networks. In a series connection of two networks, the output of the first network is equated to the input of the second network, while the input of the first network and output of the second network are declared the input and output, respectively, of the new network. In a parallel connection, the inputs of the networks are equated and declared the input of the new network; the outputs are analogously equated and the resulting pole is made the output of the entire network. Clearly, we may associate with a series connection of networks a conjunction of conduction functions, and with a parallel connection of networks a disjunction of conduction functions.

Let us give an inductive description of the class of networks that may be obtained by a parallel-series connection. A network consisting of a single contact will be said to be *elementary*. A contact network is called a *parallel-series network*, or *pi-network*, if it can be obtained from elementary networks after some number of steps by means of parallel and series connections. Clearly, we may associate with every method of constructing *pi-networks* from elementary networks a representation of conduction functions in the form of formulas containing only disjunctions, conjunctions, and negations; in addition, only the arguments are negated. Conversely, from each such formula it is possible to construct some completely defined pi-network.

9.11. Construct pi-networks from the following formulas: (a) $(x \vee y\bar{z})(xy \vee zt)$; (b) $(\bar{x}(y \vee \bar{z}) \vee \bar{t})x$.

We might ask whether every contact network is a pi-network. It turns out that this is not true.

9.12. Prove that the network (called a "bridge") in Figure 39 is not a pi-network.

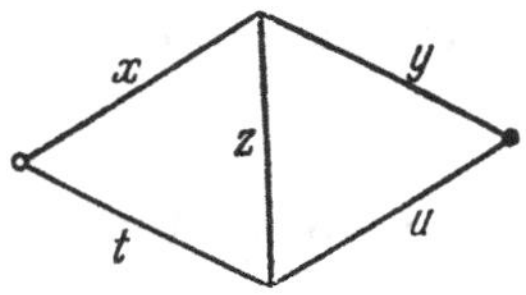

Figure 39

3. *Minimization Problem for Contact Networks.* One and the same function may be realized by different contact networks. In a realization of some function, it is natural to try to construct a contact network with either a minimal number of contacts or, at the least, one in which the number of contacts does not exceed this minimal number by too much. A network is called a *minimal network* if it contains the smallest possible number of contacts of all the networks which have the same conduction function.

There are only a few cases in which a minimal realization can be found, just as there are only a few cases in which it is possible to prove that a particular network is minimal. Let us first present some very simple examples along with background information that is sometimes helpful in establishing whether some network is minimal.

9.13. Prove that a bridge (Figure 39) is a minimal network. Attempt to formulate a sufficient network minimality condition satisfied by the example. Construct analogous examples of minimal networks.

We now present a somewhat more complicated example of a minimal network.

9.14. Prove that the network depicted in Figure 40 is a minimal network. Formulate a corresponding sufficient network minimality condition.

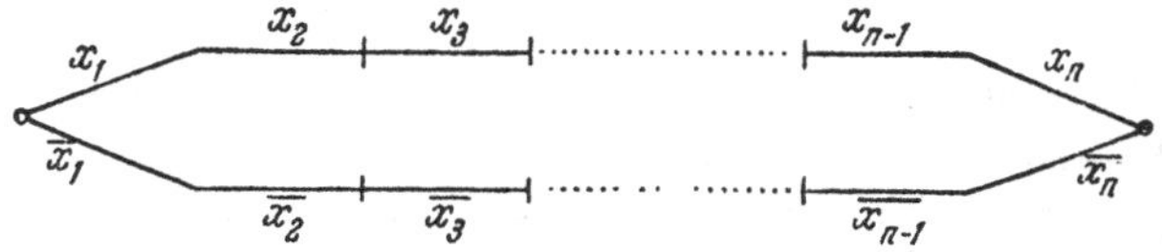

Figure 40

9.15. For the function $x_1 x_2, \ldots, x_n \vee \bar{x}_1 \bar{x}_2 \cdot \ . \ . \ \bar{x}_n$, construct a minimal network different from the network* depicted in Figure 40.

From Problem 9.15, it follows that a minimal network is not unique.

Minimal networks may be found not only in the class of all networks, but also in some narrower class of networks, for example, the class of pi-networks. We might then ask whether a minimal pi-network is minimal in the class of all networks. An answer would bring us back once again to the bridge network we have referred to several times already.

9.16. For a function which may be realized by a bridge (Figure 39), prove that there is no pi-network with the same number of contacts as the bridge.

Suppose $f(x_1, \ldots, x_n)$ is a logical function. We let $L(f)$ denote the number of contacts in the minimal network that realizes it; analogously, $L_\Pi(f)$ is the number of contacts in the minimal pi-network which realizes f. Obviously,

$$L(f) \leqslant L_\Pi(f).$$

In Problem 9.16, we have given a function for which strict inequality holds.

The maximal value of $L(f)$ for functions $f(x_1, \ldots, x_n)$ of n variables is called the *Shannon function* (analogously, $L_\Pi(n) = max\ L_\Pi(f)$, where the maximum is taken over the same set of functions). To discuss the functions $L(n)$ and $L_\Pi(n)$, we start with realizations of special function. Then we pass on to a general description of realizations of functions of n variables. We wish to estimate the maximal number of contacts needed to realize a function of n variables.

Let us first find an upper bound on $L(n)$. For this purpose, we consider some method of realization of functions by means of contact networks and find an upper bound on the

*Of course, the contacts could be represented in a trivial way in Figure 40 without changing the conduction function. But the problem speaks of an essentially different network. We will not bother stating precisely what meaning is to be assigned to these words (this may be done quite simply in the general case).

greatest number of contacts that may be required here.

9.17. Find an upper bound on $L(n)$ based on a realization of functions by means of:
(a) PDNF;
(b) PCNF;
(c) of the first two methods, the one which requires the lesser number of contacts.

We often use an inductive method of realization of functions. Considered in its simplest form, in such a method we assume that all functions of k variables have already been realized, and then realize functions of $k + 1$ variables by "expanding" them in one of the variables.

9.18. Using the method of realization of functions just given, obtain an upper bound on $L(n)$.

The methods of realization of functions considered in Problems 9.17 and 9.18 lead to pi-networks. Therefore, in these problems we have simultaneously found upper bounds on $L_{\Pi}(n)$.

Now we wish to consider a method of realization which no longer leads to pi-networks. To construct networks by means of PDNF, we must realize elementary conjunctions. Until now we have realized each elementary conjunction separately, and only subsequently connected the resulting networks in parallel. Now, however, we wish to realize all the elementary conjunctions simultaneously. We will refer to a network with a single input and k outputs as a $(1 - k)$-*pole*. A function f is said to be realized on one of the k outputs if the function is the conduction function of our network in the sense understood above whenever this particular output is made the output of the network. As a result, k logical functions are simultaneously realized at the outputs of a $(1,k)$-pole. A *universal* $(1, 2^n)$-*pole* is understood to refer to a $(1, 2^n)$-pole at whose outputs all the complete proper elementary conjunctions of n variables are realized. Such a universal multi-pole may be created by equating the inputs (but not the outputs!) of the networks (Figure 38) which realize elementary conjunctions. However, we will construct the network using a less time-consuming method (Figure 41). We hope that the method of construction is clear from the figure; a rigorous description of the method may be given by induction. If a

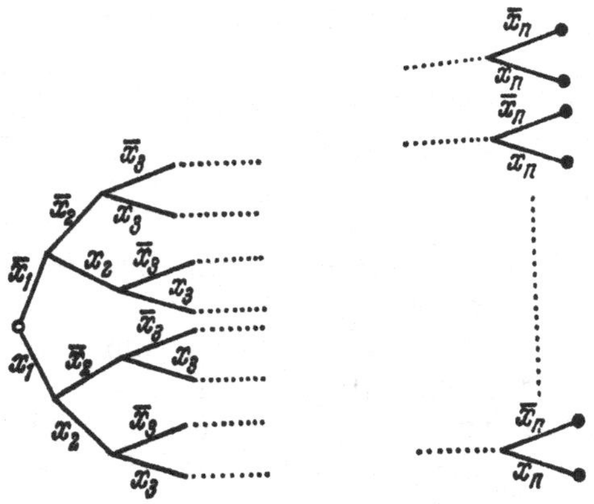

Figure 41

universal $(1,2^{k-1})$-pole has already been constructed, a universal $(1,2^k)$-pole may be obtained if the contacts x_k and $\bar{x}_k$ are adjoined to each of the 2^{k-1} outputs of the already constructed network, and if the free poles of these contacts are made the outputs of the new network (there will be k^k free poles).

9.19. Prove that the network depicted in Figure 4 is in fact a universal $(1,2^n)$-pole. Find a method of constructing a network for the logical functions which uses universal multipoles. What is the upper bound on $L(n)$ in this case?

We will say that there is zero conduction between two poles of a network if the conduction function of the network with the same graph and identically designated contacts, but with one of these poles as input and the other as output, is identically zero. Briefly, with any network connecting these poles we may associate a conjunction which is identically zero. If the conduction between any two outputs of a $(1,m)$-pole is zero, we will call it a *switching pole.*

9.20. Prove that the universal $(1,2^n)$-pole we have constructed is a switching pole.

We realize logical functions of n variables by expanding them in PDNF with respect to the last $n - k$ variables (cf. (2.16)):

$$f(x_1, \dots, x_k; x_{k+1}, \dots, x_n) = \bigvee_{(\sigma_{k+1}, \dots, \sigma_n)} \omega_{\sigma_{k+1} \dots \sigma_n}(x_1, \dots, x_k) x_{k+1}^{\sigma_{k+1}} \dots x_n^{\sigma_n}.$$

The network consists of two parts, the same for all functions f; networks for distinct functions are distinguished by the connections between these poles. The first part (M_1) constitutes a universal $(1,2^{n-k})$-pole for the variables $x_{k+1}, \dots, x_n$; it is schematically depicted in Figure 42. The second part (M_2) constitutes the set of networks of all the 2^{2^k} functions of k variables $x_1, \dots, x_k$ whose outputs have been equated. Thus M_2 may be thought of as a $(2^{2^k},1)$-pole (Figure 43).

We must stipulate which networks realize functions in M_2. From what follows, it is clear that any of the methods of realizing functions already discussed, for example the method given in Problem 9.18, can be considered. Each input of the multipole M_2 corresponds to some function of n variables.

A network S_f for f may be constructed from M_1 and M_2 in the following way. Each output of M_1 is made to correspond to some elementary conjunction $x_{k+1}^{\sigma_{k+1}}, \dots, x_n^{\sigma_n}$; note which function of k variables is a coefficient in this conjunction. We equate the input of M_2 corresponding to this function with the given output of M_1. All the outputs of M_1 may be so equated. Thus, it may turn out that several outputs of M_1 are equated to the same input of M_2.

9.21. Prove that the network S_f which thus results in fact realizes $f(x_1, \dots, x_n)$. Find an upper bound on $L(n)$.

Now we must select k for each value of n so that the resulting network S_f has the least number of contacts.

We denote by $T_{n,k}$ the number of contacts in the network. Then the greatest upper bound on the least value of $T_{n,k}$ with respect to k (for fixed n) must be found. Let us

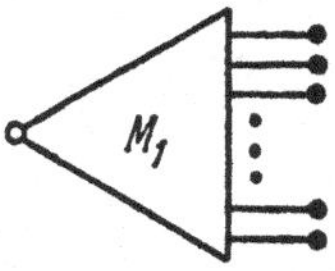

Figure 42

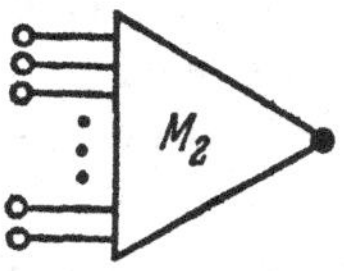

Figure 43

estimate the value of $min_k\ T_{n,k}$ for sufficiently large n, i.e., we will ignore quantities that become small as n becomes large. A more rigorous statement of the problem may be given as follows.

First let us recall some facts from analysis that may be found in any textbook on the subject. Suppose that we are given two number sequences $\{a_n\}$ and $\{b_n\}$ that tend to infinity as $n \to \infty$. We say that a_n tends to infinity *more rapidly* than b_n if

$$\lim_{n\to\infty} \frac{a_n}{b_n} = \infty .$$

If

$$\lim_{n\to\infty} \frac{a_n}{b_n} = 1,$$

the sequences $\{a_n\}$ and $\{b_n\}$ are said to be *equivalent* (written $a_n \sim b_n$); if there exist positive constants c_1 and c_2 and a natural number N such that for all $n > N$,

$$c_1 \leqslant \left|\frac{a_n}{b_n}\right| \leqslant c_2,$$

the sequences $\{a_n\}$ and $\{b_n\}$ are said to be of the same *order*. If a_n tends to infinity more rapidly than b_n, the sequences a_n and $a_n + b_n$ are equivalent.

Let us consider the sequences $\{a^{b^n}\}$ $(a > 1, b > 1)$, $\{c^n\}$ $(c > 1)$, $\{n^k\}$ $(k > 0)$, and $\{\log_p n\}$ $(p > 1)$. The sequence $\{a^{b^n}\}$ grows more rapidly than all the others, $\{c^n\}$ grows more rapidly than $\{d^n\}$ if $c > d$; $\{c^n\}$ grows more rapidly than $\{n^k\}$ for arbitrary c and k; $\{n^k\}$ grows more rapidly than $\{n^\ell\}$ if $k > \ell$; and $\{n^k\}$ grows more rapidly than $\{\log_p n\}$ for arbitrary k and p.

Note that from the above results (Problem 9.18), it follows that $L(n)$ does not grow more rapidly than $\frac{3}{2}\cdot 2^{n-1}$.

9.22. Prove that for every n, $k = k(n)$ may be selected so that as $n \to \infty$, $T_{n,k(n)}$ will be of the same order as $2^n/n$. Find an upper bound on $L(n)$ for sufficiently large n.

Until now we have been interested in an upper bound on $L(n)$. Now let us try to find a lower bound.

9.23. Is it possible to find a lower bound on $L(n)$ on the basis of the results obtained so far?

The lower bound to be determined will differ substantially from the upper bound. Let us first state precisely what we mean by a lower bound. We will start by supposing that $R(k,n)$ is the number of distinct networks with at most k contacts associated with n variables. If $k = L(n)$, it is clear that $R(L(n),n)$ cannot be less than the number of functions of n variables, i.e., 2^{2^n}, for otherwise it would be impossible to realize each function of n variables by a network with at most $L(n)$ contacts:

$$2^{2^n} \leqslant R(L(n),n).$$

From this inequality, it is also possible to obtain a lower bound on $L(n)$. Here we require an upper bound on $R(k,n)$. We will derive it in two stages. Recall that a network is a graph or set of vertices connected by edges with which the contacts are associated; we associate with the edges the symbols of the variables or their negations.

Two graphs are considered *equivalent* if a correspondence may be established between their vertices so that the number of edges connecting corresponding pairs of vertices is the same in the two graphs. For the sake of convenience, we will only consider graphs with distinct input and output vertices and without any isolated vertices. Clearly, in a minimal network for a function other than the constants, it may be assumed that there are no edges that connect some vertex to itself.

We will first estimate the number $S(k)$ of graphs with at most k edges, and then find the number $R(n,k)$.

9.24. Prove that

$$S(k) < (2k)^{2k}.$$

9.25. Prove that

$$R(k, n) < (2kn)^{2k}.$$

Thus, by the above remark and Problem 9.25, the following inequality holds:

$$2^{2^n} < (2nL(n))^{2L(n)}.$$

Using this inequality, let us try to attempt to derive a lower bound on $L(n)$. Note first that we will obtain a lower bound that increases at the same rate as the previously found upper bound. With this in mind, it is best to represent $L(n)$ in the form

$$L(n) = \frac{2^n}{2n}\alpha(n).$$

For $\alpha(n)$, we have the inequality

$$2 < (2^n\alpha(n))^{\alpha(n)/n}.$$

9.26. Find a lower bound on $\alpha(n)$. Write out the lower bound for $L(n)$.

Thus, we have found (Problems 9.22 and 9.26) that for every $\epsilon > 0$ and sufficiently large n,

$$\frac{2^n}{2n} < L(n) < (4 + \epsilon)\frac{2^n}{n}.$$

Hence, it follows in particular that as $n \to \infty$, $L(n)$ is of the same order as $2^n/n$. These inequalities were first found by Shannon in a somewhat more exact form (his lower bound has the form $(1 - \epsilon)2^n/n$). A definitive result in this area has been found by O. B. Lupanov [1], who proved that the sequence $L(n)$ is equivalent to $2^n/n$ as $n \to \infty$. In this case, the factor $4 + \epsilon$ in Shannon's upper bound must be replaced by $1 + \epsilon$.

Digression on Shannon's Upper and Lower Bounds. Shannon's greatest lower bound was found using the same method as the lower bound we have found, i.e., it may be proved that for large n and for $k = (1 - \epsilon)\cdot(2^n/n)$, the number $R(n,k)$ of networks of n variables with at most k contacts is less than the number of functions of n variables. It is remarkable that such inexact considerations lead (by Lupanov's theorem) to a greatest lower bound. Let us consider more carefully what might be the reason for this. It turns out that as $n \to \infty$, the sequences $R(L(n),n)$ and 2^{2^n} are equivalent. Hence, it it follows in turn that if $L(n)$ is the number of functions of n variables for which there is a unique realization with at most $L(n)$ contacts, then $E(n) \sim 2^{2^n}$ (in other words, for large n there is one such realization for almost all functions). Now note that a network containing precisely $L(n)$ contacts can be constructed for every function of n variables (for example, by adjoining to a minimal network dummy contacts that do not share common poles with the initial network). Then we find that for large n, precisely $L(n)$ contacts are required to realize most functions, and in addition the minimal realization is unique. More precisely, as $n \to \infty$ the number of functions of n variables that require $L(n)$ contacts for their realization is equivalent to the total number of functions of n variables.

There is an essential difference between our two methods of obtaining lower bounds on $L(n)$ (cf. Problems 9.23 and 9.26). In the first method, we found a sequence of functions f_n, where f_n depends on n variables, whereas $2n$ contacts are required to realize f_n. It would seem that a proof of the general theorem could also be attempted in an analogous fashion. However, only recently [2] has a sequence of functions f_n been found which required a nonlinearly increasing (as $n \to \infty$) number of contacts for their realization; this number increases more rapidly than any linear function of n. It has not been possible to effectively construct a sequence of functions that require $L(n)$ ($n \to \infty$) contacts for their realization.

Here we should be more precise in what we understand by the words "effectively construct." In fact, there exists an algorithm that outputs a desired function f_n for every n. But since there is a finite number of functions and a finite number of networks with at most k contacts for every n, by sorting all these functions and networks (with $k = L(n)$), it is possible to find the desired function in a finite number of steps. This we wish to exclude from consideration, so that our algorithm not become a "sorting algorithm." In quite a number of cases, it is necessary to rigorously determine when some algorithm is not a sorting algorithm, though no satisfactory answer to this question has yet been obtained. It has been conjectured that there is no algorithm other than the sorting algorithm for constructing a sequence of functions that require $L(n)$ contacts for their realization. There is indirect support for this conjecture [3]. At first glance, the hypothesis appears strange if we bear in mind that, for large n, most functions require $L(n)$ contacts for their realization. This is probably because (for large n) nearly all functions of n variables admit only equally complicated descriptions that are comparable with the process of sorting all functions of n variables. There are a few functions, however, that admit a brief description by comparison with the majority of functions, and these functions have a comparatively simple realization. Note that the sequences of functions which require few contacts for their realization, though in the minority in the case of large n, are easily constructed.

In concluding this digression, let us note that there are similar methods [4] that may be used to estimate the number of contacts required for different classes of networks:

$$L_{\Pi}(n) \sim \frac{2^n}{\ln n}.$$

Analogous estimates may be given for networks of functional elements.

4. *Realization of Linear Functions by Means of Contact Networks.* In this section we will prove, using the linear functions as an example, that our general estimates may be appreciably simplified for certain special classes of functions [5]. Thus, for every n we consider the two functions

$$P_n(x_1, \dots, x_n) = x_1 + x_2 + \dots + x_n;$$

$$Q_n(x_1, \dots, x_n) = x_1 + x_2 + \dots + x_n + 1.$$

The first of these functions is sometimes called an "odd counter," and the second an "even counter." The names derive from the fact that P_n is equal to 1 if and only if an odd number of variables is equal to 1, while Q_n is equal to 1 if and only if an even number of variables is equal to 1. Our goal is to construct contact networks for P_n and Q_n.

Digression on Switching Circuits. In our discussion of contact networks, we could just as well have used what are known as *switching circuits.* Here each contact is thought of as being in one of two states: if closed, current passes through this point in the circuit, and if open, current may not pass through this point. With each switch we associate a variable x or its negation $\bar{x}$; depending on our choice, we associate with the values $x = 1$ and $x = 0$ some state of the switch. Current is always being fed to the circuit input; there is current at the output if the conduction function is equal to one. In a light bulb circuit, we are usually dealing with a single switch (more precisely, each switch controls its own part of the network). But it is sometimes necessary to resort to switching circuits. For example, suppose that there are several switches in different parts of a room. We wish to vary the conductivity of the circuit by varying the position of one of the switches (a light must be turned on if it was not on before, or turned off it it was on). Clearly, the corresponding circuit realizes P_n or Q_n depending on the state of the circuit when the switches are turned off. That is, if the value of any one of the variables changes to its opposite, the value of the particular function changes (which is the distinctive feature of these functions; cf. Problem 4.9). The reader should have no trouble thinking of other cases in which it is natural to resort to switching circuits. We could, for example, analyze the case in which a light must be turned on if the majority of the switches are on ("majority organ"), or the case in which at least two switches must be on. Bulbs in a chandelier are turned on in sections since they have been divided into two groups, each of which is connected to a single switch; a section switch whose four possible positions correspond to the possible states of the pair of switches is sometimes used. (Do you know how such a switch is constructed? Note that three control linkages are needed in the switch.)

Let us now return to our discussion of realizations of P_n and Q_n.

9.27. Find the minimal networks that realize P_2 and Q_2. Networks for P_n and Q_n may be constructed by induction.

9.28. Construct contact networks for P_n and Q_n. Find upper and lower bounds on $L(P_n)$ and $L(Q_n)$.

Thus we have found that the linear functions may be realized by means of contact networks with linear complexity (i.e., the number of contacts required depends linearly on the number of variables).

Let us consider how to realize linear functions by means of pi-networks. We first verify that such realizations have not yet been obtained.

9.29. Prove that the networks constructed in solving Problem 9.28 are not pi-networks if $n > 2$.

Thus we can construct pi-networks only for P_2 and Q_2.

9.30. Realize the functions P_{2^k} and Q_{2^k} by means of pi-networks. Estimate the number of contacts.

9.31. For arbitrary n, realize P_n and Q_n by means of pi-networks. Find upper bounds on $L(P_n)$ and $L(Q_n)$.

Thus in the class of pi-networks we have been able to realize linear functions only by means of networks whose complexity increases as n^2. It is known [6] that $L_\Pi(P_n)$ increases nonlinearly; that is, for some $c > 0$,

$$L_\Pi(P_n) \gtrsim cn^{3/2}$$

though it is not known precisely what is the order of $L_\Pi(P_n)$. We now construct a network for arithmetic addition of binary numbers. If we are given two n-place binary numbers, their sum will be at most a $(n + 1)$-place number. An *n-place adder* is understood to refer to a $(1, n + 1)$-pole in which there are contacts associated with the variables $(x_1, \dots, x_n; y_1, \dots, y_n)$; the outputs are ordered so that the i-th rightmost digit $s_i(x_1 y)$ of the sum of the two binary numbers $(x_n x_{n-1} \dots x_1)$ and $(y_n y_{n-1} \dots y_1)$ is realized on the i-th output.

The network is constructed by induction in accordance with the ordinary method of adding numbers in a positional (in particular, a binary) number system; the digits are counted off from right to left. To perform an addition in the i-th place, it is necessary to know the i-th digits x_i and y_i of each of the two numbers and the carry p_i to this place. We must compute not only the i-th digit s_i of the sum, but also the carry p_{i+1} for the next place. Since we are dealing with the digits 0 and 1, this means determining the two logical functions that specify s_i and p_{i+1}.

9.32. Find the functions $s_i(x_i, y_i, p_i)$ and $p_{i+1}(x_i, y_i, p_i)$.

9.33. Construct the network of an n-place adder. Estimate the number of elements needed.

5. *Networks of Functional Elements for the Arithmetic Operations.* There are methods* [4, 7] of finding the above estimates for networks of functional elements similar to those obtained above for contact networks. The rate of increase of $L(n)$ is the same as for $(2^n/n)$; only the constant depends on the basis. We will not dwell on these topics, and instead construct networks from functional elements that realize the arithmetic operations of addition and subtraction. The procedure we will use for constructing such networks of functional elements is in some sense more natural than in the case of the analogous contact networks. (This is particularly clear by comparison with networks for multiplication.)

Note that there is no known effective sequence of functions that require for its realization networks of functional elements with nonlinearily increasing complexity (cf. Remark on page 197).

We will consider networks formed from zero-cycle functional elements. In estimating the number of elements, we are interested only in the rate of increase and will not try to find the exact constants. It turns out that the rate of increase of the number of elements often is independent of which basis of functional elements we are using (only the constant depends on the basis).

9.34. Construct a network of functional elements for an n-place binary adder. Find an upper bound on the number of elements required.

9.35. Construct a network of functional elements for n-place subtraction. Estimate the number of elements.

Let us now construct a network for multiplying n-place binary numbers. As in the preceding case, we start with the ordinary method of multiplying binary numbers. Recall that

---

*The reader interested in learning more about estimating the complexity of minimal networks may wish to read the survey by Yablonskii [4].

to multiply two numbers $x = (x_n x_{n-1} \dots x_1)$ and $y = (y_n y_{n-1} \dots y_1)$, it is necessary to multiply x by each y_i (xy_1), shift xy_i $(i - 1)$ units to the left (or add that many zeros on the right), and add the resulting numbers. Since $y_i = 0$ or 1 in the binary system, for all $y_i = 1$ it is sufficient to add $i - 1$ zeros to x on the right (i.e., take $2^{i-1}x$) and then add these numbers.

9.36. Construct a network of functional elements for n-place multiplication.

Now we will give [8] a more concise method of multiplication for large n. In some ways, it is analogous to the method we used to construct pi-networks for linear functions. Note, too, that the construction of a network for multiplication reduces to the construction of a network for raising a number to the power of two.

9.37. Prove that if a network containing at most $f(n)$ elements can be constructed for squaring n-place numbers, then it is possible to construct a network with less than $af(n+1) + cn$ elements which multiplies n-place numbers (a and c are independent of n).

In particular, if $f(n)$ grows more rapidly than n as $n \to \infty$ (for example, $f(n) = cn^\alpha$, $\alpha > 1$), the same (i.e., in terms of rate of increase) estimate but with a different constant may be obtained for the number of elements in the network for multiplication. Thus, we begin with a network for squaring binary numbers.

9.38. Construct a network for squaring a 2^k-place binary number that has no more than $c \cdot 3^k$ elements (c is independent of k).

9.39. Based on Problem 9.38, construct a network for squaring an n-place binary number (for arbitrary n). Estimate the number of elements. Draw a conclusion regarding the network for multiplication.

We have constructed a network for multiplication that contains at most $cn^{\log_2 3}$ elements. This result, it turns out, may be simplified [9]. For any $\alpha > 1$, there exists a network for multiplication with at most cn^α elements (c independent of n). Thus, there are methods of multiplying numbers in a positional number system that require not cn^2 elementary steps (additions and multiplications of one-digit numbers, as in ordinary arithmetic), but rather cn^α steps $(1 < \alpha < 2)$. However, let us not forget that the constants c in these estimates are not all the same, and therefore it may turn out that these new methods are less time-consuming only if n is large.

Hints

9.1. (1) Prove by induction, using the completeness of the system of functions $\{xy,\bar{x}\}$ and delay elements.

(2) $\bar{x}y$, for example (and, in general $f \notin \mathrm{R}$; cf. the hint to Problem 8.11) cannot be realized. The solution is analogous to that of Problem 8.10.

9.3. Networks (a) and (c) given in the solution of Problem 9.2 are contact networks.

9.4. For each circuit, consider the sequence of vertices through which the network passes, and prove that there is a finite number of such sequences.

9.5. Prove that if all the contacts of some circuit are closed, then an actual circuit may be found all of whose contacts are also closed.

9.7. See sec. 2.6 for an algorithm that transforms formulas to DNF. The resulting DNF may be simplified by means of formulas (2.20) - (2.25).

9.8. Use formulas (2.20), (2.22), and (2.23).

9.9. First construct networks that realize elementary disjunctions.

9.11. Each formula specifies a method of constructing a function from variables and negations of variables by means of conjunctions and disjunctions. Each network must be constructed by induction, based on the foregoing construction.

9.12. Prove that the network cannot be divided into two networks connected in series, nor into two networks connected in parallel.

9.13. Note that none of the variables in the conduction function are actual.

9.14. Prove that in every network that realizes the given function, we may associate with each variable at least one positive contact and one negative contact.

9.15. The network in Figure 40 corresponds to a PDNF. The desired network may be found from some CNF.

9.16. Prove that the function cannot be represented by a formula by means of conjunctions and disjunctions in which each variable is encountered precisely once. More precisely, the function cannot be represented in the form of a conjunction or disjunction of functions without common variables.

9.17. (a), (b) $L(n) \leqslant n2^n$; (c) $L(n) \leqslant n2n^{n-1}$.

9.18. $L(n) \leqslant 3 \cdot 2^{n-1} - 2$; prove that if c_k is the number of contacts that may be required in the method of realization of functions of k variables indicated here, then $c_{k+1} \leqslant 2c_k + 2$.

9.19. That a $(1,2^n)$-pole is a universal pole may be proved by induction. To realize the function, we need only equate the outputs of the multipole created from the corresponding elementary conjunctions. The following estimate is then obtained.

$$L(n) \leqslant 2^{n+1} - 2.$$

9.20. In any circuit connecting the outputs, there is one positive contact and one negative contact for some variable. The proof can be easily carried out by induction.

9.21. To prove that there are no redundant circuits with non-zero conduction, we use the fact that the universal multipole M_1 is a switching circuit.

For any $k \leqslant n$, we have

$$L(n) \leqslant T_{n,k} = 2 \cdot 2^{n-k} + 2 \cdot 2^k \cdot 2^{2^k}.$$

9.22. $k(n)$ must be selected so that the second term in $T_{n,k}$ (cf. hint to Problem 9.21) grows more slowly than the first. The first term must be made as small as possible. To satisfy these two requirements, it is sufficient to select (for any $\beta > 0$) the greatest integer not exceeding $\log_2 n - \beta$ as our $k(n)$.

As a result, we find that for any > 0 and large enough n,

$$L(n) < (4 + \quad) \frac{2^n}{n}.$$

9.23. Use the result of Problem 9.14.

9.24. Find an upper bound on the number of vertices in a graph with k edges; renumber the vertices in the graph and note that a graph is determined if the number of edges that connect each pair of vertices is specified.

9.25. Find the number of networks for n variables all of which have the same graph (with k contacts). Then use the estimate of the preceding problem.

9.26. $\alpha(n) > 1$; $L(n) > \dfrac{2^n}{2n}$.

9.27. Use a PDNF or PCNF; cf. Problem 9.14.

9.28. We have

$$P_{n+1}(x_1, \dots, x_n, x_{n+1}) = P_2(P_n(x_1, \dots, x_n), x_{n+1});$$

$$Q_{n+1}(x_1, \dots, x_n, x_{n+1}) = Q_2(P_n(x_1, \dots, x_n), x_{n+1}).$$

Further, $P_n = \overline{Q}_n$.

Construct by induction a (1,2)-pole that realizes P_n and Q_n; use the networks of the preceding problem. For $n > 1$, we have

$$2n \leqslant L(P_n) \leqslant 4(n-1);$$

$$2n \leqslant L(Q_n) \leqslant 4(n-1).$$

For the lower bound, use Problem 9.14.

9.29. For the sake of simplicity, consider as an example the network for $P_3(x,y,z)$. Prove that it cannot be represented in the form of a parallel or series connection of two (non-empty) networks. This fact may be proved by analogy with the proof of Problem 9.12. The corresponding property of the networks for P_n and Q_n may be proved by induction.

9.30. We have

$$P_{2m}(x_1, \dots, x_m; y_1, \dots, y_m)$$
$$= P_2(P_m(x_1, \dots, x_m), P_m(y_1, \dots, y_m));$$
$$Q_{2m}(x_1, \dots, x_m; y_1, \dots, y_m)$$
$$= Q_2(P_m(x_1, \dots, x_m), P_m(y_1, \dots, y_m)).$$

By means of these formulas, it is possible to realize P_{2m} and Q_{2m} using pi-networks if P_m and Q_m have already been realized. With this in mind, 4^k contacts are sufficient for realizing P_{2^k} and Q_{2^k}.

9.31. Suppose that $2^k < n < 2^{k+1}$ (the case $n = 2^k$ has already been analyzed), and let $n - 2^k = m$.

First Method. Construct the network for $P_{2^{k+1}}$ and transform it using the fact that $x_p = 0$ for $p > n$. An obvious upper bound results:

$$L_\Pi(P_n) \leqslant 4n^2$$

or the least upper bound

$$L_\Pi(P_n) \leqslant 2n^2.$$

Second Method. We have

$$P_n(x_1, \dots, x_{2^k}; y_1, \dots, y_m) = P_{2^k}(P_2(x_1,y_1); \dots ;$$
$$P_2(x_m,y_m); x_{m+1}, \dots, x_{2^k}).$$

Find the number of contacts $k(n)$ in such a network and its upper bound:

$$\frac{k(n)-n^2}{n^2}.$$

Here we may replace n by a continuous variable and find the greatest value of the resulting function using methods from differential calculus, obtaining as a final result

$$L_{\Pi}(P_n) \leqslant \frac{9}{8}n^2.$$

9.32. $s_i(x_i, y_i, p_i) = x_i + y_i + p_i$; $p_{i+1}(x_i, y_i, p_i) = x_iy_i \vee x_ip_i \vee y_ip_i$.

9.33. The construction is performed by induction. Construct a $(1, i + 2)$-pole T_i on whose outputs the digits of the sum $s_1, \dots, s_i$ are realized, carry over p_{i+1} to the $(i + 1)$-th place, and let $q_{i+1} = \overline{p_{i+1}}$. The number of contacts in the adder grows linearly.

9.34. For every digit, construct networks that realize s_i and p_{i+1}, and connect these networks in a suitable fashion. The number of elements in this network is less than cn for some constant c (independent of n).

9.35. Suppose that we wish to subtract one binary number $x = (x_nx_{n-1}\dots x_1)$ from another binary number $s = (s_ns_{n-1}\dots s_1)$; we will assume that $s_n = 1$. The network may be constructed based on the ordinary method of digit-by-digit subtraction (from right to left). Recall that in the binary system the "tens" digit is the number two; therefore, when we subtract if there is a unit in the $(i + 1)$-th place, we must add two units to the i-th place. Here it is clear that we have no need for more than one unit.* (Of course, this remark relates only to the case in which we are subtracting one number, rather than several.) In constructing the network, technically it is more convenient to require (unlike the ordinary subtraction method) that a unit occupy the $(i + 1)$-th place, to

*Strictly speaking, this assertion must be proved by induction on n.

place, to store this unit, and to then perform the operation on the i-th digit independent of the value of s_{i+1} (i.e., whether it is possible to subtract a unit from s_{i+1} or whether it is necessary to more further to the left until a digit is reached from which it is in fact possible to subtract a unit); then, moving the $(i + 1)$-th digit again, we disregard digits further to the left. Thus once we have considered the i-th digit, we know the value of s_i and x_i, and also whether it is necessary to subtract a unit from the i-th digit for the preceding digits. Let us introduce the quantity q_i equal to 1 if a unit must be added, and equal to 0 otherwise.

From s_i we must subtract x_i and q_i. If $s_i < x_i + q_i$, we put a unit in the $(i + 1)$-th place; then s_i is increased by 2 and the subtraction may be performed; further, we obtain a quantity y_i equal to 1 or 0. Thus, there are two functions to determine: $y_i(s_i, x_i, q_i)$ and $q_{i+1}(s_i, x_i, q_i)$, i.e., the i-th digit of the difference y_i and the quantity q_{i+1}, which was borrowed from the $(i + 1)$-th place. We have

$$y_i(x_i, x_i, q_i) = s_i + x_i + q_i;$$

$$q_{i+1}(s_i, x_i, q_i) = q_i x_i \vee \bar{s}_i(x_i \quad q_i).$$

The construction of the network is analogous to the construction of an adder network. The number of contacts is not greater than cn. If $q_{n+1} = 1$, then $s < x$.

Subtraction can also be reduced to addition using the following properties of binary numbers. Suppose that $\bar{x} = (\overline{x_n x_{n-1} \cdots x_1})$; then

$$x + \bar{x} = (\underbrace{1\ 1 \ldots 1}_{n}) = 10^{n+1} - 1.$$

9.36. The product of two n-digit numbers contains at most $2n$ digits ($x < 2^n$, $y < 2^n \Rightarrow xy < 2^{2n}$). We denote the product xy by $p = (p_{2n}p_{2n-1} \cdots p_2p_1)$. Thus,

$$p = y_1x + y_2x \cdot 2 + \ldots + y_ix \cdot 2^{i-1} + \ldots + y_nx \cdot 2^{n-1}.$$

Recall that in the binary system the number 2^k is written as a unit followed by k zeros. We let $p^{(j)}$ denote the sum of the first j terms in the expression for p. Since the number of digits in each succeeding term is at most one unit greater than the number of digits in the preceding terms, $p^{(j)}$ contains at most $n + j$ digits. All succeeding terms in the sum for p have zeros as their final digits, therefore the last j digits of $p^{(j)}$ coincides with the digits in the same places in p, i.e., $p^{(j)}$ has the form

$$p^{(j)} = (t_n^{(j)} \ldots t_1^{(j)} p_j p_{j-1} \ldots p_1).$$

We have

$$p^{(j+1)} = p^{(j)} + y_{j+1} x \cdot 2^j =$$

$$= (t_n^{(j+1)} \ldots t_1^{(j+1)} p_{j+1} p_j \ldots p_1).$$

Comparing these formulas, we see that

$$t^{(j)} + y_{j+1} x = (t_n^{(j+1)} \ldots t_1^{(j+1)} p_{j+1}).$$

(Here we have set $t^{(j)} = (t_n^{(j)} t_{n-1}^{(j)} \ldots t_1^{(j)})$). As a result we have found a way of constructing the network by induction; on the $(j + 1)$-th step, starting with two n-digit numbers $t^{(j)}$ and x and a one-digit number y_{j+1}, find the $(j + 1)$-th digit of the product p_{j+1} and the number $t^{(j+1)}$. The number of elements in the circuit does not exceed cn^2.

9.37. Use the formula $(a + b)^2 = a^2 + b^2 + 2ab$ and Problem 9.35.

9.38. Prove that if $d(m)$ elements suffice for squaring an m-digit number, then $3d(m) + cm$ elements suffice for squaring a $2m$-digit number, where c is independent of m. To prove the assertion, represent a $2m$-digit number in the form $x \cdot 2^m + y$, where x and y are m-digit numbers. The required formula may be proved by induction on k.

9.39. Use the network for squaring a 2^k-digit number, where $n \leqslant 2^k$. The number of elements may be estimated by the inequality

$$d(n) \leqslant cn^{\log_2 3}.$$

Solutions

9.1. (1) Every function may be represented in the form of a composition of conjunctions and a negation. The basis of the induction is a function which coincides with its argument (x). The function is obviously realizeable.

The inductive transition consists of two parts. Suppose that the final operation in the representation of f in the form of such a composition is to apply the negation: $f = \overline{\omega}$. The function ω may be represented in the form of a composition in fewer steps than f. Let us assume that ω has already been represented in the form of a circuit with delay ν. Then by connecting the output of this circuit to the relay winding depicted in Figure 31, we finally find a network that realizes f with delay $\nu + 1$.

Now suppose that the final operation in the composition is the conjunction $f = \omega_1\omega_2$, while the functions ω_1 and ω_2 are realized with delays ν_1 and ν_2, respectively. By connecting the output of one of them in series with delay elements, we find that the network that realizes ω_2 has delay ν one cycle greater than the network that realizes ω_1. Now feeding the output of the network for ω_1 to the relay winding, and the output of the network for ω_2 to the contact of a make contact relay, we obtain a network that realizes $\omega_1\omega_2$ with delay $\nu + 1$. The proof is complete.

(2) Suppose that we are given a network S that realizes $\overline{x}y$ with delay ν, where ν is even. At time t, we feed the signals $\{x = x_0, y = 1\}$. At time $t + \nu$, we obtain $\overline{x}_0$ at the output, independent of the signals at the inputs of S at times other than t. At time $t + k$ (k is any integer) we feed the signals $\{x = x_0, y = 1\}$ if k is even, and $\{x = \overline{x}_0, y = 1\}$ if k is odd. By induction, it may be proved that in any conductor there is either a signal independent of x_0 or the same signal as at the input of x. (Under the assumption that the signals at the relay inputs possess the desired property, it can be checked that the signal at the relay output also possesses this property.) Therefore, the signal at the output of S is either independent of x_0 or is x_0 (ν is even). We have arrived at a contradiction. Similarly, if ν is odd, feeding $x = 0$ we find that S cannot realize y. Thus, if we are given only a break contact relay, it is impossible to realize even $\overline{x}y$.

9.2. Consider the networks in Figure 44.

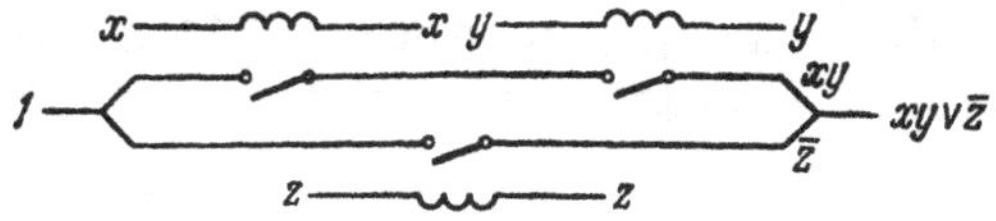

(a) One-cycle network

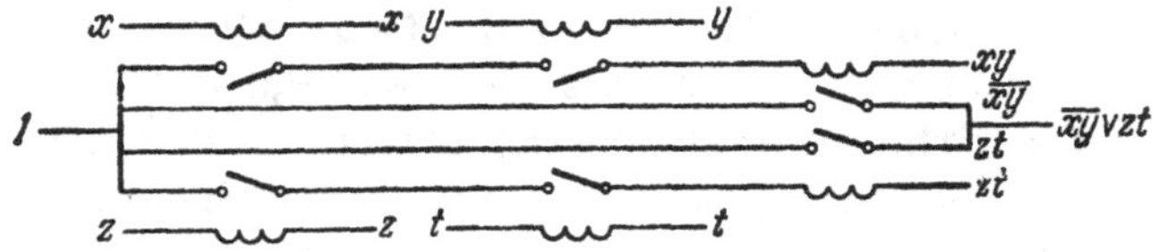

(b) Two-cycle network

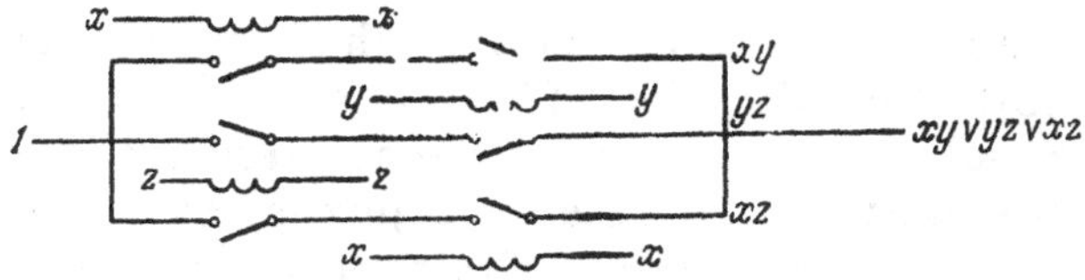

(c) One-cycle network

Figure 44

9.3. Consider Figure 45. It is clear how much more compact is the notation used in Chapter 2.

9.4. We write out in succession the vertices through which the actual circuit passes. None of these vertices may repeat. It is clear that the number of sequences of nonrepeating vertices is finite. Though in general a sequence of vertices does not uniquely determine a circuit (there can be several contacts with the same poles), there are only a finite number of circuits with the same sequence of vertices.

Remark. There is another way of defining actual circuits as circuits with nonrepeating vertices. This definition is not

(a) (b)

Figure 45

equivalent to the one given above. The proof that the number of actual circuits is finite then becomes somewhat simpler.

9.5. Suppose that we are given a circuit in which some vertex is encountered twice. We discard all the contacts which are encountered between two passes through this vertex. It is clear that in this case we again obtain a circuit, further, if all the contacts of the initial circuit are closed, all the contacts of the newly obtained circuit are also closed. Put differently, the conjunction corresponding to the initial circuit absorbs the conjunction for the resulting circuit (cf. Definition 5.3(c)). Thus, by successively contracting the circuit we obtain an actual circuit that is conductive if the initial circuit is conductive. There is current at the output if all the contacts in at least one actual circuit are closed, and this is equivalent to the assertion just proved. It would be possible to start with the definition of an actual circuit presented in the remark appended to the solution of the preceding problem.

9.6. (a) $xyt \vee tyt \vee xz \vee tz \vee \overline{yxt} \vee \overline{yx}yz = yt \vee xz \vee tz \vee \overline{yx}t$.

(b) $xy\overline{x} \vee xt\overline{x} \vee xy\overline{t}\,y\overline{x} \vee xty\overline{t\,x} \vee xy\overline{x} \vee x\overline{t\,x} \vee x\overline{t}\,yt\overline{x} \vee xyty\overline{x} = 0$.

(c) $xy \vee tu \vee xzu \vee tzy$.

9.7. (a) $y \vee z \to x\overline{y} = \overline{y \vee z} \vee \overline{xy} = \overline{yz} \vee x\overline{y}$ (cf. Figure 46a).

(b) $\overline{z\overline{y}} \sim yx = (\overline{zy} \vee yx)(z\overline{y} \vee \overline{yx}) = (\overline{z} \vee y \vee yx)(z\overline{y} \vee \overline{y} \vee \overline{x}) = (z \vee y)(\overline{y} \vee \overline{x}) = \overline{zy} \vee \overline{zx} \vee y\overline{x}$ (cf. Figure 46b).

(c) $x + y + z = xyz \vee xy\overline{z} \vee x\overline{y}z \vee \overline{x}yz$ (cf. Figure 46c).

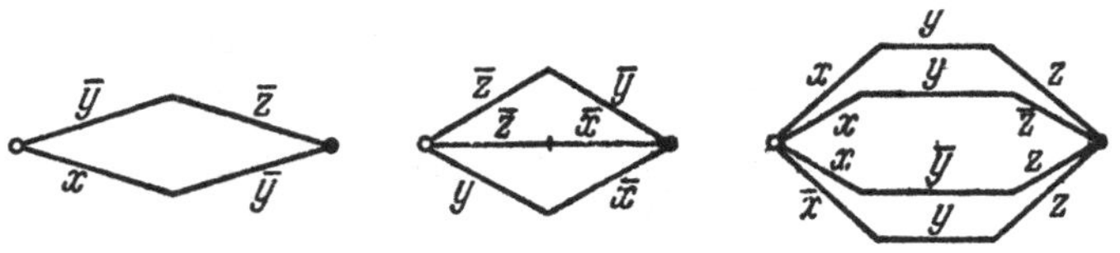

(a) (b) (c)

Figure 46

9.8. (1) Formula (2.20): In constructing the conduction function, we disregard circuits containing other circuits (in fact, we already used this rule to equate the actual circuits; an actual circuit cannot be absorbed by another actual circuit).

(2) Formulas (2.22) and (2.23): If some contact is a circuit, the contact opposite it cannot be included in any of the other circuits.

Based on these formulas, more complicated rules may be formulated. Using (2.20) - (2.25), it is often possible to simplify networks without varying their conduction functions. We will not be discussing this point here.

9.9. With every elementary disjunction $x_1^{\sigma_1} \vee \ldots \vee x_k^{\sigma_k}$ we may associate the network depicted in Figure 47. We then connect these networks in series for all the elementary disjunctions occurring in CNF (Figure 48). This is done in such a way that the input of a succeeding network coincides with the output of its preceding network. The input of the first and output of the last of these networks become the input and output, respectively, of the resulting network.

9.10. (a) $(y \vee z) \rightarrow x\bar{y} = \overline{yz} \vee x\bar{y} = \bar{y}(\bar{z} \vee x)$ (Figure 49a).

(b) $z\bar{y} \sim yx = (\bar{z} \vee y)(\bar{y} \vee \bar{x})$ (Figure 49b).

(c) $x + y + z = (x \vee y \vee z)(\bar{x} \vee \bar{y} \vee z) \,\&\, (\bar{x} \vee y \vee \bar{z})(x \vee \bar{y} \vee \bar{z})$ (Figure 49c).

9.11. (a) We begin with the elementary networks (Figure 50). We require two copies of each of the first two contacts. Now, the conjunctions $\bar{y}z$, xy, and zt may be realized by means of successive connections (Figure 51). Then, using parallel connections we realize the functions $x \vee y\bar{z}$ and $xy \vee$

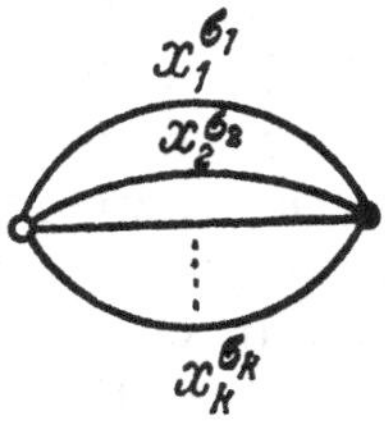

Figure 47

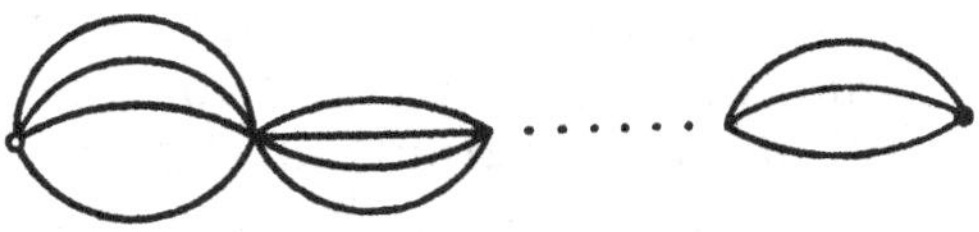

Figure 48

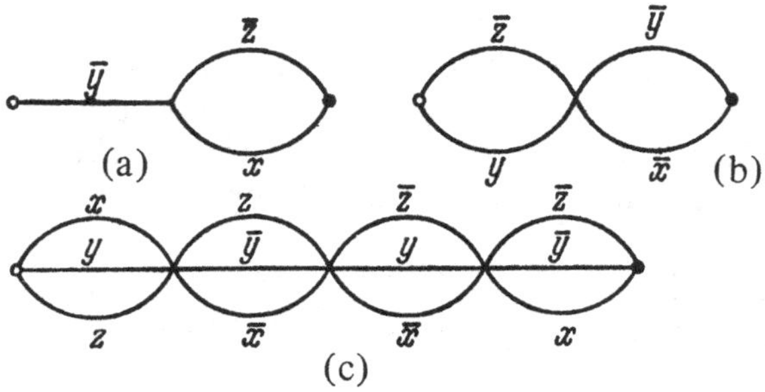

Figure 49

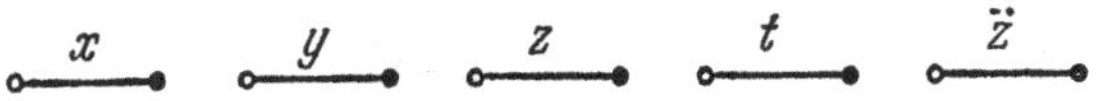

Figure 50

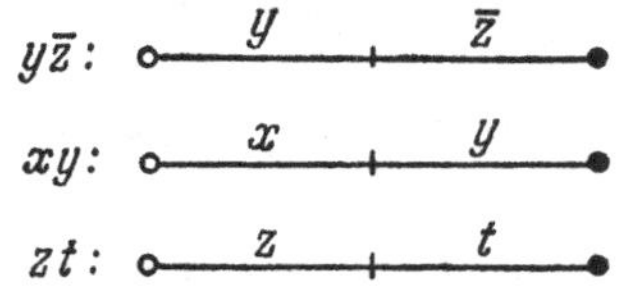

Figure 51

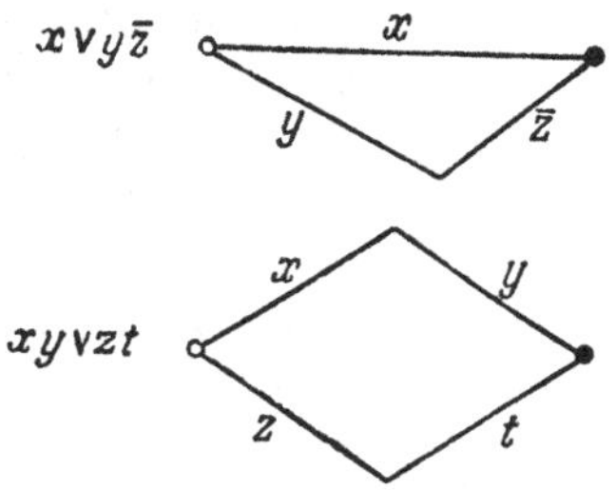

Figure 52

zt (Figure 52). Finally, we successively connect these networks (Figure 53).

(b) cf. Figure 54.

9.12. Since the network is not an elementary network, for it to be a pi-network it must be possible to divide it into two sub-networks connected either in series or in parallel. If the two networks are connected in series, all the poles in the resulting network (other than the pole through which the connection is made) either has no contacts in common with the input of the entire network or no contacts in common with the network output, or shares a contact in common only with the input or only with the output. It is clear that if any of the internal poles in Figure 39 is made a connecting sub-network, the remaining pole will share a common contact

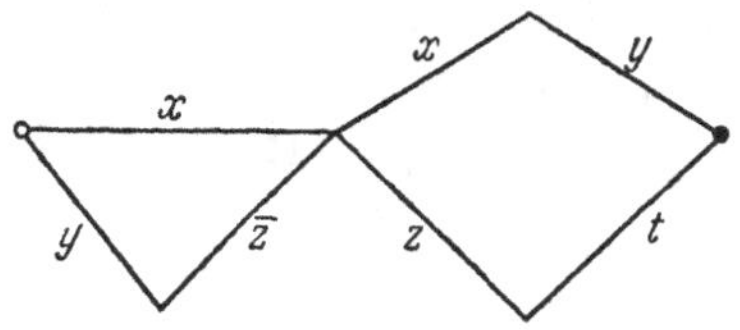

Figure 53

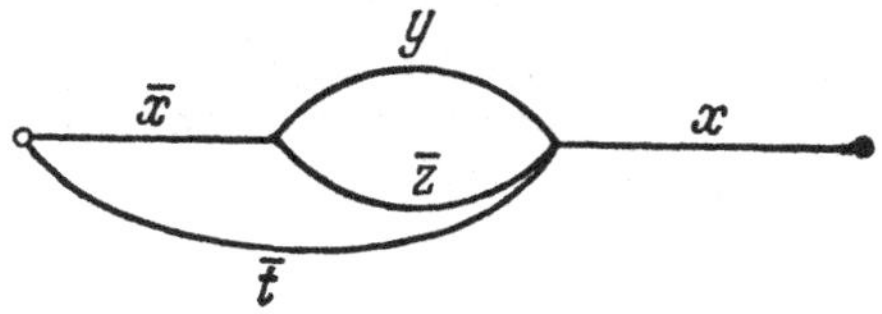

Figure 54

with both the input as well as the output of the network. Therefore, our network cannot be obtained by a series connection of two networks.

If the network is a parallel connection of two networks this will mean that, in particular, its contacts and poles may be divided into two parts such that either one of the parts contains only contacts directly connecting the input to the output (and in that case there are no poles in this part other than the input and output) or poles that occur in different parts of the network and are neither the input nor output do not have common contacts. The first possibility cannot be realized in our network, since it has no contacts that directly connect the input and output. The second possibility also cannot be realized, since our network has a total of two internal networks, which also share a common contact. Thus, the network also cannot be obtained by a parallel connection of two non-empty networks. That is, it is not a pi-network.

9.13. If some variable occurs as an actual variable in a function, the network that realizes this function must contain at least one contact connected to this variable (this is self-evident). In Problem 9.6c, we found the conduction function for a bridge: $xy \vee tu \vee xzu \vee tzy$. All the variables in the network are actual; if $t = z = u = 0$ and $y = 1$, the value of the function depends on the value of x (that the variables y, t, and u are actual may be proved in an analogous fashion); if $x = u = 1$, everything depends on the value of z. Since every variable in this network corresponds to precisely one contact, the network is minimal.

General Principle. If each of the contacts in a network is associated with distinct variables, and if all these variables occur as actual variables in the conduction function, the network is minimal.

9.14. Suppose that in some network there are only positive contacts associated with a variable x_1. Then for all $x_2, \dots, x_n$, the conduction function $f(x_1,x_2, \dots, x_n)$ satisfies the inequality

$$f(1, x_2, \dots, x_n) \geqslant f(0, x_2, \dots, x_n)$$

(If some of the contacts in the network are closed, the conduction in the network cannot vanish.) Analogously, if only break contacts are associated with x_1,

$$f(1, x_2, \dots, x_n) \leqslant f(0, x_2, \dots, x_n).$$

Now suppose that there exists a sequence $(\alpha_2, \dots, \alpha_n)$ such that

$$f(1, \alpha_2, \dots, \alpha_n) < f(0, \alpha_2, \dots, \alpha_n).$$

Then we say that f is non-increasing with respect to x_1; in every network that realizes f, there must be a negative contact associated with the variable x_1. Analogously, if there exists a sequence $(\alpha_2, \dots, \alpha_n)$ for which

$$f(1, \beta_2, \dots, \beta_n) > f(0, \beta_2, \dots, \beta_n)$$

(in this case, we say that f is non-increasing with respect to x_1), then in every network that realizes f there must be a positive contact associated with x_1. But if f is non-increasing and non-decreasing with respect to x_1, in every network for f there must be both positive and negative contacts associated with x_1.

Let us now consider the function $x_1x_2 \dots x_n \vee \bar{x}_1\bar{x}_2 \vee \bar{x}_n$, which is the conduction function for the network depicted in Figure 40. Setting $\alpha_2 = \alpha_3 = \dots = \alpha_n = 0$, we see that it is non-increasing with respect to x_1, and setting $\beta_2 = \beta_3 = \dots = \beta_n = 1$, we see that it is non-decreasing with respect to the same variable. Since the function is symmetric in all its variables, this property is possessed by all the other variables as well. And since in our network we have associated with each variable a single positive contact and a single negative contact, we find that our network is minimal.

A sufficient network minimality condition is as follows. If at most one positive contact and one negative contact are associated with each variable in a network, these variables

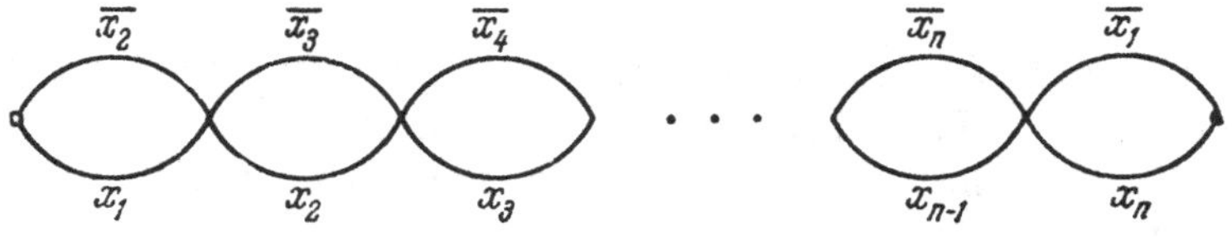

Figure 55

are all actual variables for the conduction function; moreover, if the function is non-increasing and non-decreasing with respect to the variables supplied with both contacts, the network is minimal.

9.15. The function may be represented in the following way in CNF:

$$(x_1 \vee \bar{x}_2)(x_2 \vee \bar{x}_3)(x_3 \vee \bar{x}_4)\dots(x_{n-1} \vee \bar{x}_n)(x_n \vee \bar{x}_1).$$

A minimal network (Figure 55) may be found using this representation.

9.16. If there would be such a pi-network, with each variable we could associate a single contact and, by virtue of the solution of Problem 9.14, this contact would be positive. By means of a pi-network, it would then be possible to construct a formula in which each variable occurs precisely once, using the operations of conjunction and disjunction alone. Suppose the final operation is disjunction. Then the function $f(x,y,z,t,u) = xy \vee tu \vee xzu \vee tzy$ may be represented in the form $f = f_1 \vee f_2$, where f_1 and f_2 do not have any variables in common and are not constants. Suppose that the variable z occurs in f_1. We set $y = t = 0$. Then f is transformed into $\omega(x, z, u) = xzu$. Further, f_2 in that case becomes identically zero, since there are no values of x and u for which $\omega = 1$ independent of the value of z, whereas f_2 is independent of z. Thus, after this substitution f_1 turns into xzu, which means that x and u occur in f_1 as actual variables. Analogously, setting $x = u = 0$, we find that f_1 must depend actually on y and t. As a result, all five variables occur in f_1 as actual variables, and we have arrived at a contradiction.

We leave it to the reader to similarly prove that f cannot be represented in the form $f_1 f_2$, where f_1 and f_2 do not have any variables in common and are not constants.

9.17. (a) To realize an elementary conjunction, we require n contacts. The number of elementary conjunctions in a PDNF does not exceed 2^n. That is, the total number of contacts does not exceed $n2^n$.*

(b) This problem can be considered analogously.

(c) The number of terms in a PDNF is equal to the number of sequences on which the function is equal to 1, while the number of terms in a PCNF is equal to the number of sequences on which it is equal to 0. One of these numbers is at most 2^{n-1}. Applying the method of (a) or (b) as necessary, we obtain a network with at most $n2^{n-1}$ contacts.

9.18. We expand the function $f(x_1, x_2, \dots, x_k, x_{k+1})$ with respect to the variable x_{k+1} (cf. (2.16)):

$$f = x_{k+1}\omega(x_1, \dots, x_k) \vee \overline{x_{k+1}}\psi(x_1, \dots, x_k).$$

If ω and ψ have already been realized, f can be realized, as is shown in Figure 56. Here —[ω]— and —[ω]— denote the networks which realize ω and ψ; their outputs have been equated. If at most c_k contacts are required to realize functions of k variables (in particular, ω and ψ), at most $2c_k + 2$ contacts will be required to realize f, i.e., $c_{k+1} \leqslant 2c_k + 2$. If we now bear in mind that $c_1 = 1$ (a single contact suffices for functions of one variable), we find that

$$c_k \leqslant 2^{k-1} + 2^{k-1} + 2^{k-2} + 2^{k-3} + \dots + 2^2 + 2$$

$$= 3 \cdot 2^{k-1} - 2.$$

Thus, $L(n) \leqslant 3 \cdot 2^{n-1} - 2$.

9.19. In Figure 41 we saw that the contacts in a network may be divided into "levels" in a natural way, and that in

*It is often convenient to exclude the case of constants, since, for example, in realizing 0 by this method it is necessary to consider a network that contans no contacts at all: O.

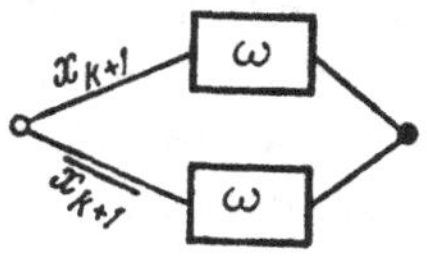

Figure 56

each level contacts may be found that may be associated with some variable. From the figure, it is also clear that only a single actual circuit leads to each output; the circuit contains precisely one contact from each level; with each such circuit we may associate some complete proper elementary conjunction, distinct conjunctions corresponding to distinct outputs. Thus, with each elementary conjunction we may associate some output.

A rigorous proof may be carried out by induction. Suppose that it has already been proved that a $(1, 2^k)$-pole is a universal pole; let us now prove that a $(1, 2^{k+1})$-pole is a universal pole. From the induction hypothesis, it follows that with each output of a $(1, 2^k)$-pole we may associate a unique actual circuit that constitutes a complete elementary conjunction of k variables, distinct conjunctions corresponding to distinct outputs. Each output of a $(1, 2^{k+1})$-pole may be associated with a unique output of a $(1, 2^k)$-pole, whereas two outputs of a $(1, 2^{k+1})$-pole may be associated with a single output of a $(1, 2^k)$-pole, one with the contact x_{k+1} and the other with the contact $\overline{x_{k+1}}$. Hence it follows that each output of a $(1, 2^{k+1})$-pole realizes its own complete elementary conjunction.

2. Suppose we wish to realize the function $f(x_1, \dots, x_n)$. Let us represent it in PDNF. In a universal $(1, 2^n)$-pole we equate the outputs on which the elementary conjunctions occurring in the PDNF are realized. The pole obtained as a result is associated with the output of the network. It is clear that the resulting network realizes $f(x_1, \dots, x_n)$.

The number of contacts in this network is equal to the number of contacts in a universal $(1, 2^n)$-pole, i.e.,

$$2 + 4 + 8 + \dots + 2^n = 2^{n+1} - 2.$$

We obtain the estimate

$$L(n) \leqslant 2^{n+1} - 2.$$

Thus, we have found an estimate that is more approximate than the one in Problem 9.18, though the networks we have used are not pi-networks. Below, however, we will be able to simplify substantially the estimate for $L(n)$ by means of a universal multipole.

9.20. It is intuitively clear that as we move along a circuit that connects the outputs of one part of a pole to the other, we encounter two opposite contacts at the point where motion changes direction. A rigorous proof may be carried out by induction. Suppose that the assertion we wish to prove is true for a $(1, 2^k)$-pole. Let us consider a circuit connecting the outputs of a $(1, 2^{k+1})$-pole. It is clear that the first and last contacts in this circuit connect these outputs to the outputs of a $(1, 2^k)$-pole. Discarding these contacts, we obtain a circuit that connects the outputs of a $(1, 2^k)$-pole. By the induction hypothesis, the conduction between these outputs is equal to zero if they are distinct (the case in which they coincide is self-evident); therefore, we may associate with the resulting circuit, hence with the initial circuit as well, elementary conjunctions that are identically zero.

9.21. The multipoles M_1 and M_2 are connected in such a way that each output of M_1 may be equated to a unique input of M_2, whereas several outputs of M_1 may be equated to a single input of M_2; further, there are some inputs of M_2 with which no output of M_1 may be equated. The poles which are obtained after equating the corresponding outputs of M_1 and inputs of M_2 are called the *connecting poles* of S_f.

Suppose that $F(x_1, \dots, x_n)$ is the function that is realized by the resulting network. We wish to prove that $F(x_1, \dots, X_n) = f(x_1, \dots, x_n)$. Note first that $F(x_1, \dots, x_n) \geqslant f(x_1, \dots, x_n)$. This nearly self-evident fact may be proved by representing each of the functions $\omega_{\sigma_{k+1} \cdots \sigma_n}(x_1, \dots, x_k)$ in DNF using their realizations in the multipole M_2 (i.e., in the form of a disjunction of conjunctions corresponding to the actual circuits). As a result, we have a representation of f in DNF. With each elementary conjunction occurring in it,

$$x_{m_1}^{\sigma_1} \cdots x_{m_\ell}^{\sigma_\ell} \qquad x_{k+1}^{\sigma_{k+1}} \cdots x_n^{\sigma_n} \quad (m_\ell \leqslant k),$$

we may associate some circuit in our network S_f; this circuit is a series connection of the circuit corresponding to the conjunction of $x_{k+1}^{\sigma_{k+1}} \ldots x_n^{\sigma_n}$ in the $(1, 2^{n-k})$-pole M_1 and the circuit in M_2 corresponding to $x_{m_1}^{\sigma_1} \ldots x_{m_l}^{\sigma_l}$ (that it exists follows from the choice of the DNF for f). Thus, if $f = 1$ on some sequence, $F = 1$ on this sequence too.

These circuits may be distinguished from all the actual circuits of S_f in such a way that they pass only once through the connecting pole. In fact, each circuit with this property may be decomposed into an actual circuit in M_1, i.e., into a circuit that realizes some conjunction $x_{k+1}^{\sigma_{k+1}} \ldots x_n^{\sigma_n}$ (cf. solution of Problem 9.19), and into an actual circuit in M_2 that starts at the input of M_2 corresponding to the function $\omega_{\sigma_{k+1} \ldots \sigma_n}$ (by the construction of S_f); the conjunction

$$x_{m_1}^{\sigma_1} \ldots x_{m_l}^{\sigma_l} \qquad (m_j \leqslant k)$$

corresponding to the second circuit occurs in the DNF we have constructed for $\omega_{\sigma_{k+1} \ldots \sigma_n}$.

It remains for us to prove that all the other actual circuits in S_f have zero conduction. Suppose that some circuit in S_f passes at least twice through the connecting poles. Since the circuit is an actual circuit, these poles are distinct. Let us consider the segment of the circuit between two successive passes through the connecting poles. This segment lies entirely either in M_1 or in M_2. In the first case, the conduction in the segment (that is, in the entire circuit) is zero since M_1 is a switching circuit. In the second case, by the construction of M_2 the actual circuit that connects the inputs of M_2 must pass through the output of M_2 (which coincides with the output of S_f). But then the initial circuit would have to contain the output of S_f as an interior pole, which would contradict the fact that the output is an actual output (the initial circuit would have to pass twice through the output of S_k).

Let us estimate how many contacts there are in the network. By Problem 9.19, the number of contacts in M_1 is less than $2 \cdot 2^{n-k}$ (we have given a somewhat rough estimate);

by Problem 9.18, there are at most $2 \cdot 2^k \cdot 2^{2^k}$ contacts in M_2. Thus, for any $k \leqslant n$ we have

$$L(n) \leqslant T_{n,k} = 2 \cdot 2^{n-k} + 2 \cdot 2^k \cdot 2^{2^k}.$$

9.22. Let us first present a number of points that would make it possible to find $k(n)$ and guess the order of $T_{n,k(n)}$. We have

$$T_{n,k} = 2 \cdot 2^{n-k} + 2 \cdot 2^k \cdot 2^{2^k}.$$

Suppose that $k(n) \geqslant \log_2 n$. Then the second term is greater than $2n \cdot 2^n$, i.e., for large n, $T_{n,k(n)}$ yields an estimate for $L(n)$ which is worse than that found in Problems 9.18 and 9.19. Now suppose that $k = k(n) = \log_2 n - \alpha(n)$, where $\alpha(n) \geqslant \beta > 0$. Then

$$2^k \leqslant \frac{n}{2^\beta} = \delta n, \quad \delta = 2^{-\beta} < 1.$$

It is clear that the second term now grows more slowly than the first term, since 2^n grows more rapidly than $2^{2^k} \leqslant (2^\delta)^n$. Thus, everything depends on the first term. Now we wish to find out how great k can be, since 2^k occurs in the denominator. In fact, k may be made the greatest integer less than $\log_2 n - \beta$. Then

$$\log_2 n - \beta - 1 < k(n) \leqslant \log_2 n - \beta,$$

i.e.,

$$\frac{n}{2^{1+\beta}} < 2^k;$$

$$2 \cdot 2^{n-k} = \frac{2 \cdot 2^n}{2^k} < 4 \cdot 2^\beta \cdot \frac{2^n}{n} = (4 + \varepsilon)\frac{2^n}{n};$$

$$\varepsilon = 4(2^\beta - 1).$$

Since the second term of $T_{n,k(n)}$ grows more slowly than the first term, with this choice we find that $T_{n,k(n)}$ is of the same order of magnitude as $2^n/2$. For $L(n)$, we find that (since $\beta > 0$ may be made as small as desired) for any $\varepsilon > 0$,

$$L(n) \lesssim (4 + \varepsilon)\frac{2^n}{n},$$

where the symbol $\lesssim$ means that the inequality holds for sufficiently large n ($n > N$, where N depends on ε).

It is clear that for large n, the resulting estimate is better than those we have been able to obtain up to now.

9.23. By Problem 9.14 there is a function of n variables whose realization requires $2n$ contacts. Therefore,

$$L(n) \geqslant 2n.$$

9.24. In a graph with k edges, the number of vertices cannot exceed $2k$, since each edge has two vertices, and there are no isolated vertices. Let us number the vertices in our graph and stipulate that the input is numbered 1, and the output numbered 2. Now a graph is uniquely determined if we can show which vertices connect each of the edges. With each edge we may associate a pair of natural numbers (p, q), which are the numbers associated in turn with the pair of the vertices connecting this edge; $p, q \leqslant 2k$ (the relative order of magnitude of p and q is not essential). The number of such pairs is less than $(2k)^2$. There are no more than k edges in a graph; to determine the graph, we need specify no more than k such pairs; some of these pairs may coincide, since distinct edges may connect identical vertices. These pairs may be selected in fewer than $((2k)^2)^k = (2k)^{2k}$ ways. (Permutations with repetitions; cf. Chapter 2; in fact, the order of the edges is not essential, and therefore combinations with repetitions could also be used, though this would be sufficient for the estimate.)* We could have obtained a more precise estimate, though this is not necessary.

9.25. To obtain a network from a graph, either the symbol of some variable or the symbol for the negation of the variable must be associated with each edge. If there are k edges in the graph and if there are n variables, $(2n)^k$ networks may be obtained from this graph, since there are $2n$ ways of selecting a symbol for one edge. (We are again using permutations with repetitions). Now taking

*If there are fewer than k edges, the set of edges may be chosen so as to contain k elements; for example, we may formally add the pair (1,1), which cannot correspond to any edge.

into account the results of the preceding problem (number of graphs with k edges), we find that the number of networks with k contacts for n variables is given as $R(k,n) < (2n)^k(2^k)^{2k} \leqslant (2kn)^{2k}$ if $n > 1$. The case $n = 1$ could be considered separately, though it is of no interest.

9.26. We have

$$2 < 2^{\alpha(n)} \cdot \alpha(n)^{\alpha(n)/n}.$$

If $\alpha(n) \leqslant 1$, the first factor on the right side will not exceed 2, and the second factor 1. Therefore $\alpha(n) > 1$, that is

$$L(n) > \frac{2^n}{n}.$$

Note that, unlike the previously obtained upper bound, this estimate is valid for all $n > 1$ (when $n = 1$, the inequality turns into an equality), and not only for large enough n.

9.27. We have $Q_2(x_1,x_2) = x_1x_2 \vee \bar{x}_1\bar{x}_2$. The corresponding network is depicted in Figure 57. Further $P_2(x_1,x_2) = x_1\bar{x}_2 \vee \bar{x}_1x_2$; P_2 is realized by the network depicted in Figure 58. That the first of these networks is minimal was proved in Problem 9.14; the minimality of the second network may be proved analogously (and can also be derived from the minimality of the first network). The PCNF for these functions also leads (Figure 59) to minimal networks (cf. Problem 9.15).

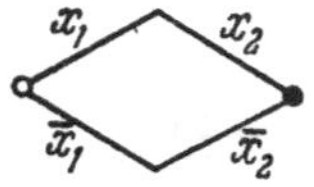

Figure 57

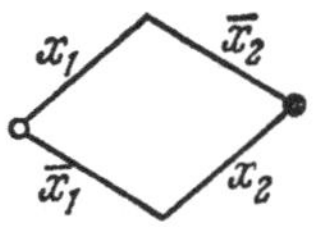

Figure 58

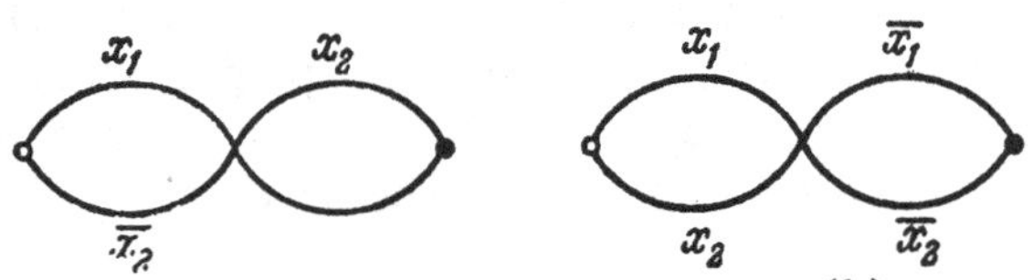

Figure 59

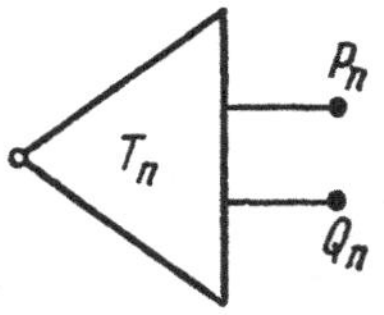

Figure 60

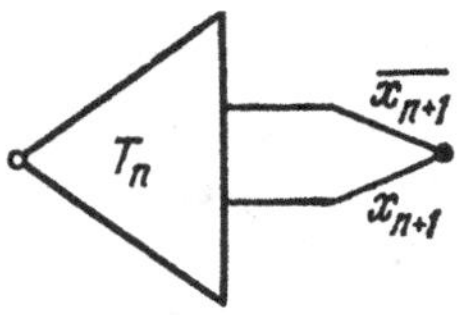

Figure 61

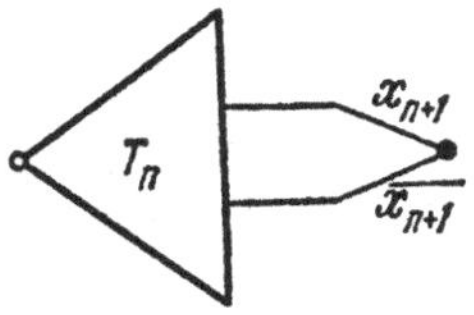

Figure 62

9.28. Suppose that we are given a (1, 2)-pole T_n (Figure 60) at whose outputs the functions $P_n(x_1, \dots, x_n)$ and $Q_n(x_1, \dots, x_n)$ are realized. By the inductive formulas for P_n and Q_n presented in the hint and the fact that $P_n = \overline{Q_n}$, using the networks constructed in the preceding problems (Figures 57 and 58) we then find that $P_{n+1}(x_1, \dots, x_n, x_{n+1})$ is realized by the network depicted in Figure 61, and Q_{n+1} by the network depicted in Figure 62. Combining these networks, we find that the (1,2)-pole shown in Figure 63 may be taken as T_{n+1}. From the network, it is clear that a (1, 2)-pole T may be constructed with four more contacts than in T_n. If we now take into account the fact that a network with two contacts (Figure 64) may be interpreted as a (1, 2)-pole T_1, we then find by induction that $2 + 4(n - 1)$ contacts are sufficient for constructing T_n. Further, both P_n and Q_n may be obtained from T_{n-1} by the addition of two contacts (Figures 61 and 62), so that $4(n - 1)$ contacts suffice for realizing them:

$$L(P_n) \leqslant 4(n - 1); \quad L(Q_n) \leqslant 4(n - 1).$$

To obtain a lower bound, we need only bear in mind (cf. solution of Problem 9.14) that a linear function is non-increasing and non-decreasing with respect to each of its (actual) variables, i.e., in the network that realizes it each variable may be associated with at least two contacts (x_i and $\overline{x}_i$):

$$2n \leqslant L(P_n); \quad 2n \leqslant L(Q_n).$$

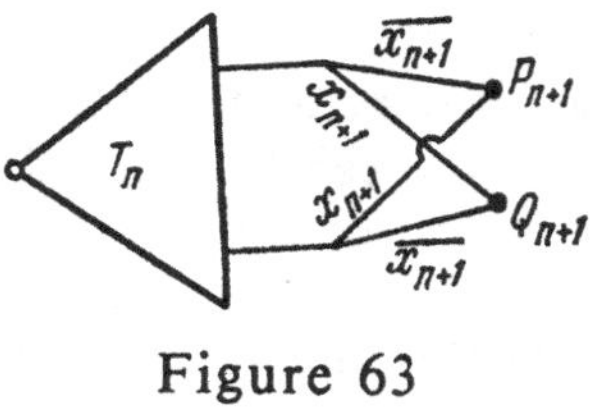

Figure 63

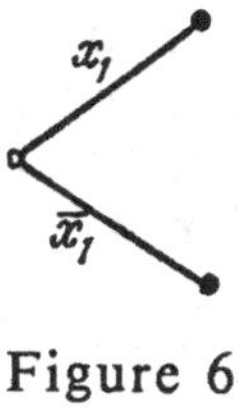

Figure 64

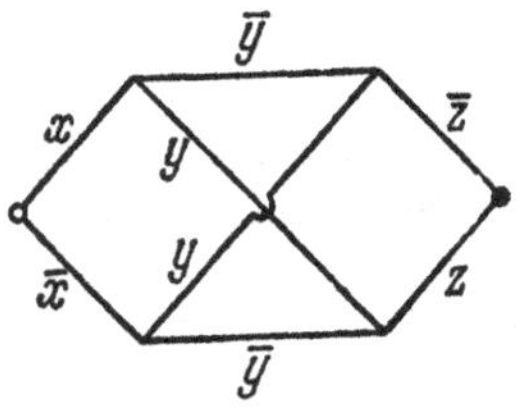

Figure 65

9.29. Let us consider the network (Figure 65) which realizes $P_3(x,y,x)$. If this network is a series connection of two sub-networks, the vertices of the first sub-network (which contains the network input) cannot be connected to the network output by means of a circuit that does not pass through the connecting pole. (Analogously, the vertices of the second sub-network cannot be connected to the network input by by-passing the connecting pole.) It is clear that no matter which pole is made the connecting pole in our network, the remaining poles can be connected together by means of circuits that do not pass through this pole. Thus our network cannot be represented in the form of a series connection of two networks.

Based on the foregoing considerations, we may formulate the following necessary criterion of a network that is in the form of a series connection of two sub-networks: There must exist a vertex through which any circuit connecting the network input to its output passes. (This vertex is the connecting pole.) That such a vertex is lacking may be easily proved by induction for all the networks constructed in Problem 9.28.

From the solution of Problem 9.12, it follows that if a network is a parallel connection of two sub-networks, there is either a contact that connects the input to the output, or all the vertices (other than the input and output) may be divided into two groups in such a way that the vertices of different groups cannot be connected by a circuit that passes through the input or output. In Figure 65, the input and output are not connected by a contact, whereas all the other vertices may be connected by a circuit that does not pass through the input or the output. That this is so may be easily proved by induction for all the networks constructed in Problem 9.28 for $n > 2$, i.e., none of these networks may be represented in the form of a parallel connection of two non-empty networks.

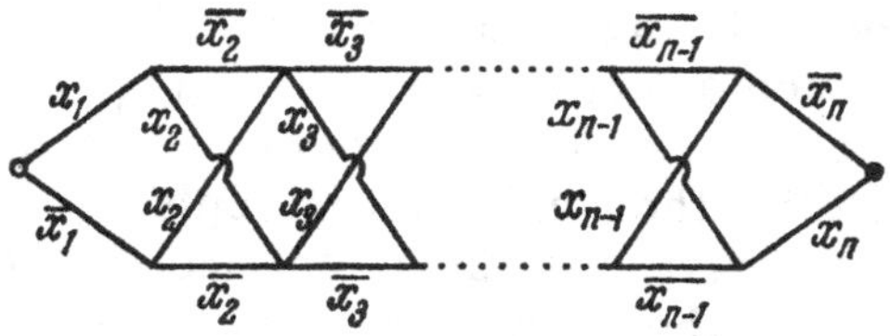

Figure 66

For the sake of convenience, we present the network for P_n (Figure 66).

9.30. Suppose that we are given two pi-networks that realize P_m and Q_m (Figure 67). Then by the formulas presented in the hint to the problem and by Problem 9.27, the network depicted in Figure 68, in which the contacts $x_1, \dots, x_m$ occur in the two networks $P_m(x)$ and $Q_m(x)$, while the contacts $y_1, \dots, y_m$ occur in the two networks $P_m(y)$ and $Q_m(y)$, realizes the function $P_{2m}(x_1, \dots, x_m; y_1, \dots, y_m)$. Analogously, the network in Figure 69 realizes $Q_{2m}(x_1, \dots, x_m; y_1, \dots, y_m)$.

Thus, pi-networks may be constructed by induction for P_{2^k} and Q_{2^k} by doubling the number of variables at each step. Also, at each step the number of contacts increases four-fold. Thus, there are 4^k contacts in all. In other words,

$$L_\Pi(P_n) \leqslant n^2, \quad L_\Pi(Q_n) \leqslant n^2$$

if $n = 2^k$.

9.31. Suppose that $2^k < n < 2^{k+1}$; $n = 2^k + m$, $0 < m < 2^k$.

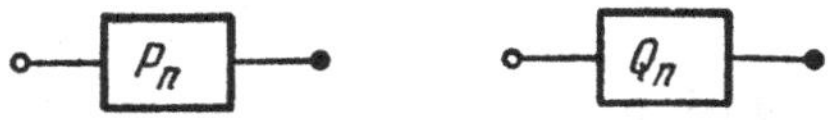

Figure 67

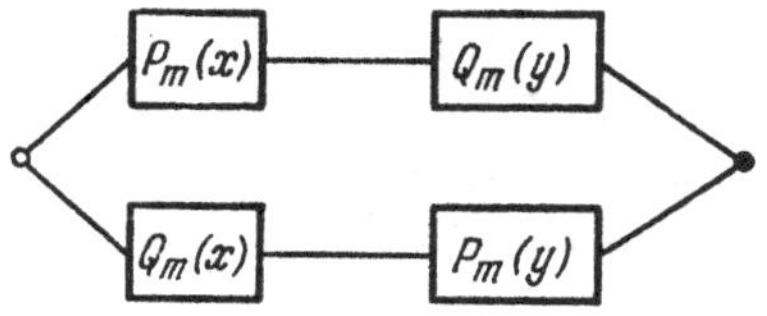

Figure 68

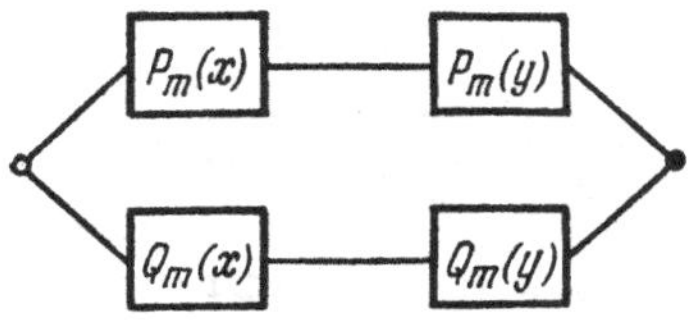

Figure 69

First Method. Using the method indicated in Problem 9.30, we construct a network for $P_{2k+1}(x_1, \ldots, x_n, x_{n+1}, \ldots, x_{2k+1})$. Now let us make a general remark. If in some function f, one of the variables is replaced by the additive identity, all positive contacts connected to y must be eliminated from the network that realizes f (without changing the poles of the contacts), while the negative contacts must be eliminated by equating the vertices they connect. For example, the network depicted in Figure 71 may be obtained from the network depicted in Figure 70. By induction on the construction of a pi-network, it can be easily proved that under such a transformation a pi-network is transformed into a pi-network; here we must consider networks in which the input and output coincide (the conduction function for such networks is the multiplicative identity). This accords with our understanding that 0 replaces an arbitrary variable in a formula containing only disjunctions, conjunctions, and negations (of arguments only); note how such formulas are related to pi-networks (cf. sec. 9.2).

In our case, we substitute $x_p = 0$ if $p > n$. As a result we obtain the network for P_n from the network for $P_{2^{k+1}}$. Let us estimate the number of contacts in the resulting network. This network, first of all, has fewer contacts than does the network for $P_{2^{k+1}}$, i.e., this number is less than 4^{k+1}, and since $n > 2^k$, $4^{k+1} < 4n^2$, i.e.,

$$L_{\Pi}(P_n) < 4n^2.$$

Let us now find the least upper bound. In the network for $P_{2^{k+1}}$, all the variables are assigned the same number of contacts, a fact which may be proved, for example, by induction, i.e., each variable is assigned $4^{k+1}/2^{k+1} = 2^{k+1}$

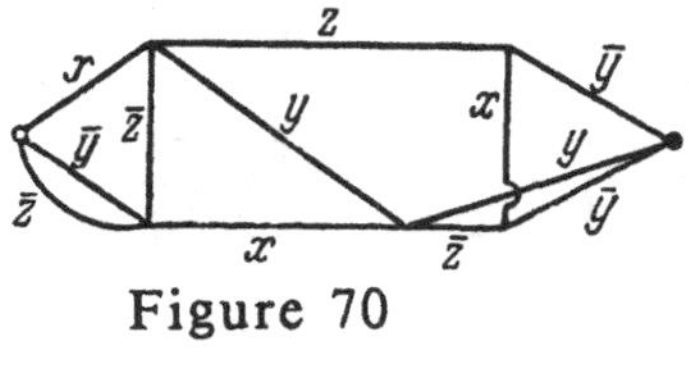

Figure 70

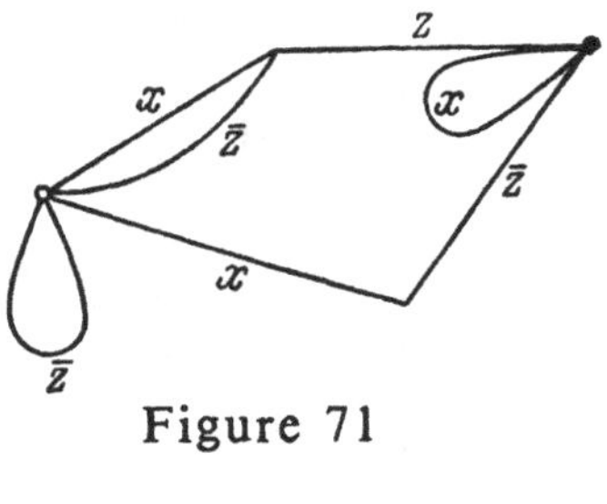

Figure 71

contacts. After the network has been transformed, it has left only those contacts that are associated with the variables x_1, ... , x_n, i.e., the resulting network will have $n2^{k+1}$ contacts. Since $2^k < n$, we find that

$$L_{\Pi}(P_n) < 2n^2.$$

The function Q_n may be considered analogously.

Second method. For the sake of convenience, we denote the variables in P_n thus:

$$x_1, \dots, x_m, x_{m+1}, \dots, x_{2^k}, y_1, \dots, y_m.$$

We realize P_n using the formulas presented in the hint for this problem and the networks for P_2 and P_{2^k}. By the reasoning given in the first method, we require $\frac{1}{2} \cdot 2^k$ copies of each of the networks $P_2(x_p, y_p)$ and $Q_2(x_p, y_p)$ $(p \leqslant m)$, with 2^k contacts (half positive, half negative) being associated with each of the variables x^p $(p > m)$. In the resulting network, there will be 2^{k+1} contacts associated with each of the variables x_p and y_p $(p \leqslant m)$, and 2^k contacts associated with each x_p $(p > m)$. The total number of contacts is then

$$R(n) = 2m \cdot 2^{k+1} + (2^k - m)2^k = 4^k + 3m \cdot 2^k.$$

$R(n)$ can be compared to $n^2 = 4^k + 2m \cdot 2^k + m^2$ in a natural way. Consider

$$\frac{R(n)-n^2}{n^2} = \frac{m \cdot 2^k - m^2}{4^k + 2m \cdot 2^k + m^2} \quad (0 < m < 2^k).$$

and the function

$$f_a(x) = \frac{ax - x^2}{(a+x)^2} \quad (0 \leqslant x \leqslant a).$$

We have

$$\frac{R(n)-n^2}{n^2} = f_{2^k}(m).$$

The function $f_a(x) > 0$ on the segment $[0,a]$ and vanishes at the endpoints of the segment. Let us find the maximum value of $f_a(x)$ on $[0,a]$. We have

$$f'_a(x) = \frac{(a-2x)(a+x)+2(x^2-ax)}{a+x} = \frac{-3ax+a^2}{a+x}.$$

The maximum value is attained when $x = a/3$; it is independent of a and equal to $1/8$, i.e.

$$f_a(x) \leqslant \frac{1}{8} \quad \text{if} \quad 0 \leqslant x \leqslant a.$$

That is,

$$\frac{R(n)-n^2}{n^2} \leqslant \frac{1}{8}.$$

Hence, it follows in turn that

$$R(n) \leqslant \frac{9}{8} n^2,$$

and also that

$$L_\Pi(P_n) \leqslant \frac{9}{8} n^2.$$

9.32. (a) Since we are dealing with binary numbers, to obtain s_i we must add x_i, y_i, and p_i modulo 2, i.e.,

$$s_i(x_i, y_i, p_i) = x_i + y_i + p_i \quad (i \leqslant n); \quad s_{n+1} = p_{n+1}.$$

(b) In the binary system, there is a forward carry to the next place whenever we add together the three digits in the same place in each of three addends if there are units in the same place in at least two of the addends. Only forward carry to the next place may occur. Thus,

$$p_{i+1}(x_i, y_i, p_i) = x_i y_i \vee x_i p_i \vee y_i p_i, \quad p_1 = 0.$$

9.33. We first find the networks for s_i and p_{i+1} under the assumption that p_i is realized by a single contact (Figure 72). That is, we must have not only p_i, but $\overline{p}_i$ as well. Since $x_i y_i \vee x_i p_i \vee y_i p_i$ is a self-dual function (cf. Chapter 3, Problem 3.3), to realize $p_{i+1}(x_i, y_i, p_i)$ we need only replace all the

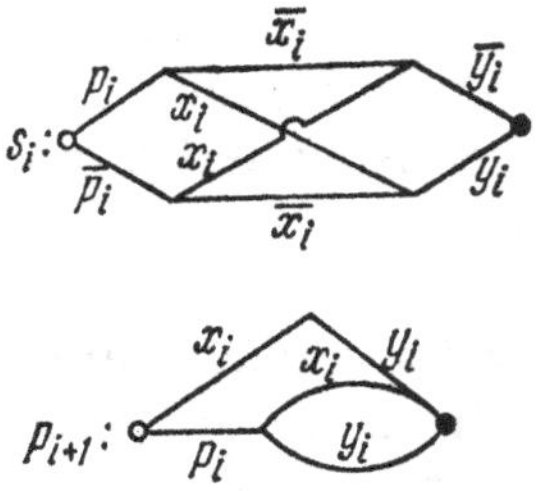

Figure 72

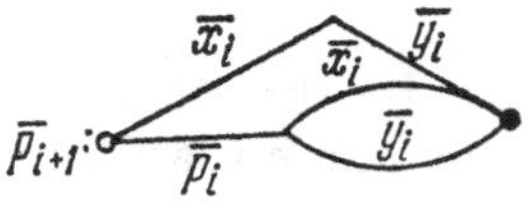

Figure 73

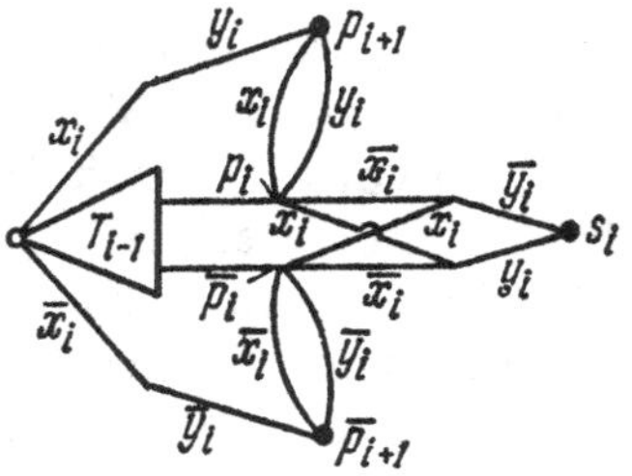

Figure 74

contacts in the network for p_{i+1} (Figure 72) by negative contacts (Figure 73). We will start with these three networks. Suppose a (1, i+1)-pole T_{i-1} at whose outputs $s_1, s_2, \dots, s_{i-1}, p_i, \overline{p}_i$ are realized has already been constructed. We construct the analogous (1, i+2)-pole T_i. The outputs of T_{i-1} corresponding to $s_1, \dots, s_{i-1}$ also serve as outputs of T_i and are not used in the further constructions and, therefore, they are not shown in the figure. We must extend the network so as to obtain three more outputs: s_i, p_{i+1}, and $\overline{p_{i+1}}$. Here we will use the outputs of T_{i-1} on which p_i and $\overline{p}_i$ are realized. We obtain the network depicted in Figure 74. The network has been constructed. (Recall once again that the outputs for $s_1, \dots, s_{i-1}$ are taken from T_{i-1}; they are not shown in the figure.) In constructing T_n from T_{n-1}, the part that realizes $\overline{p_{n+1}}$ cannot be constructed, and the output p_{n+1} must be denoted by s_{n+1}. With this, the construction of the network concludes.

Let us estimate the number of contacts. To derive T_i from T_{i-1} we add 14 contacts at each step. Therefore, the required number of contacts does not exceed 14n. In fact, it is somewhat less, since $p_0 = 0$, and on the first step eight contacts suffice; on the last step 10 contacts suffice, since it is not necessary to realize $\overline{p_{n+1}}$, i.e., $14n - 10$ contacts suffice, though this is not a significant difference for large n. It is more important that we have constructed a network exhibiting linear complexity as $n \to \infty$.

Remark. In the process of constructing an n-place adder, we have constructed adders with a lesser number of places. As the number of places is increased, the network we wish to construct may be altered without changing the part already constructed.

9.34. Let us construct networks that realize s_i and p_{i+1} as functions of p_i, x_i, and y_i (Problem 9.32). We may assume that we have been given a network R_i with three inputs and two outputs (Figure 75). The design of this network is independent of i. Now we make n copies of these networks and connect them as shown in Figure 76. Note that $p_1 = 0$ and $s_{n+1} = p_{n+1}$ (cf. solution of Problem 9.33). As a result, we obtain a network with $2n$ inputs $(x_1, \dots, x_n; y_1, \dots, y_n)$ and $n+1$ outputs $(s_{n+1}, s_n, \dots, s_1)$. If c contacts are needed to construct a single network R_i (c, obviously, depends on the basis of functional elements used), the total number of elements is equal to cn. Thus, n-place adders may be realized by means of networks exhibiting linear complexity in n. Note that as in the case of contact networks, to increase the number of places we only have to expand the network and don't have to transform it. Note, too, that an adder network composed of functional elements is simpler than a contact network.

9.35. *First Method.* The first steps of the proof have been given in the hint. Note that since we must add 2 to carry a digit from the $(i+1)$-th place to the i-th place, and since the operations are performed modulo 2, $1 \equiv -1$ (mod

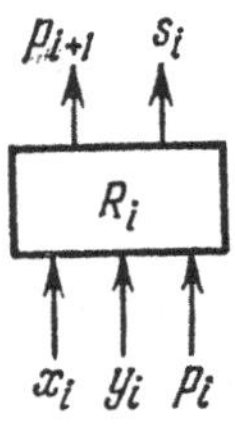

Figure 75

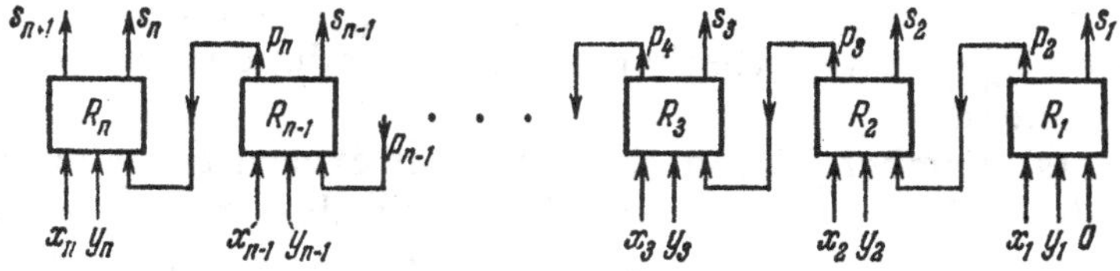

Figure 76

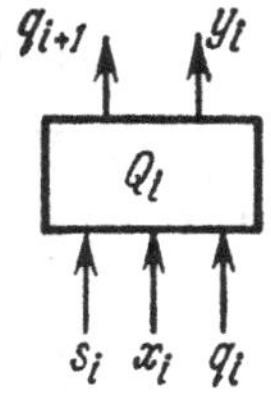

Figure 77

2), to obtain y_i we need only add s_i, x_i, and q_i modulo 2, i.e.,

$$y_i = s_i + x_i + q_i \,.$$

Further, it is necessary to carry the $(i+1)$-th place, if either $x_i = q_i = 1$ (since even if $s_i = 1$, we must subtract 2), or if $s_i = 0$, whereas $x_i = 1$ or $q_i = 1$ (from 0, it is necessary to subtract 1); subtraction is possible in the remaining cases. Therefore,

$$q_{i+1} = x_i q_i \vee \bar{s}_i(x_i \vee q_i).$$

Let us construct a network that realizes y_i and q_{i+1} (Figure 77). The design of the network Q_i is independent of i. These networks are connected in the same way as in the construction of the adder in Problem 9.34 (Figure 78).

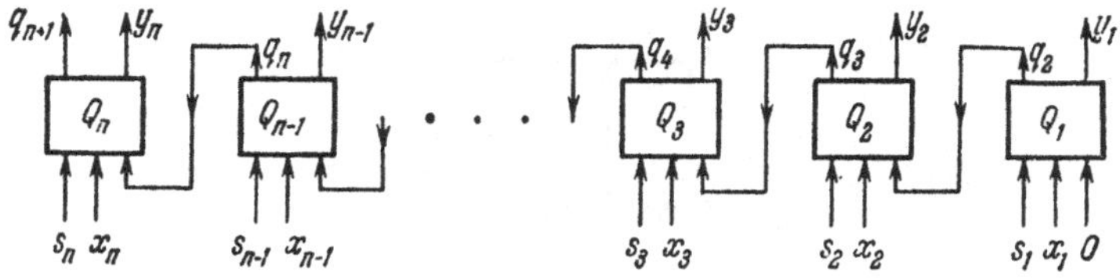

Figure 78

Here $q_i = 0$ and $q_{n+1} = 0$ if $s \geqslant x$, and $q_{n+1} = 1$ if $s < x$, i.e., the output q_{n+1} states whether or not the minuend is less than the subtrahend.

The total number of contacts does not exceed cn, where c is the number of elements in Q_i. (But this does not mean that the constants for the adder are the same as the constants for the circuit we are constructing; in general, we will often denote distinct constants by the same letter.)

Second Method. $\bar{x}$ must be added to s, and then 1. We denote the resulting network by $y = (y_{n+2}\, y_{n+1} \dots y_1)$. Here if $s \geqslant x$, then $y_{n+2} = 1$ or $y_{n+1} = 1$. Further, from y we must subtract 10^{n+1}. Suppose that $s \geqslant x$. For this purpose, it is necessary to replace y_{n+1} by 0 (if $y_{n+1} = 1$) without changing the other digits, or set $y'_{n+2} = 0$ and $y'_{n+1} = 1$ (if $y_{n+1} = 0$) again without changing the other digits. We will not bother sketching the corresponding network (which will also have linear complexity).

9.36. Let us construct a network P_{i+1} with $2n+1$ inputs $t_1^{(i)}, \dots, t_n^{(i)};\ x_1, \dots, x_n;\ y_{i+1}$, at whose output the successive digits of the sum $t^{(i)} + y_{i+1}x = (t_n^{(i+1)} \dots t_1^{(i+1)} p_{i+1})$ are realized (Figure 79). This network may be easily obtained from an n-place adder; the conjunctions $y_{i+1}x_1, y_{i+1}x_2, \dots, y_{i+1}x_n$ are fed to the inputs corresponding to the digits of the first addend (and the digits $t^{(i)}$ to the other inputs). As before, the number of elements in this network admits a linear estimate, since n networks must be added to the adder for conjunctions of two variables (if there are a elements in one such circuit, another na elements are needed).

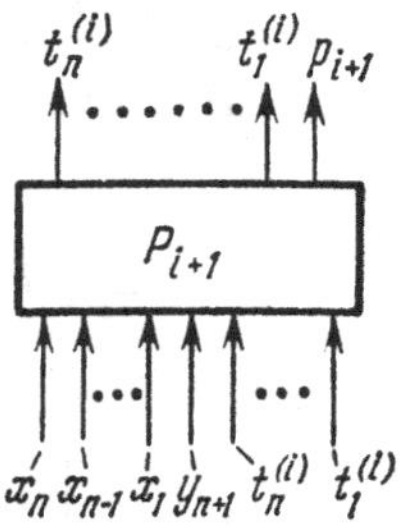

Figure 79

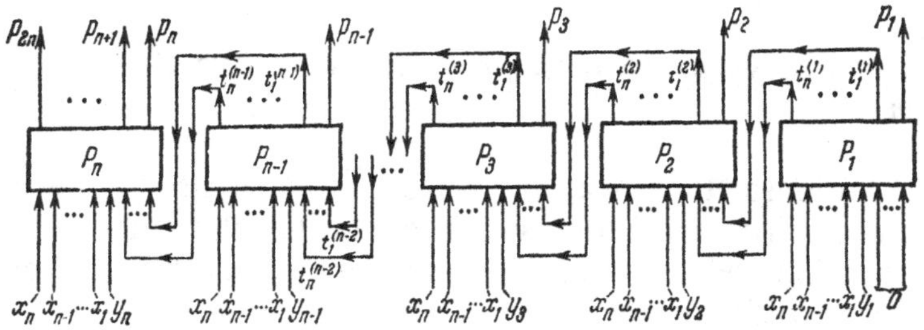

Figure 80

Let us now connect the networks P_i, as shown in Figure 80. Here we must bear in mind that $t^{(0)} = (0, 0, \dots, 0)$ and that $t^{(n)} = (p_{2n}, p_{2n-1}, \dots, p_{n+1})$. So as not to complicate the drawing, the inputs $x_1, \dots, x_n$ of the different P_i are not equated. As a result, we have constructed a network with $2n$ outputs $(p_{2n}, \dots, p_i, \dots, p_1)$ and $2n$ inputs $(x_n, \dots, x_1; y_n, \dots, y_1)$ that realizes the product xy.

Since all the networks P_i are identical and since they may be realized by cn elements, the total number of contacts does not exceed cn^2. Thus we are able to construct a multiplier network with quadratic complexity.

Remark. The reader may easily verify that a contact network for multiplication constructed according to the same principle will require a substantially greater number of contacts (in order of magnitude as $n \to \infty$).

9.37. We have $ab = \frac{1}{2}[(a+b)^2 - a^2 - b^2]$. As we have already remarked, the number $a+b$ has at most $n+1$ digits. Therefore, at most $f(n+1) + 2f(n) \leq 3f(n+1)$ elements are needed to construct the networks for $(a+b)^2$, a^2, and b^2; another $c_1 n$ elements are used in a preliminary computation of the sum (by Problem 9.34). Further, to compute $2ab$ the network of Problem 9.35 must be used twice for numbers containing up to $2(n+1)$ digits (i.e., as many digits as there are in the square of an $(n+1)$-digit number). For this, once again no more than $c_2 n$ elements are needed. In the binary system, dividing an even number of 2 is carried out by discarding the last digit (0). As a result, the total number of elements does not exceed

$$3f(n+1) + c_3 n.$$

If $f(n) = cn^\alpha$ $(\alpha > 1)$, by increasing the constant c we may replace $n+1$ by n and discard the linear term.

Remark. In solving the next problem, we prove that $f(n+1) \leqslant f(n) + cn$.

9.38. Suppose that we are able to construct a network consisting of $d(m)$ elements which computes the square of an m-digit number. We now square the number $x \cdot 2^m + y$, where x and y are m-digit numbers:

$$(x \cdot 2^m + y)^2 = x^2 \cdot 2^{2m} + 2xy \cdot 2^m + y^2.$$

On the right-hand side is the product xy for which we are still unable to construct the desired network. As in the solution of the preceding problem, we replace $2xy$ by $(x+y)^2 - x^2 - y^2$:

$$(x \cdot 2^m + y)^2 = x^2 \cdot 2^{2m} + (x+y)^2 \cdot 2^m - x^2 \cdot 2^m - y^2 \cdot 2^m + y^2.$$

To compute x^2 and y^2, we need $2d(m)$ elements. The number $x+y$ may contain $m+1$ digits, but by the formula $(2z+u)^2 = z^2 \cdot 2^2 + zu \cdot 2^2 + u^2$, where u is a one-digit number (0 or 1), $d(m+1) \leqslant d(m) + c_1 m$. Thus, to compute three required squares we need

$$3d(m) + c_2 m$$

elements. Then, it is necessary to add and subtract numbers containing at most $4m$ digits. Since the total number of required operations is independent of m, by Problems 9.34 and 9.35 $c_2 m$ elements are sufficient for this purpose. As a result, we find that

$$d(2m) \leqslant 3d(m) + c_3 m,$$

where c_3 is independent of m.

Based on this formula, we find the following estimate for $d(2^k)$:

$$d(2^k) \leqslant 3d(2^{k-1}) + c_3 2^{k-1}.$$

Since $d(1) = 1$ $(x^2 = x)$, by induction we find that

$$d(2^k) \leqslant 3^k + c_3(3^{k-1} + 3^{k-2} \cdot 2 + \ldots + 3^{k-i-1} 2^i + \ldots + 2^{k-1}).$$

Extracting 3^k on the right side as a factor yields

$$d(2^k) \leqslant 3^k\left[1 + \frac{c_3}{3}\left[1 + \frac{2}{3} + \dots + \left(\frac{2}{3}\right)^i + \dots + \left(\frac{2}{3}\right)^{k-1}\right]\right].$$

The expression within the brackets is the sum of a geometric progression with ratio 2/3; it does not exceed the sum of the corresponding infinite progression

$$\frac{1}{1 - \frac{2}{3}} = 3.$$

Thus

$$d(2^k) \leqslant c \cdot 3^k.$$

9.39. Suppose that $n \neq 2^m$ and let k be the smallest number such that $n \leqslant 2^k$, $k = [\log_2 n] + 1$. To find the square of a 2^k-place number, we use the network constructed in Problem 9.38. This may be done by letting the leftmost digits be zero. (Once again, it may be necessary to require that a finite number of elements suffice to realize the additive identity.) The number of elements in the network is less than

$$c3^k \leqslant 3c \cdot 3^{[\log_2 n]} \leqslant c_1 3^{\log_2 n} = c_1 \cdot 2^{\log_2 3 \cdot \log_2 n} = c_1 n^{\log_2 3}.$$

Thus, a network of no more than $cn^{\log_2 3}$ elements may be constructed. This estimate is better than the one we found earlier, since $\log_2 3 < 2$. In order of magnitude, this estimate also holds for multiplication (by Problem 9.34).

Chapter 10
ELEMENTS OF PROBABILISTIC LOGIC

1. *Probabilistic Boolean Algebra.* We will retain here the *formal* approach to probability theory, on occasion clarifying the relationship between our axiomatic theory and real "probabilistic" problems. We will not be using any results from probability theory, assuming them to be already known. Nevertheless, some familiarity with the elements of probability theory would be helpful (cf. Refs. 1-4).

Suppose that $\mathfrak{U}$ is a regular* Boolean algebra (sec. 2.3, 2.4, Definitions 2.5 and 2.7); we call its elements *events*.

Definition 10.1. The function $P(x)$ defined on $\mathfrak{U}$ is called a *probability measure* (or simply a *probability*) if it satisfies the following axioms**:

(1) $0 \leqslant P(x)$ for all $x \in \mathfrak{U}$;
(2) $P(\mathbf{1}) = 1$;

---

*We could limit ourselves to the additional axioms $x \vee \bar{x} = \mathbf{1}$ and $x\bar{x} = \mathbf{0}$, which hold in regular Boolean algebras (cf. solution of Problem 2.22); each of these axioms is a consequence of the other one together with the rest of the axioms of Boolean algebra.

**In this chapter $\mathbf{1}$ and $\mathbf{0}$ are designated elements in a Boolean algebra, while 1 and 0 are real numbers.

(3) if $xy = 0$, $P(x \vee y) = P(x) + P(y)$.

A regular Boolean algebra $\mathfrak{U}$ together with the probability measure P is called a *probabilistic Boolean algebra* $\{\mathfrak{U}, P\}$.

That the algebras discussed in the problems are regular will follow from the fact that they may be considered (to within isomorphism) as (Boolean) algebras of sets (cf. Problem 2.23).

But we are not interested in precisely which method is used to associate the values of P_i with the elements (events) of an algebra ; below we present concrete examples. Probability theory considers the probability of compound events, i.e., events that may be obtained from certain initial events (whose probabilities are already known) by repeated application of the operations of disjunction, conjunction, and negation. Because probabilistic Boolean algebras are regular, compound events may be obtained from initial events by applying the Boolean operations (Definition 2.8). It turns out that from the very simple properties of probability stated in the axioms alone, quite a bit can be concluded regarding the probabilities of compound events without knowing anything about the method used to determine the probabilities of the simple events.

Definition 10.2. Two events x and y whose conjunction coincides with 0 are said to be *disjoint.*

The meaning of this definition is that x and y cannot occur simultaneously; this is in accord with the ordinary understanding that 0 represents an event that can never occur, and analogously 1 represents an event that always occurs.

10.1. Prove that:

(a) $P(\overline{x}) = 1 - P(x)$ for all $x \in \mathfrak{U}$;
(b) $P(x) \leqslant 1$ for all $x \in \mathfrak{U}$;
(c) $P(0) = 0$;
(d) if $x \rightarrow y = 1$ (the event x implies the event y), then $P(x) \leqslant P(y)$;
(e) $P(x \vee y) \leqslant P(x) + P(y)$ for any $x,y \in \mathfrak{U}$ (cf. Axiom 3);

(f) $P(x \vee y) = P(x) + P(y) - P(xy)$;

(g) $P(x_1 \vee x_2 \vee \ldots \vee x_k) \leqslant P(x_1) + P(x_2) + \ldots + P(x_k)$ for all $x_1, \ldots, x_k$;

(h) $P(x_1 \vee x_2 \vee \ldots \vee x_k) = P(x_1) + P(x_2) + \ldots + P(x_k)$ if $x_i x_j = \mathbf{0}$ for all $i \neq j$;

(i) $P(xy) \leqslant P(x)$ for any x, y;

(j) $P(xy) \geqslant P(x) + P(y) - 1$ for any x, y.

These formulas all admit simple meaningful interpretations. For example, if one event x implies another event y, the probability of x cannot exceed the probability of y (Axiom d); the probability that at least one of two events x and y will occur does not exceed the sum of their probabilities and is equal to this probability only when the probability that x and y will occur together is equal to zero, in particular when x and y are disjoint events (Axioms e, f); the probability that some events will occur together does not exceed the probability that each will occur (Axiom i), and so on.

As will be clear from what follows, the probability of a compound event is not, in general, uniquely determined by the probabilities of its constituent simple events.

2. *Conditional Probability.*

Definition 10.3. Suppose that a is a designated element of $\mathfrak{U}$, where $P(a) \neq 0$. The function

$$P_a(x) = \frac{P(ax)}{P(a)} \qquad (x \in \mathfrak{U})$$

is called the *conditional probability* of the event x *relative to* the event a (or *under the condition* a).

10.2. Prove that the function $P_a(x)$ is a probability measure on $\mathfrak{U}$, i.e., that is satisfies Axioms (1)-(3).

10.3. Express $P_y(x)$ in terms of $P(x)$, $P(y)$, and $P_x(y)$.

Definition 10.3. A system of events $a_1, a_2, \ldots, a_n$ is called a *complete system of disjoint events* if these events are pairwise disjoint and if

$$a_1 \vee a_2 \vee \ldots \vee a_n = \mathbf{1}.$$

10.4. Suppose that $a_1, a_2, \dots, a_n$ is a complete system of disjoint events.

(a) Prove that

$$P(x) = P(a_1)P_{a_1}(x) + P(a_2)P_{a_2}(x) + \dots + P(a_n)P_{a_n}(x).$$

(b) Express $P_x(a_i)$ in terms of $P_{a_j}(x)$ and $P(a_j)$ $(1 \leqslant j \leqslant n)$.

The formulas of Problem 10.4 are very well known. Formula (a) is called the *composite probability formula*; it is used to compute an unconditional probability if the conditional probabilities relative to a complete system of disjoint events with given probabilities are known. Formula (b) is called *Bayes' formula* (or "hypothesis probability" formula). It corresponds to the following word problem. Suppose that there are n mutually disjoint "hypotheses" that exhaust all the possibilities regarding some event. Suppose that the probabilities of these hypotheses and the probability that some event x will occur (depending on which hypothesis is valid) are known. What may be said about the probabilities of the hypotheses a_i once the event x has already occurred? (For more detailed information about the meaning and possible applications of Bayes' formula, we refer the reader to Ref. 2.)

Definition 10.5. If $P_y(x) = P(x)$, the event x is said to be *independent* of the event y.

Since the equality $P_y(x) = P(x)$ is obviously equivalent to the equality

$$P(xy) = P(x)P(y).$$

which may also be considered a definition of the independence of the events x and y,* the relation of independence of events is symmetric. If x is independent of y, then y is independent of x. (In this case, the events x and y are also said to be (mutually) *independent*.)

10.5. If two events x and y are independent, x and $\overline{y}$ are also independent.

*In this definition, we can no longer exclude the possibility of events with zero probability.

3. *Probabilities Defined on Finite Free Boolean Algebras.* Suppose that $\mathfrak{A}_n = \mathfrak{A}(a_1, \dots, a_n)$ is a free Boolean algebra generated by a finite number of elements $a_1, \dots, a_n$. The term, "free Boolean algebra," denotes that there are no relations between the elements of $\mathfrak{A}_n$ that are not consequences of the axioms of Boolean algebra. Then $\mathfrak{A}_n$ consists of 2^n elements, each of which has the form $f(a_1, \dots, a_n)$, where f is a logical function of n variables (for example, represented in PDNF), and the elements corresponding to non-equivalent functions do not coincide.* In particular, $f = 0$ corresponds to $\mathbf{0}$, and $f = 1$ corresponds to $\mathbf{1}$. Now let us discuss how to define a probability measure on $\mathfrak{A}_n$. We specify the value of P on a_i thus: $P(a_i) = p_i$.

10.6. Prove that if all the p_i take the value 0 or 1, a probability on $\mathfrak{A}_n$ is uniquely defined, and P also takes only the values 0 or 1.

Remark. In general, a homomorphism of a Boolean algebra $\mathfrak{A}$ into the algebra $\{0, 1\}$ induces on a probability measure P that takes the values 0, 1 (Axioms (1)-(3) are satisfied). If $\mathfrak{A}$ is a propositional algebra, the assignment of a probability measure P may be interpreted as assigning to the proposition numbers that specify the probabilities of their truth values. As we will see below, this point of view makes sense in certain problems, though of course it is scarcely possible to assign probabilities to the truth values of arbitrary real propositions in a reasonable fashion (cf. [5]).

Let us consider how to define a probability on $\mathfrak{A}_n = \mathfrak{A}(a_1, \dots, a_n)$ in the general case. It may be assumed that every function f may be represented in PDNF. Events corresponding to the elementary conjunctions $a_1^{\sigma_1} \dots a_n^{\sigma_n}$ are disjoint and form a complete system of disjoint events. Every element of $\mathfrak{A}(a_1, \dots, a_n)$ is the union of a number of such events. Then by Axiom 3, the probability $P(x)$ on $\mathfrak{A}_n$ is uniquely determined by specifying the probabilities

---

*The homomorphisms of $\mathfrak{A}_n$ into $\{0,1\}$ are uniquely determined by specifying the images of the elements $a_1, \dots, a_n$, which may be selected arbitrarily. There are 2^n such homomorphisms, and these homomorphisms separate any two pairs of elements of $\mathfrak{A}_n$, i.e., $\mathfrak{A}_n$ is a regular Boolean algebra.

$$p_{\sigma_1 \dots \sigma_n} = P(a_1^{\sigma_1} \dots a_n^{\sigma_n}).$$

Now let us consider what conditions (other than non-negativity) must be satisfied by the numbers $p_{\sigma_1 \dots \sigma_n}$. It is first clear that their sum must equal 1 (Axiom 2); second, since the probabilities of the events a_i have been specified in advance, the following conditions must hold:

$$\sum_{\sigma_j=1} p_{\sigma_1 \dots \sigma_i \dots \sigma_n} = p_i \quad (i = 1, 2, \dots, n).$$

On the left is the probability of the event a_i expressed in terms of

$$p_{\sigma_1 \dots \sigma_n} (a_i = \bigvee_{\sigma_i=1} a_1^{\sigma_1} \dots a_n^{\sigma_n}).$$

It is clear that all the conditions are now satisfied. Thus, besides the positivity condition, $n+1$ linear conditions are imposed on the 2^n numbers $p_{\sigma_1 \dots \sigma_n}$.

If the non-negativity condition were not imposed on the numbers $p_{\sigma_1 \dots \sigma_n}$, then, clearly, there would exist infinitely many solutions of the resulting system of linear equations if $n > 2$ (a plane of dimension $2^n - n - 1$ in 2^n-dimensional space). However, since the coefficients and right sides of the linear equations are non-negative, it can also be easily verified that the analogous result holds for non-negative solutions.

If it is assumed that the events $a_1, \dots, a_n$ are independent, then by specifying the probabilities p_i we are uniquely determining a probability on $\mathfrak{U}_n = \mathfrak{U}(a_1, \dots, a_n)$. Let us make these remarks more rigorous, and state precisely what we mean when we say that several events are independent (this concept has so far been defined only for two events).

Definition 10.6. The events $a_1, \dots, a_n$ are said to be *independent* if for any distinct $1 \leqslant i_1, \dots, i_k \leqslant n$,

$$P(a_{i_1} \dots a_{i_k}) = P(a_{i_1} \dots P(a_{i_k}).$$

Otherwise, $a_1, \dots, a_n$ are said to be *dependent*. Note that this condition cannot be replaced by the pairwise independence of the events:

$$P(a_i a_j) = P(a_i)P(a_j) \quad (1 \leqslant i \neq j \leqslant n).$$

10.7. Give an example of pairwise independent, but dependent events.

10.8. Prove that a probability measure on $\mathfrak{U}(a_1, \dots, a_n)$ is uniquely determined by the probabilities p_i if the events $a_1, \dots, a_n$ are independent.

4. *Systems of Elementary Events.* Now we may consider more special cases. Based on the foregoing -- at least in the case of a Boolean algebra $\mathfrak{B}$ consisting of a finite number of elements -- to specify a probability it is natural to identify a complete system of disjoint events such that every event in $\mathfrak{U}$ is a disjunction of some of these events.

Definition 10.7. A complete system E of disjoint events from a finite Boolean algebra $\mathfrak{B}$ is called a *system of elemenary events* if every event in $\mathfrak{B}$ is a disjunction of events in $\mathfrak{U}$.

To specify a probability on $\mathfrak{U}$, we need only specify this probability on elementary events. These probabilities must be non-negative, and must sum to 1. There are no other conditions. The probabilities of the elementary events, by contrast, are specified based on a real situation.

Suppose, for example, we are tossing a die in the form of a cube on which the digits from 1 to 6 are inscribed. We are interested in learning what digit is inscribed on the face, which comes out on top. There are six outcomes to this trial; we let e_i $(1 \leqslant i \leqslant 6)$ denote the event in which the digit i is inscribed on the face that comes out on top. It is clear that these events form a complete system of disjoint events, since at least one of them must occur, and two of them cannot occur simultaneously. Further, if the die is regularly shaped and is "tossed honestly," it is natural to consider that all the events e_i are *equally probable*, i.e., we set $P(e_i) = 1/6 \quad (1 \leqslant i \leqslant 6)$.

Now let us consider the following distinct non-elementary events:

(a) an even number appears;
(b) an odd number appears;
(c) a number divisible by 3 appears;
(d) a number not greater than 4 appears.

10.9. Find the probabilities of the events a, b, c, and d. Are any of them independent?

Thus, we see that if some trial is underaken, the outcomes of the trial may be made the elementary events. If we start with the conditions of the trial and assign equal probabilities to all the outcomes in a natural way, the probability of any one event will equal the ratio of the number of outcomes in which it occurs to the total number of outcomes. In this rather obvious example (of equally probably outcomes), it is natural to consider what it means to say that any two events *a* and *b* are independent. Specifically, it means that the number of outcomes in which *a* occurs is the same whether expressed as a proportion of all the outcomes or as a proportion of the outcomes in which *b* occurs.

Bear in mind that a set of elementary events is not introduced, in general, in a uniquely defined way. For example, in the above problem we might try making *a* and *b* the elementary events (they are disjoint and form a complete system); however, in that case events *c* and *d* could not be represented in the form of their disjunction. In other words, to say that certain events are "elementary" means that any event we are interested in must be representable as a disjunction of these elementary events. If we take a single ball from an urn in which there are 6 white and 4 black balls, we could mentally number the balls and declare that events in which we have removed some designated ball are the elementary events. Then these events may all be considered equally probable (with probability 1/10). But if we are interested only in events associated with the colors of the balls, events in which a white or black ball has been removed may be declared the elementary events by assigning to them the probability 3/5 or 2/5, respectively. Let us consider several examples.

10.10. Two dice are tossed independently. What is the probability of the following events:

(a) identical numbers appear on the top face;
(b) the sum of the resulting numbers is even;
(c) the resulting numbers are mutually prime (do not have common positive divisors other than 1).

The example presented in the solution to Problem 10.7 may be given the following interpretation. Suppose that we are given a tetrahedron on whose faces the numbers 2, 3, 5 and 30 have been inscribed. The face on which it presently rests is chosen randomly (a die in the form of a tetrahedron may be tossed). The events a_1, a_2, and a_3 assert that a number divisible by 2, 3, and 5, respectively, have been inscribed on the base.

10.7". Verify that a_1, a_2, and a_3 are dependent, though also pairwise independent, events.

In the problems below, it is convenient to use the concept of conditional probability.

10.11. Suppose that we are given three discs. Both sides of one of the discs are colored red, both sides of another disc are colored blue, while one side of the third disc is colored red and the other blue. After the discs have been jumbled, one of them is selected and one of its sides is shown. It is necessary to guess the color of the other side. What strategy should be followed in this game?

10.12. The letters of the word "algebra" are written on separate cards. After the cards have been shuffled, three of them are chosen in succession. What is the probability that (a) the word "beg" is obtained; (b) the word "rag" is obtained?

We have presented somewhat more or less traditional problems that are usually found in textbooks of probability theory. In the books cited above, the reader may find many other problems. (Of course the reader could think up some of them independently.) Let us present some results.

We considered the case in which a finite number of elementary events $E = \{e_1, \dots, e_n\}$ is chosen and on which a probability $0 \leqslant P(e_i) \leqslant 1$, $P(e_1) + \dots + P(e_k) = 1$ is defined. Event a may then be interpreted as a subset of the set E (collection of elementary events whose disjunction is a). With the negation of an event, we associate the complement of the corresponding set with respect to all of E; with the conjunction of events, an intersection of sets; and with the disjunction of events, their union. Thus, in this case a Boolean algebra of events is isomorphic to a Boolean algebra of subsets of the set of elementary events. The probability of an event is equal to the sum of the probabilities of the elementary events. We have thereby arrived at another possible axiomatic approach to probability theory. It is different from the approach presented earlier in that here we are dealing not with abstract Boolean algebras, but rather with Boolean algebras of subsets of some set. In fact, we already know that every regular Boolean algebra is isomorphic to such an algebra (Problem 2.25).

Though we have been discussing a finite set of elementary events E, we could also speak of all the subsets of E. But if E is infinite, the function P cannot be defined on all subsets without violating Axioms 1-3; in particular, P is non-zero on only a finite number of elements of E (single-element subsets). Therefore, if the set of elementary events is infinite, we may consider some Boolean algebra of its subsets $\mathfrak{U}(E)$ on which $P(x)$, $x \in \mathfrak{U}(E)$, is defined (x is not a point in E, but rather a *subset* of E!) and satisfies Axioms 1-3; P is sometimes called the *probability measure* on E (more precisely, on $\mathfrak{U}(E)$).

Digression on Probabilistic Measure. If E is the collection of all the points in a square of unit area, $\mathfrak{U}(E)$ can be the collection of figures for which area is defined (we will not rigorously state what this means; it is important that $\mathfrak{U}(E)$ be a Boolean algebra), and $P(x)$ can be the area of x. It can be easily verified that in this case $P(x)$ satisfies Axioms 1-3. Now let us consider the following probabilistic problem. An object whose dimensions are of no importance (a "point"!) is tossed randomly into the above square. We may assume that the target itself has the shape of a square and that a sighting beam emanates from it. It is then natural to assume that the probability of hitting some target region in the square is proportional to its area.*

In the case of finite sets E, it is substantially more

*Note that this interpretation is used in modern computational mathematics to calculate areas of curvilinear figures, in what is known as the Monte-Carlo method.

advantageous to define P on *elements* of E instead of directly on the Boolean algebra. It is precisely in this case that P, understood as a function of E, need only satisfy simple and easily checked conditions, whereas it is a far more complicated matter to verify the axioms. As we have already remarked, in the case of infinite sets E, P cannot be defined on elements of E. We may then ask about those collections of subsets $\mathfrak{B}(E) \subset \mathfrak{U}(E)$ on which $P(x)$ may be defined without any special constraints such that $P(x)$ may be uniquely continued to $\mathfrak{U}(E)$. This problem has been studied in great depth in measure theory.

5. *Random Variables.*

Definition 10.8. Suppose that we are given a probability P defined on the Boolean algebra $\mathfrak{U}$. If there is some finite complete system of disjoint events $a_1, \dots, a_k$, $a_i \neq 0$, and if with every a_i we may associate a number $\xi(a_i)$, the random variable ξ is said to be defined on $\mathfrak{U}$. The system of events $\{a_i\}$ is called the *characteristic system* of events for ξ.

Suppose that we are given some other complete system of disjoint events $\tilde{a}_1, \dots, \tilde{a}_\ell$, $\tilde{a}_j \neq 0$, where for every $\tilde{a}_j$ there exists an event a_i such that $\tilde{a}_j \to a_i = 1$ (then every a_i is the disjunction of several $\tilde{a}_j$). The system of events $\{\tilde{a}_j\}$ is said to be *subordinate* to the system $\{a_i\}$. Note that for distinct i and k, it cannot happen that $\tilde{a}_j \to a_i = 1$ and $a_j \to a_k = 1$, since in that case $\tilde{a}_j \to a_i a_k = 1$ and $\tilde{a}_j \to \mathbf{0} = 1$ (by the fact that a_i and a_k are disjoint), i.e., $\tilde{a}_j$ would coincide with $\mathbf{0}$, which contradicts the hypothesis. Then a random variable ξ may be carried over may be into a system of events $\{\tilde{a}_j\}$: $\tilde{\xi}(\tilde{a}_j) = \xi(a_i)$ if $\tilde{a}_j \to a_i = 1$ (by the above remark, this definition is consistent). The variable $\tilde{\xi}$ is said to be *induced* by the variable ξ on $\{\tilde{a}_j\}$. In the language of set theory, we are dealing with a partitioning of the space of elementary events into non-empty disjoint sets $a_1, \dots, a_k$. With every a_i we associate a number $\xi(a_i)$. If there is a finer partitioning $\tilde{a}_1, \dots, \tilde{a}_\ell$, then with every $\tilde{a}_j$ we associate the value of ξ on the set a_i containing $\tilde{a}_j$. If E is a finite set, ξ may be defined, in particular, for a partitioning of E into elementary events. On the other hand, a *maximal* partitioning of E may be considered, i.e., a partitioning into sets (events) such that ξ assumes distinct values on distinct a_i. (It is necessary to consider the disjunction of all events on which ξ takes the same value.)

In the case of an infinite set E, a random variable ξ may be interpreted as a function on E that takes a finite number of values such that for every value of ξ the set of points E in which this value is taken belongs to the Boolean algebra $\mathfrak{U}(E)$.*

Definition 10.9. Suppose that $\{a_1, a_2, \ldots, a_k\}$ is a characteristic system of events of the random variable ξ. The number

$$M\xi = \xi(a_1)P(a_1) + \ldots + \xi(a_k)P(a_k),$$

is called the *mathematical expectation*, or mean value of the random variable ξ.

If all the events a_i were equally probable, $M\xi$ would coincide with the arithmetic mean of ξ. In the general case, however, every value occurs with weight equal to its probability.

Let us prove certain properties of the mathematical expectation.

10.13. The mathematical expectation $M\tilde{\xi}$ of an induced random variable $\tilde{\xi}$ for a system of events $\{\tilde{a}_j\}$ subordinate to the system $\{a_i\}$ coincides with $M\xi$.

Henceforth, we will not distinguish random variables obtained by introducing a subordinate system of events. By Problem 10.13, $M\xi$ is independent of the choice of the characteristic system of events. In particular, it is always possible to compute $M\xi$ starting with a maximal characteristic system of events.

Definition 10.10. Suppose that we are given two random variables ξ and η, the first associated with the system of events $\{a_1, \ldots, a_n\}$ and the second with the system $\{b_1, \ldots, b_\ell\}$. We consider the system of events $\{a_ib_i\}$, $1 \leqslant i \leqslant k$, $1 \leqslant j \leqslant \ell$;

*In probability theory, an important role is played by random variables that take an infinite number of values; we will not be discussing this topic here.

these events are disjoint and form a complete system. (Prove it!). The random variable

$$(\xi+\eta)(a_ib_j) = \xi(a_ib_j) + \eta(a_ib_j) = \xi(a_i) + \eta(b_j),$$

is called the *sum* of the random variables ξ and η. We will now use the fact that the system of events $\{a_ib_i\}$ is subordinate to each of the systems $\{a_i\}$ and $\{b_j\}$.

10.14. Prove that $M(\xi+\eta) = M\xi + M\eta$.

A random variable may be multiplied by a scalar factor. Clearly,

$$M(c\xi) = cM\xi.$$

where c is a number.

If $\{a_i\}$ consists of the single event 1 and if $\xi(1) = c$, the random variable ξ is called a constant and denoted c. (It corresponds to a constant function on E.) It is clear that

$$Mc = c,$$

where c is a constant random variable.

10.15. Prove that $M(\xi - M\xi) = 0$.

Here we have subtracted a constant random variable equal to the constant $M\xi$ from the random variable ξ.

Let us present examples of random variables. In the case of a finite Boolean algebra, a random variable, as we already noted, may also be considered a function on a set of elementary events.

As an example, let us consider the events associated with the tossing of a die (Problem 10.9). The number which appears may be interpreted as a random variable, i.e., with the elementary event e_i we associate the number $\xi(e_i) = i$ ($1 \leqslant i \leqslant 6$). Since ξ has distinct values on all the e_i, $\{e_i\}$ is a maximal characteristic system.

10.16. Find the mathematical expectation of the above random variable.

10.17. In the space of events of Problem 10.10, consider the following random variables: ξ, the number of even numbered faces that occur among the numbers i and j, and η, the sum of all the numbers that appear. Find $M\xi$ and $M\eta$.

10.18. There are n boxes numbered 1 to n and n counters numbered the same way. The counters are shuffled, after which precisely one of them is placed in each box. Find the mathematical expectation of the random variable ξ, the quantity of counters such that the number of each counter is the same as number of the box in which it is found.

In the next problem we present an important result from probability theory, called the *Chebyshev inequality.*

10.19. Prove that

$$P(|\ \xi\ |) \geqslant \delta) \leqslant \frac{M(\xi^2)}{\delta^2}.$$

for every positive number δ. On the left is the probability that the value of the random variable ξ is not less (in modulus) than δ; the square of a random variable is determined in an obvious way: the value of ξ for each event is squared.

We introduce one more characteristic of a random variable.

Definition 10.11. The number

$$D\xi = M[(\xi - M\xi)^2]$$

is called the *variance* of the random variable ξ. The variance states by how much the value of a random variable deviates from its mathematical expectation.

By means of the variance, it is possible to rewrite the Chebyshev inequality thus:

$$P(|\xi - M\xi| \geqslant \delta) \leqslant \frac{D\xi}{\delta^2}.$$

To derive this inequality, we need only replace ξ by $\xi - M\xi$ in the expression for the Chebyshev inequality given above. Thus, by means of the variance it is possible to estimate the probability that the deviation of a random variable from its mathematical expectation exceeds some designated number.

Let us discuss some properties of the variance. From its definition, it follows at once that

$$D(c\xi) = c^2 D\xi$$

for any number c. Before introducing the variance of a sum of random variables, let us discuss the mathematical expectation of a product of random variables. (How these two problems are related is explained below.)

Definition 10.12. The random variable with characteristic system of events $\{a_i b_j\}$ $(1 \leqslant i \leqslant k,\ 1 \leqslant j \leqslant \ell)$ defined by the condition

$$\xi\eta(a_i b_j) = \xi(a_i)\eta(b_j).$$

is called the *product* of the random variables ξ and η with characteristic systems of events $\{a_1, \ldots, a_k\}$ and $\{b_1, \ldots, b_\ell\}$, respectively. It can be easily verified that introducing two subordinate systems of events $\{\tilde{a}_i\}$ and $\{\tilde{b}_j\}$ leads to a subordinate system of events $\{\tilde{a}_i\tilde{b}_j\}$ for $\{a_i b_j\}$. This shows that our definition is well-formed.

It turns out that, in general, $M(\xi\eta) \neq M\xi \cdot M\eta$. To verify this result, consider the following example. If we are given some event a, we may construct the random variable I_a with characteristic system of events $\{a, \bar{a}\}$ thus:

$$I_a(a) = 1, \quad I_a(\bar{a}) = 0.$$

The quantity I_a is called the *indicator* of the event a. We have

$$MI_a = P(a).$$

If we are given two events a and b, the product of their *indicators* is equal to the indicator of their conjunction:

$$I_a I_b = I_{ab}.$$

We have found that the mathematical expectation of the product $I_a I_b$ is equal to the product of the mathematical expectations of I_a and I_b only when the two events a and b are independent. An analogous situation holds for arbitrary random variables.

Definition 10.13. The random variables ξ and η with characteristic systems of events $\{a_i\}$ and $\{b_j\}$ are said to be *independent* if the events a_i and b_j are independent for any i and j.

10.20. Prove that

$$M(\xi\eta) = M\xi\ M\eta.$$

for two independent random variables ξ and η.

Note that independence is only a sufficient, but not a necessary, condition for the validity of this formula.

10.21. Prove that

$$D(\xi + \eta) = D\xi + D\eta.$$

for two independent random variables ξ and η.

To compute the variance, it is often convenient to use the formula derived in the next problem.

10.22. Prove that

$$D\xi = M(\xi^2) - (M\xi)^2.$$

10.23. Compute the variances of the random variables presented in Problems 10.16 - 10.18.

6. *Bernoulli trials.* Let us consider more carefully the case in which a probabilistic (free) Boolean algebra $\{\mathfrak{U}_n, p\}$ is generated by n independent events $a_1, \dots, a_n$ with identical probabilities ($P(a_i) = p$). It may be assumed that a trial (experiment) with two outcomes has been conducted n times independently; the probability of one outcome (we will say it is positive) is equal to p, and the probability of the other (negative) outcome is equal to $q = 1-p$. Such a series of trials is referred to as *Bernoulli trials* (or a *Bernoulli chain*).

The event a_i asserts that there is a positive outcome in the i-th trial. We are dealing with Bernoulli trials if we toss a coin independently and wish to find out whether a head (positive outcome) or tail (negative outcome) occurs; here $p = q = 1/2$. If we toss a die and let a 6 represent a positive outcome, then we have a Bernoulli chain with $p = 1/6$ and $q = 5/6$. If in producing a batch of the same type of item, there are no systematic causes of defective items, then successive trials of the items for the presence of defects will constitute Bernoulli trials.

We call the algebra $\{\mathfrak{U}_n, p\}$ a *Bernoulli algebra.* Its elements are logical functions $f(a_1, \dots, a_n)$ of $a_1, \dots, a_n$. The events $a_1^{\sigma_1} \dots a_n^{\sigma_n}$ are the elementary events. By Problem 10.8, the probabilities of these events

$$p_{\sigma_1 \dots \sigma_n} = p^m q^{n-m} = p^m(1 - p)^{n-m},$$

where m is the number of units in the sequence $(\sigma_1, \sigma_2, \dots, \sigma_n)$. Then

$$P(f(a_1, \dots, a_n)) = h_f(p) = \sum_{m=0}^{n} c_m(f) p^m (1-p)^{n-m}, \tag{10.1}$$

where $c_m(f)$ is the number of sequences containing m units on which f is equal to unit. Clearly,

$$0 \leqslant c_m(f) \leqslant C_n^m,$$

where C_n^m is the total number of sequences with m units; it is equal to the number of combinations of n elements taken m at a time (it is known that $C_n^m = n!/m!(n-m)!$). Let us consider the random variable ξ_n (number of positive outcomes) on $\mathfrak{U}_n$. We define it on the elementary events:

$$\xi_n\left[a_1^{\sigma_1} \dots a_n^{\sigma_n}\right] = m(\sigma),$$

where $m(\sigma)$ is the number of units in the sequence σ.

10.24. Find $M\xi_n$ and $D\xi_n$.

Along with the random variable ξ_n, we may also consider the random variable ζ_n, or mean number, or frequency, of positive outcomes

$$\zeta_n = \frac{\xi_n}{n} = \frac{m(\sigma)}{n}.$$

By Problem 10.24, we have

$$M\zeta_n = p; \quad D\zeta_n = \frac{pq}{n}.$$

We apply the Chebyshev inequality to the random variable ζ_n:

$$P(|\zeta_n - p| \geqslant d) \leqslant \frac{pq}{n\delta}.$$

Thus, we have an estimate for the probability that the mean number of positive outcomes deviates from the probability of this outcome p by a quantity greater than δ. Since $q = 1-p$ ($0 \leqslant p \leqslant 1$), $pq \leqslant 1/4$, and we have found an estimate which is independent of p:

$$P(|\zeta_n - p| \geqslant \delta) \leqslant \frac{1}{4n\delta}. \qquad (10.2)$$

Let us study the algebraic meaning of the latter inequality. For this purpose, we must compute $P(|\zeta_n - p| \geqslant \delta)$.

10.25. Find $P(|\zeta_n - p| \geqslant \delta)$.

We take limits in inequality (10.2) as $n \to \infty$, obtaining

$$\lim_{n\to\infty} P(|\zeta_n - p| \geqslant \delta) = 0 \qquad (10.3)$$

or, what is the same thing,

$$\lim_{n\to\infty} P(|\zeta_n - p| < \delta) = 1.$$

In other words, for large enough n the frequency of positive outcomes is arbitrarily close to p with probability arbitrarily close to 1. To give a rigorous meaning to these remarks, we introduce the following

Definition 10.14. A sequence of random variables $\xi_1, \ldots, \xi_n, \ldots$ *approaches in probability* the random variable ξ as $n \to \infty$ (i.e., $\lim_n \mathrm{pr}\ \xi_n = \xi$) if for any $\varepsilon_1 > 0$ and $\varepsilon_2 > 0$, there exists a number N such that for all $n > N$,

$$P(|\xi_n - \xi| < \varepsilon_1) > 1 - \varepsilon_2 .$$

We have proved that $\lim_{n\to\infty}\text{pr}\ \zeta_n = p$, where p is a constant random variable. This is the simplest form of what is known as the *law of large numbers.*

By this law, probabilities may be assigned to the outcomes of an experiment that may be repeated independently an infinite number of times in accordance with the frequencies of the outcomes. Our assertion exhibits a typical feature found in certain theorems of probability theory. In these theorems an assertion is made with probability close to one which may also be stated in nonprobabilistic terms. Applications of probability theory to real-life problems are based on a series of results of this type.

7. *Bernshtein [Bernstein] Polynomials.* We now use the results already derived for Bernoulli trials to prove Weierstrass' theorem for uniform approximation of functions by polynomials (proof due to S. N. Bernshtein).

Definition 10.15. Suppose that $\omega(p)$ is a function continuous on the segment [0,1]. The polynomial

$$B_n(\omega;p) = \sum_{m=0}^{n} \omega\left(\frac{m}{n}\right) C_n^m\, p^m(1 - p)^{n-m}$$

is called the n-th *Bernshtein polynomial* of the function $\omega(p)$. It is a polynomial of degree n (in p).

10.26. Prove that the sequence $B_n(\omega;p)$ converges uniformly to $\omega(p)$ as $n \to \infty$.

We have already proved that every continuous function $\omega(p)$ on the segment [0,1] may be approximated uniformly by polynomials with any degree of accuracy, and have also given an actual method of constructing the approximating sequence of polynomials $(B_n(\omega;p))$. Let us now add one more stipulation to this assertion. Suppose that the function $\omega(p)$ takes integral values at the end-points of the segment [0,1]. For this case, we will prove that $\omega(p)$ may be approximated uniformly on [0,1] by polynomials with integral coefficients.

From $\omega(p)$, we construct a sequence of polynomials

$$C_n(\omega;p) = \sum_{m=0}^{n} \left[\omega\left(\frac{m}{n}\right) C_n^m\right] p^m(1-p)^{n-m},$$

where $[x]$ is the integral part of the number x. In our case, $[\omega(0)C_n^0] = \omega(0)$ and $[\omega(1)C_n^n] = \omega(1)$, since $\omega(0)$ and $\omega(1)$ are integers.

10.27. Prove that the sequence of polynomials $C_n(\omega;p)$ converges uniformly to $\omega(p)$.

Suppose that the function $\omega(p)$ satisfies the condition $0 \leq \omega(p) \leq 1$; $\omega(0)$ and $\omega(1)$ are equal to 0 or 1. Then $C_n(\omega;p)$ has the form

$$\sum_{m=0}^{n} a_m p^m(1-p)^{n-m}, \tag{10.4}$$

where a_m are integers; $0 \leq a_m \leq C_n^m$. As we have already seen (cf. (10.1)), a polynomial of this type may be interpreted as the probability $h_f(p)$ of the event $f(a_1,a_2, \ldots, a_n)$ (f is a logical function) in the Bernoulli algebra $(\mathfrak{U}_n,p)$. (The probabilities of a_i are equal to p.) Recall that a_m is the number of sequences containing m units on which f is equal to 1. We will call such a polynomial the *Bernshtein polynomial corresponding to the logical function f.** (The function f is not uniquely defined.) Thus, we have proved that every function ω, $0 \leq \omega(p) \leq 1$, defined on the segment [0,1] and equal to 0 or 1 at the endpoints of the segment may be approximated uniformly by Bernshtein polynomials that correspond to the logical functions.

10.28. Prove that every function $\omega(p)$ defined on [0,1] which satisfies the above conditions as well as the relation

$$\omega(1-p) = 1-\omega(p), \quad (0 \leq p \leq 1), \tag{10.5}$$

can be approximated uniformly by Bernshtein polynomials corresponding to the self-dual logical functions.

---

*Bernshtein polynomials corresponding to the logical functions (see, for example, the result of Problem 10.28) have been discussed by P. S. Novikov (unpublished lectures).

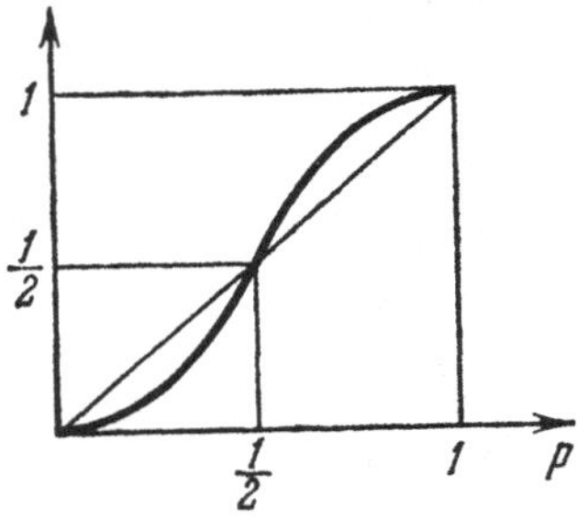

Figure 81

Geometrically, (10.5) asserts that the graph of $\omega(p)$ is symmetric relative to the point (1/2, 1/2) (Figure 81).

8. *Bernshtein Polynomials Associated With the Logical Functions.* Let us now discuss Bernshtein polynomials associated with the logical functions and learn how to approximate continuous functions by means of Bernshtein polynomials. We will start by considering the realization of the logical functions by means of circuits composed of functional elements and contact networks.

Definition 10.16. A *Bernoulli channel with parameter p* is to a device at whose output the signal 1 appears with probability p, and the signal 0 with probability $1-p$; signals at distinct moments of time are independent.

Let us consider some network S consisting of zero-cycle functional elements that realize the function $f(x_1, \dots, x_n)$. If there are independent Bernoulli channels with the same parameters p at the inputs of the network, it may be assumed that units occur independently at the inputs of the network with probability p. Then a unit occurs at the network output with probability $h_j(p)$, where $h_j(p)$ is the Bernshtein polynomial corresponding to f (cf. (10.1)). Thus, a network consisting of zero-cycle functional elements is a device that transforms Bernoulli channels (starting with channels with parameters p, we may obtain a channel with parameter $h_f(p)$). The polynomial h_f is called the *law of transformation* of Bernoulli channels into the network S. From what we have proved, it follows that, given an arbitrary function ω continuous on the segment [0,1] that takes the values 0 and 1 at the endpoints of the segment, networks may be constructed consisting of functional elements that transform Bernoulli channels according to the law $h_f(p)$, which differs as little as desired (uniformly in p) from $\omega(p)$. Analogously, we may consider proper multi-cycle networks of functional elements.

There is one drawback to the use of proper networks of functional elements for transforming Bernoulli channels; it is then necessary to have several independent channels with parameters p so as to obtain a channel with parameter $h_j(p)$.* A somewhat different starting point using automata without feedback is more convenient. Suppose that we are given such an automaton realized, for example, by a (non-regular) network of one-cycle functional elements with a single input (Figure 82). Suppose that the output state of the automaton $\mathfrak{U}_\nu$ at time t depends on the input state at preceding moments of time $s^1 = s(t-1), \dots, s^\nu = s(t-\nu)$ and that the output signal is equal to $f(s^1, \dots, s^\nu)$ (i.e., the delay in $\mathfrak{U}_\nu$ is equal to 1, and the index is ν). In Chapter 8 (Problem 8.15), it was proved that any automaton without feedback can be represented by a network (which is not, in general, proper) of one-cycle functional elements. If the input of the network (automaton) $\mathfrak{U}_\nu$ is connected to a Bernoulli channel, at every moment of time there is a **1** at its output with probability $h_f(p)$ and a **0** with probability $1 - h_f(p)$. If there are signals at the output of $\mathfrak{U}_\nu$ every ν cycles, they are independent. Thus, it may be assumed that $\mathfrak{U}_\nu$ transforms one Bernoulli channel with parameter p into another Bernoulli channel with parameter $h_f(p)$ (because of the variation in the scale of time, the new unit of time is ν times the old scale). Once again, any law of transformation specified by a continuous function $\omega(p)$ ($\omega(0)$ and $\omega(1)$ are equal to 0 or 1) may be approximated to any

Figure 82

*It may be assumed that channels associated with distinct network inputs are characterized by distinct parameters.

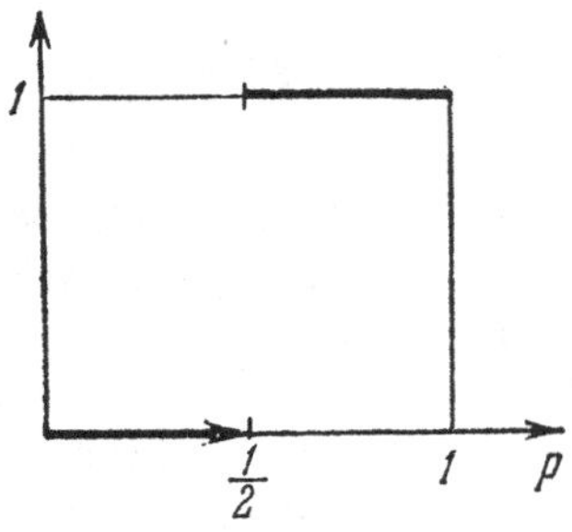

Figure 83

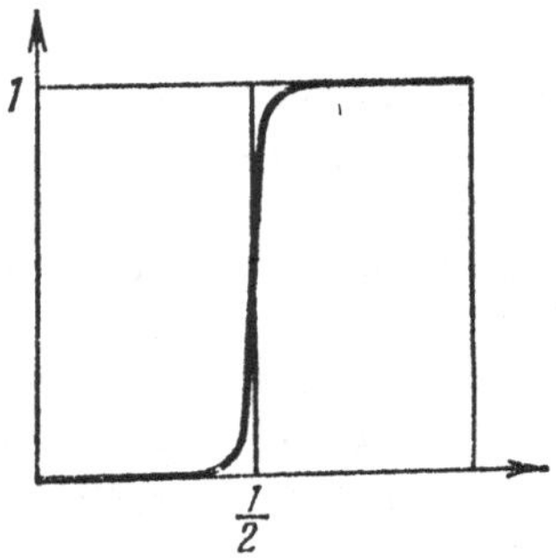

Figure 84

degree of accuracy by means of an appropriately chosen automaton.

In particular, let us consider the function $\chi(p)$ equal to 0 if $p < 1/2$ and to 1 if $p > 1/2$ (Figure 83), and approximate it by a continuous function $\omega(p)$ (Figure 84), i.e., except for a small neighborhood of 1/2, to the left of the point 1/2, ω is everywhere close to 0, and to the right of 1/2, close to 1. If we consider an automaton that transforms Bernoulli channels according to the law $h_f(p)$, where $h_f(p)$ approximates $\chi(p)$ sufficiently closely,* then $h_f(p)$ will possess the properties of $\omega(p)$ cited above, and we have found an automaton that operates in the following way. If **0** (correspondingly, **1**) is fed to the automaton's input with probability $p < 1/2$ sufficiently greater (or lesser) than 1/2, **0** (correspondingly, **1**) appears at the output with probability that differs insignificantly from 1. In other words, we are dealing with a special type of amplifier: if a signal is fed to the input with error probability p sufficiently less than 1/2, at the output the error probability differs insignificantly from 0. Such a decrease in the error probability is due, roughly speaking, to the fact that the output signal depends not on a single input signal, but rather on some number ν of preceding signals (it is said that "the statistics on ν signals have been collected"), and as the range of deviation of the probability from zero decreases, more signals are required. It is clear that once a suitable automaton $\mathfrak{U}_\nu$ is chosen, an interval can be selected

*It can be easily proved that, for sufficiently large k, f can be a function of 2k+1 arguments equal to 1 if most of its arguments are equal to 1.

around 1/2 and the deviation of the error probability from 0 can be made arbitrarily small. Similar considerations are used below in other problems.

Now let us consider a problem from automaton theory already referred to in Chapter 8; the problem (behavior of an automaton in a random environment) has been discussed by M. L. Tseitlin [6]. We are presented with the following situation. If **0** appears at the output of an automaton, then with probability p, **0** appears at its input at the next moment of time, and **1** arrives with probability $1-p$ (the environment the automaton finds itself in "penalizes" it with probability p, and "rewards" it with probability $1-p$). If the output signal is **1**, the opposite pattern appears: **1** appears at the input with probability p, and **0** with probability $1-p$. The number p is fixed, but not known in advance. It is necessary to construct an automaton that would be penalized with a possibly lesser probability. It is clear that with probability close to **1**, the automaton produces **0** if $p < 1/2$, and produces **1** is $p > 1/2$.

Let us begin by considering the simplest one-cycle automaton (Figure 85), in which the element ω realizes the function $\omega(x,y) = x+y = x\bar{y} \vee \bar{x}y$. We place the automaton in a random environment and feed a signal to the input y that specifies a penalty or reward. Suppose that P is the probability that **0** appears at the output of the automaton. It is clear that

$$P(1 - p) + (1 - P)p = P,$$

since **0** appears at the output if it appeared at the output on the preceding cycle and if the automaton was *not* penalized (here, the conditional probability is equal to $1-p$) or if the automaton produced a **1** on the preceding cycle and *was* penalized. In other words, the strategy of the automaton is to alter its behavior when penalized and not alter it when rewarded. From the above equality, it follows that

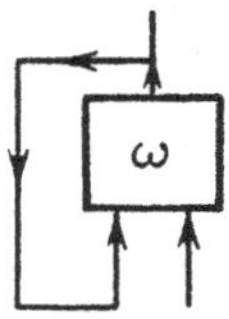

Figure 85

Figure 86

$$P = 1 - p,$$

i.e., the automaton's strategy is "efficient", more precisely, it is more likely to produce the signal for which it is less likely to be penalized (that is to say, there is a greater probability for it to produce the signal which there is a lesser probability it would be penalized for); specifically, if $p < 1/2$, the signal is **0**, and if $p > 1/2$ it is **1**.

However, we would like to have an automaton that produces, with probability close to one, the signal which it is less likely to be penalized for. For this purpose, let us consider the above automaton without feedback $\mathfrak{U}_\nu$ which transforms a Bernoulli channel by the law $h_f(p)$, where f is a function of ν variables. The graph of $h_f(p)$ has the form shown in Figure 84; it approximates $\chi(p)$ (Figure 83) outside a small neighborhood of the point 1/2. Let us consider another automaton without feedback $\mathfrak{B}_\nu$ (Figure 86) in which the output signal at time t has the form

$$\overline{s_1(t-1)} \vee \overline{s_1(t-2)} \vee \ldots \vee \overline{s_1(t-\nu-3)} \vee s_2(t-1),$$

where $s_1(t)$ and $s_2(t)$ are the signals to the inputs s_1 and s_2 at the corresponding moments of time. Thus, there is a zero signal at the output of $\mathfrak{B}_\nu$ only if there was a unit signal at the first input at the preceding $\nu + 3$ moments of time and a zero signal to the second input the preceding moment. Consider the automaton (now with feedback) in Figure 87. Here the one-cycle element ω realizes the function $x+y$; a penalty or reward signal is fed to the input of $\mathfrak{U}_\nu$. From a comparison with the automaton depicted in Figure 85, it is clear that the output signal of the automaton (element ω) changes if there was a zero signal at the output of $\mathfrak{U}_\nu$ at the preceding moment of time (suppose this moment is t), i.e., if there was a unit signal at the output of $\mathfrak{U}_\nu$ from time $(t-\nu-3)$ to time $(t-1)$ (that is, the signal at the output does not change from the $(t-\nu-2)$-th to the t-th moment) and if there was a zero signal at the output of $\mathfrak{U}_\nu$ at time $t-1$. The signal at the

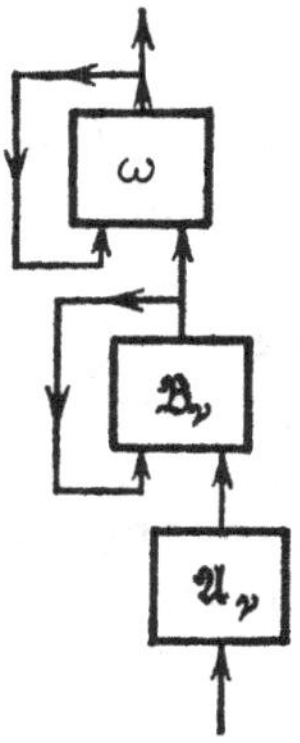

Figure 87

output of $\mathfrak{U}_\nu$ corresponds to the penalty-reward signal obtained in response to the automaton's output signals from time $(t-\nu-2)$ to time $(t-2)$. But as we have already remarked, at these times the automaton's output signal must be the same as at time t, i.e., under these conditions the probability that $\mathfrak{U}_\nu$ has a zero output signal is equal to $h_f(p)$ or $h_f(1-p)$, depending on whether the automaton's output signal at time t is equal to **0** or **1**. As a result, if $h_f(p)$ approximates $\chi(p)$ sufficiently closely, then with probability close to one the automaton's output signal changes if it is penalized with probability greater than 1/2, and does not change if this probability is less than 1/2. For a sufficiently lengthy period of time, $\mathfrak{B}_\nu$ does not change its output signal. During this period of time, $\mathfrak{U}_\nu$ determines how often this signal acts as a penalty. $\mathfrak{U}_\nu$ may be chosen so that this question can be answered sufficiently reliably (here ν is large). We will limit ourselves to these theoretical considerations without bothering with the detailed computations.

We have constructed an example of a sequence of automata that behaves asymptotically optimally in a random environment. Other designs of such sequences of automata are known [6]. Usually, the discussion is not limited to the case of a symmetric environment, as was done above (the probabilities of the penalties for the output signals **0** and **1** sum to one).

9. *Unreliable Relays.* Let us consider a number of similar problems related to contact networks.

Definition 10.17. A *probabilistic relay with parameters* (a,b) is a relay whose contact is closed with probability a whenever there is a current through the coil and whose contact is closed with probability b whenever there is no current.

In the case of the parameters (1,0) we have a make-contact relay, and in the case (0,1) a break-contact relay. Otherwise, it may be assumed that we are dealing with an unreliable relay which (with some probability) is inoperable in every state. By constructing networks from such contacts, we can compute the probabilities that a network is closed for particular sequences of states of the coils of the circuit's relays -- under the assumption that all the relays comprising the network operate independently. We will limit our discussion to the case in which all the contacts are associated with the same coil and the parameters (a,b) are characteristics of the contacts only, i.e., it is assumed that errors occur in the contacts independently, and are not due to the coil. To represent polynomials of a single variable, we assume that we are given only two types of contacts, those with parameters $(1-p,p)$, $p < 1/2$, the "make contacts," and those with parameters $(p, 1-p)$, the "break contacts" (p is the same for all the contacts). Then a contact network formed from such contacts* that realizes some self-dual logical function $f(x_1, \ldots, x_n)$ may be considered a compound relay with parameters $(h_f(1-p) = 1-h_f(p), h_f(p))$, where $h_f(p)$ is the Bernshtein polynomial associated with the function f; identity holds for h_f by virtue of the self-duality of f.

By analogy with the method of transforming Bernoulli channels, we may speak of transforming probabilistic relays (contacts). Here, once again, any continuous law of transformation may be approximated uniformly by means of (self-dual) contact networks. But if the law of transformation is given by the discontinuous function shown in Figure 83, it can be approximated uniformly everywhere outside any interval containing the point 1/2 (Figure 84). If p is interpreted here as the probability that the contact is inoperable, the error probability of the compound contact $h_f(p) < p$, and for an appropriate choice of f this probability may be made arbitrarily small. Thus, arbitrarily reliable contacts may be obtained from unreliable contacts.

---

*We formally assign to all the contacts the symbols of the different variables $x_1, \ldots, x_n$ or their negations $\overline{x}_1, \ldots, \overline{x}_n$, though all the contacts are associated with the same coil.

Until now we have been forced to use both make and break (insufficiently reliable) contacts in our method of constructing reliable make contacts, for example. We might naturally wonder whether arbitrarily reliable make contacts can be obtained from unreliable make contacts alone. Since networks of make contacts are monotone functions, this question is related in a natural way to the study of Bernshtein polynomials associated with monotone functions.

10.29. Prove that for $0 < p < 1$, the inequality

$$h_f'(p) \geqslant \frac{h_f(p)(1 - h_f(p))}{p(1 - p)}, \tag{10.6}$$

holds for the Bernshtein polynomial $h_f(p)$ associated with the monotone logical function $f(x_1, \dots, x_n)$, and that equality is possible only if $h_f(p) \equiv 0$, $h_f(p) \equiv 1$, or $h_f(p) = p$.

From this result (due to Moore and Shannon [7]), it follows that, in particular, $h_f(p)$, which is a function of a real variable, is monotone if f is a monotone logical function, since $h_f \geqslant 0$, i.e., h_f is non-decreasing. This result may be easily established directly.

10.30. Prove that if $h_f(p) = 0$, 1, or p, then $f = 0$, $f = 1$, or $f(x_1, \dots, x_n) = x_i$, respectively.

10.31. The graph of the Bernshtein polynomial $h_f(p)$ corresponding to a monotone logical function crosses the line $h(p) = p$ $(0 < p < 1)$ at one point at most.

Problem 10.30 gives us extremely important information about $h_f(p)$. If $f \neq$ const, $h_f(0) = 0$ and $h_f(1) = 1$. In addition, since we are interested in self-dual functions, let us suppose that $f \neq x_i$. Since for every self-dual function f, the graph of $h_f(p)$ crosses the line $h(p) = p$ at the point $(1/2, 1/2)$, in the case of a monotone self-dual function $f \neq x_i$ this point of intersection is unique, i.e., the graph of $h_f(p)$ in this case has the form shown in Figure 88. It is essential that when $p < 1/2$, the graph of $h_f(p)$ lies below the line $h(p) = p$; this follows from the fact that by (10.6), $h_f'(1/2) > 1$ (in our case, strict inequality holds).

Let us now return to the question we began with. It is clear that if we create a network corresponding to f from

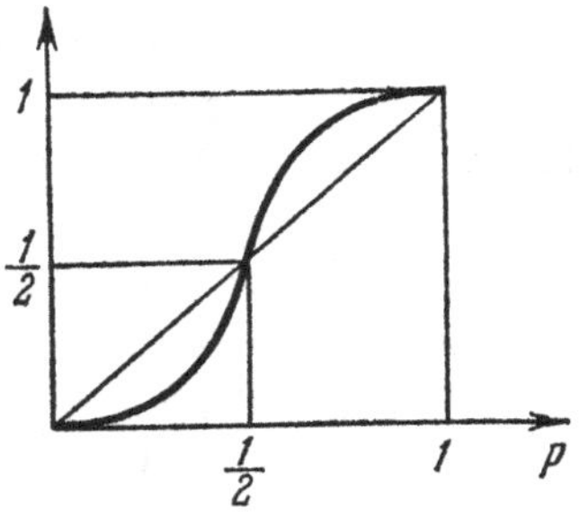

Figure 88

make contacts with parameters $(1-h_f(p), h_f(p))$, we will end up with a compound contact with parameters $(1-h_f(p), h_f(p))$. Since $h_f(p) < 0$ if $0 < p < 1/2$, we have obtained a more reliable contact. To prove that it is possible to construct an arbitrarily reliable contact, it is sufficient to show that the function χ of Figure 83 may be approximated as closely as desired outside an interval around the point 1/2 by functions $h_f(p)$ corresponding to monotone self-dual functions f. By making a compound contact with parameters $(1-h_f(p), h_f(p))$ $(0 < p < 1/2)$ the elementary contact, we could instead form the network for f from compound contacts. As a result, we obtain a contact whose error probability is equal to $h_f(h_f(p))$. Continuing on with this process, we obtain a sequence of contacts with error probabilities

$$p_k = h_f(h_f(...(h_f(h_f(p)))...)).$$

The procedure for successively obtaining p is an ordinary iterational process (Figure 89). Clearly, $p_{k+1} < p_k$, i.e., we are dealing with a monotonically decreasing sequence of positive

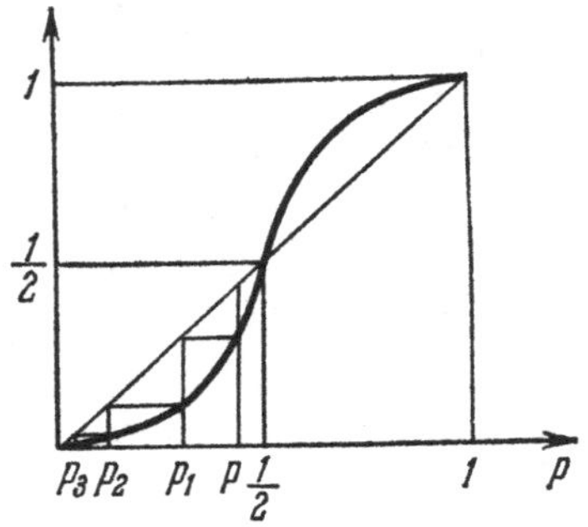

Figure 89

numbers. Its limit must satisfy the equality $h_f(a) = a$, hence $a = 0$. Thus, for every $0 < p < 1/2$, an arbitrarily reliable compound contact can be constructed. We have used here only the fact that $h_f(p) < p$ if $0 < p < 1/2$ (and not the monotonicity of $h_f(p)$). Since h_f is monotone, p may everywhere be considered the upper bound of the contact's error probability, and not its exact probability.

Let us make one more remark regarding inequality (10.6). This inequality is a necessary condition for the existence of a monotone function among the logical functions corresponding to the Bernshtein polynomials. It can be easily proved that the condition is not necessary. It has been shown [8] that there exists a function of two variables $F(x,y)$ $(0 \leqslant x, y \leqslant 1)$ (which may be constructed explicitly) such that a monotone logical function may be associated with $h_f(p)$ if and only if

$$h_f'(p) \geqslant F(p, h_f(p)) \qquad (0 < p < 1).$$

In this case $F(x,y) \geqslant y(1-y)/x(1-x)$, which implies the Moore-Shannon inequality (10.6).

10. *Completeness of Unreliable Functional Elements.* Let us consider the reliability of networks formed from (zero-cycle) functional elements [9]. We assume that the functional elements occurring in the network may operate incorrectly (independently of each other) with some probability. More precisely, we are given a sequence* of logical functions $\Phi = \{\omega_1, \dots, \omega_n\}$, each of which is supplied with an unlimited number of functional elements** that realize it; further, the error probability of these elements does not exceed ε_i $(0 \leqslant \varepsilon_i < 1/2)$ for any sequence of input signals. Thus, we are not establishing the error probability of a functional element, but only finding an upper bound on this probability. This point of view is more natural, since in a real situation exact probabilities may be difficult to determine, and, moreover, may differ for different functional elements that realize the

*We are limiting our discussion to finite systems of functions Φ, though in the problems we are considering infinite systems are also of interest (cf. [9]).

**We will denote them by the same letters.

same logical function; the error probability of an element may differ for different sequences of input signals, and so on. Since the elements in a network may operate incorrectly independent of each other, once we know the value of ε_i the error probability of the network may be estimated. We are interested in learning when it is true that a logical function F may be realized arbitrarily reliably by means of a network consisting of elements ω_i, i.e., when does there exist (for every $\gamma > 0$) a network consisting of elements of Φ that realizes F with error probability less than γ. If an arbitrarily reliable realization is possible for any logical function, the system of functional elements Φ is said to be *h-complete.* For certain cases we will find h-completeness conditions expressed in terms of functionally closed classes. Let us first explain why h-completeness is a stronger requirement than the ordinary completeness requirement.

10.32. Prove that if a system of functional elements Φ is h-complete, then the system of functions realized by them is complete (in the ordinary sense; cf. Chapter 6).

Note that an h-complete (finite!)* system must contain absolutely reliable elements. This is implied by the following assertion (von Neumann's lemma).

10.33. Prove that the error probability of a network cannot be less than the error probability ε of its output element if $\varepsilon < 1/2$.

If there were no absolutely reliable elements in a finite system of functional elements Φ, we could not construct networks that realize functions with error probability less than the smallest error probability of the elements of Φ.

We might naturally ask whether there are h-complete systems that do not contain any complete sub-systems of absolutely reliable elements. We will give here an important method of constructing such systems based on the properties of an element, called a "mixer"; this element realizes the function $m(x,y,z) = xy \vee yz \vee xz$.

*An infinite system of elements containing arbitrarily reliable and complete (in the ordinary sense) subsystems is, as is easily seen, h-complete.

10.34. Prove that every system of functional elements Φ that contains an absolutely reliable mixer such that its corresponding system of functions is complete is *h*-complete.

The use of a mixer to increase reliability is a quite natural strategy. A mixer may be thought of as "voting" for all the signals arriving at its inputs; from its output signal it is possible to decide which signals there were more of. It is clear that if such a "majority organ" operates absolutely reliably, we will obtain as a result a network that is more reliable than any of the initial networks.

10.35. Prove that the above assertion remains true if the mixer is replaced by an element that realizes an arbitrary self-dual monotone function that does not coincide with any variable whatsoever.

As we have seen in Chapter 7, problems involving functionally closed classes are simplified in the case of extended composition, i.e., if there are constants present. In the case of networks consisting of unreliable elements, the corresponding assertion presupposes the existence of absolutely reliable elements that realize the constants. Under this assumption we will first consider *h*-completeness, applying the results of Chapter 7 on functionally closed classes to extended composition.

10.36. Find necessary and sufficient conditions of *h*-completeness for a finite system of unreliable functional elements $\Phi = \{f_1, \dots, f_k\}$ containing the absolutely reliable constants 0 and 1.

Note one important consequence of the above results. It turns out that a system of absolutely reliable functional elements containing elements that realize both constants and the functions $\omega_1 \notin L$, $\omega_2 \notin D^{01}$, and $\omega_3 \notin K^{01}$ is a *universally reliable system* in the sense that an *h*-complete system may be obtained by extending such a system to a complete system that satisfies the conditions of Post's theorem by means of arbitrary elements, the upper bounds of whose error probabilities ε_i are less than 1/2. In other words, the *h*-completeness conditions for a system of elements containing reliable constants (more precisely, conditions imposed on the absolutely reliable part of the system) are independent of the range of error probabilities ε_i of the unreliable elements only if $\varepsilon_i < 1/2$. If we do not assume the existence of reliable constants, the conditions imposed on the absolutely reliable part depend very greatly on ε_i. Below we present a rigorous statement of this result, but now we wish to discuss a simpler problem dealing with the conditions imposed on a universal reliable system of elements, i.e., a system of elements R whose extension to a complete system by means of arbitrary elements with range of error probabilities less than 1/2 yields an h-complete system R. Until now we have discussed necessary and sufficient conditions for the universal reliability of a system of elements containing constants. In addition, from Problem 10.34 it follows that a mixer $m(x,y,z)$ and, in general, any self-dual monotone function that does not coincide with an argument, forms a universal reliable system. Let us first formulate a sufficient condition.

A system of absolutely reliable functional elements is *universally reliable* if the functions $\omega_1 \notin \mathbf{L}$, $\omega_2 \notin \mathbf{D}^{01}$, $\omega_3 \notin \mathbf{K}^{01}$, $\omega_4 \notin \mathbf{F}^{(2)}$, and $\omega_5 \notin \mathbf{G}^{(2)}$ are among the functions it realizes. We will require certain additional facts supplementing the results derived in Chapter 7 on functionally closed classes. (Of course, the table of classes (Figure 6) could be used, but if possible we would like to avoid using results that have not been proved here.) More precisely, we wish to find the $\mathbf{S}_{01}$-pre-complete functionally closed classes, where $\mathbf{S}_{01}$ is the class of self-dual functions that preserve the additive identity (and thereby automatically preserve the multiplicative identity).

10.37. Prove that the following systems of functions are bases in $\mathbf{S}_{01}$: (a) $xy+yz+xz+y+z$; (b) $\{xy+yz+xz,\ x+y+z\}$.

10.38. Find the $\mathbf{S}_{01}$-pre-complete functionally closed classes.

10.39. Prove that the class **MS** of self-dual monotone functions is in $\mathbf{F}^{(2)}$, and also in $\mathbf{G}^{(2)}$.

10.40. Prove that the universal reliability conditions formulated above for a system of elements are sufficient.

Obviously, these conditions are not necessary, since, for example, the system of functions consisting of a single mixer $m(x,y,z)$ does not satisfy them (by Problem 10.39).

Let us now formulate necessary and sufficient conditions for the universal reliability of a system of functional elements R. These conditions may be obtained if $\mathbf{F}^{(2)}$ and $\mathbf{G}^{(2)}$ are replaced in the sufficient condition given above by $\mathbf{F}^{(3)}$ and $\mathbf{G}^{(3)}$, respectively. Thus, our system R of absolutely reliable functional elements is universally reliable if and only if the functions $\omega_1 \notin \mathbf{L}$, $\omega_2 \notin \mathbf{D}^{01}$, $\omega_3 \notin \mathbf{K}^{01}$, $\omega_4 \notin \mathbf{G}^{(3)}$, and $\omega_5 \notin \mathbf{F}^{(3)}$ are among the functions it realizes.

10.41. Using the table of functionally closed Post classes, prove that the above condition is sufficient.

The proof that this condition is necessary is too complicated to be presented here.

Let us pass on to our general result on h-completeness conditions of a system of elements [9]. Suppose that we are given some system Φ of functional elements consisting of a system $R = \{f_1, \ldots, f_k, \ldots\}$ of absolutely reliable elements and a system $T = \{g_1, \ldots, g_m, \ldots\}$ of unreliable elements; ε_i is the upper bound of the error probability of the element g_i, $0 < \varepsilon_i < 1/2$.

Suppose that

$$\kappa(T) = \inf_{T} \varepsilon_i .$$

We let $\kappa(T) > 0$. In the case of finite systems T, the least upper bound coincides with the smallest of all the ε_i; in this case the above condition is always satisfied.

Suppose, further, that $T(\nu)$ is a set of elements $g_i \in T$ such that $\varepsilon_i < \nu$. For every system of elements $T' \subset T$, we set

$$\kappa(T') = \inf_{g_i \in T'} \varepsilon_i .$$

Now we can formulate the h-completeness criterion [9].

A system of elements $\Phi = R \cup T$ ($\kappa(T) > 0$) is h-complete if and only if:

(1) Φ is complete in the ordinary sense;
(2) R contains:
  (a) a nonlinear function;
  (b) a function that does not belong to $\mathbf{D}^{01}$;
  (c) a function that does not belong to $\mathbf{K}^{01}$;

(3) There exists a system of integers $\mu_1 > \mu_2 > \dots > \mu_t > 1$ such that if $T_i = T(1/\mu_i)$, $R_1 = R$, $R_{i+1} = R \cup T_i$; $\kappa_0 = \kappa(T)$, $\kappa_i = \kappa(T \setminus T_i)$, then

(a) $\kappa_{i-1} < 1/\mu_i$,

(b) R_i contains a function that does not belong to the class $F^{(\mu_i+1)}$;

(c) R_i contains a function that does not belong to the class $G^{(\mu_i+1)}$;

(d) The system $R_{t+1} = R \cup T_t$ is complete in the ordinary sense.

To prove sufficiency, it is convenient to use the function*

$$S_n^{\ell}(x_1, \dots, x_n) = \begin{cases} 0 & \text{if} \quad x_1 + \dots + x_n < \ell \\ 1 & \text{if} \quad x_1 + \dots + x_n \geq \ell \end{cases}$$

i.e., S_n^{ℓ} is equal to zero on sequences containing fewer than ℓ units. These functions generalize the concept of a mixer; in fact, $m(x,y,z) = S_3^2(x,y,z)$. It is clear that $S_{\mu k+1}^{k+1}$ belongs to $F^{(\mu)}$ for any k (any μ sequences of length $\mu k+1$ containing no more than k units share a common additive identity). The following result will play a central role in what follows:

10.42. Prove that

$$\lim_{k \to \infty} h_{f_k}(p) = \begin{cases} 1 & \text{if} \quad p > 1/\mu \\ 0 & \text{if} \quad p < 1/\mu \end{cases}$$

where $f_k = S_{\mu k+1}^{k+1}$ and h_{f_k} is the Bernshtein polynomial associated with f_k (cf. (10.1)).

10.43. Prove that the above h-completeness condition is sufficient.

---

*Here we are using ordinary addition (not addition modulo 2).

Hints

10.1. (a) $x \vee \bar{x} = 1$, $x\bar{x} = 0$;
(b) Use (a) and Axiom 1;
(c) Use Axiom 2 and formula (a);
(d) Prove that if $x \to y = 1$, then $y = x \vee \bar{x}y$; in this case, the events x and $\bar{x}y$ are disjoint;
(e) $x \vee y = x \vee \bar{x}y$; $\bar{x}y \to y = 1$;
(f) Compare (e) and the formula $\bar{x}y \vee xy = y$; note that $\bar{x}y$ and xy are disjoint;
(g), (h) Prove by induction.
(i) $x = xy \vee x\bar{y}$;
(j) Compare (f) and (b).

In proving relations for the form $P(x) \leqslant P(y)$, it is best to represent y in the form of a disjunction of x and some other event z, where x and z are disjoint, after which the additivity and positivity of the probability (Axioms 1 and 3) may be applied.

10.2. Axiom 1 is self-evident; Axiom 2 follows from the fact that $a \cdot 1 = a$; Axiom 3 follows from the fact that if $xy = 0$, then $ax \cdot ay = 0$.

10.3. Use the formula

$$P(xy) = P(x)P_x(y),$$

which follows directly from the definition of conditional probability.

10.4. (a) $x = a_1x \vee a_2x \vee \dots \vee a_nx$; the events a_ix are disjoint.
(b) Use the definition of conditional probability and part (a).

10.5. $x = x\bar{y} \vee xy$.

10.6. It is sufficient to verify that if the probability is equal to 0 or 1 on some elements, it is uniquely defined and takes one of these values on all the elements that may be obtained from the initial elements by means of the basic operations of Boolean algebra; further, this value agrees with the truth table. From the latter remark, it follows that the definition of probability by the method indicated in the problem is consistent for any $p_i = 0$ or 1.

10.7. Consider the algebra $\mathfrak{U}(a_1, a_2, a_3)$ generated by the events a_1, a_2, and a_3 whose probabilities p_i are equal to 1/2. The numbers $p_{\sigma_1\sigma_2\sigma_3}$ are specified in the following way:

$$p_{111} = p_{100} = p_{010} = p_{001} = \frac{1}{4},$$

$$p_{000} = p_{011} = p_{101} = p_{110} = 0.$$

Prove that a_1, a_2, and a_3 are dependent, though pairwise independent.

10.8. Based on the independence of $a_1, a_2, \ldots, a_n$, compute the probabilities $p_{\sigma_1 \ldots \sigma_n}$.

10.9. Represent the events in the form of a disjunction of elementary events.

10.10. Introduce a system of elementary events in the following way: pairs of numbers (i, j) $(1 \leqslant i, j \leqslant 6)$ which have appeared. *Answer:* $P(a) = 1/6$; $P(b) = 1/2$; $P(c) = 23/36$.

10.11. At first glance it might seem possible to prove that the opposite side can, with equal probability, be colored with the color we see and also with the opposite color. But, in fact, this is not so. Introduce a system of elementary events and compute the corresponding conditional probabilities.

10.12. The letters in the word "algebra" could be numbered (in which case the letter "a" is numbered twice) by the digits 1, 2, ... , 7, and ordered triplets of digits -- among which there are no repetitions of the numbers assigned to the selected letters -- could then be made the elementary events. We could also compute the conditional probabilities that the next letter is the one desired under the condition that the preceding letters were already chosen correctly. In this case, for every conditional probability it is best to select an appropriate system of elementary events.

10.16. $M\xi = \frac{7}{2}$.

10.17. $M\xi = 1$; $M\eta = 7$. It is best to use the formula for the mathematical expectation of a sum of random variables (especially for computing $M\eta$).

10.18. Introduce a system of elementary events. Use the formula for the mathematical expectation of a sum of random variables. *Answer:* $M\xi = 1$.

10.19. In the sum of $M(\xi^2)$, discard terms corresponding to events for which $|\xi|$ is greater than ε.

10.21. Use the formula

$$D(\xi+\eta) = D\xi + D\eta + 2M((\xi - M\xi)(\eta - M\eta)),$$

as well as the fact that subtracting arbitrary constants from independent random variables results again in independent random variables.

10.23. (1) If ξ is the random variable from Problem 10.16, then $D\xi = 2\frac{11}{12}$.

(2) For the quantities in Problem 10.17, $D\xi = 1/2$ and $D\eta = 5\frac{5}{6}$. Use the formula for the variance of a sum of independent random variables.

(3) The variables $\xi_1, \ldots, \xi_n$ into which the random variable ξ was partitioned in solving Problem 10.18 are not independent. However, in computing $D\xi$ it is nevertheless best to use the representation $\xi = \xi_1 + \ldots + \xi_n$. Compute $M(\xi^2)$. We have $D\xi = 2$.

10.24. Represent ξ_n in the form $\xi_n = \xi^{(1)} + \ldots + \xi^{(n)}$, where $\xi^{(n)} = 1$ if the i-th trial has a positive outcome, and $\xi^{(i)} = 0$ otherwise. *Answer:* $M\xi_n = np$; $D\xi_n = npq$.

10.25. Find the probability that there are m positive outcomes in n trials.

10.26. Consider the inequality

$$|\omega(p) - B(\omega;p)| \leqslant \sum_{m=0}^{n} \left|\omega(p) - \omega\left[\frac{m}{n}\right]\right| C_n^m\, p^m(1-p)^{n-m}.$$

Divide the last sum into two parts:

$$\sum_{|\frac{m}{n}-p|<\delta} \quad \text{and} \quad \sum_{|\frac{m}{n}-p|\geqslant\delta} .$$

Select $\delta > 0$ so that the first sum is less than ε. (This is possible by virtue of the uniform continuity of $\omega(p)$.) Then select N so that the second sum is less than ε if $n > N$ for the value of δ chosen (use Problem 10.25).

10.27. By Problem 10.27, it is sufficient to prove that the difference, $|B_n(\omega;p) - C_n(\omega;p)|$ tends to zero uniformly. This difference may be majorized in modulus by the sum

$$S_n(p) = \sum_{m=1}^{n-1} p^m(1-p)^{n-m}.$$

Terms with $m = 0$ and $m = n$ do not occur, since $f(0)$ and $f(1)$ are integers. It is sufficient to prove that this sum (note that it is a geometric progression) tends to zero uniformly on [0,1] as $n \to \infty$. This may be verified by finding the maximum of the sum $S_n(p)$ and computing the limit of the sequence of maxima. But it is simpler to prove that $S_n(p)$ is a monotonically decreasing sequence of continuous functions that approaches the additive identity (hence the sequence converges uniformly to zero).

10.28. Prove that the polynomial (10.4) corresponds to a self-dual logical function if and only if

$$a_m = a_{n-m}.$$

To construct the approximating sequence, consider the polynomials

$$D_n(\omega;p) = \sum a_m^{(n)} p^m(1-p)^{n-m},$$

where

$$a_m^{(n)} = \begin{cases} [\omega(\frac{m}{n})\, C_n^m] & \text{if} \quad m < n/2 \\ [\omega(\frac{m}{n})\, C_n^m] + 1 & \text{if} \quad m > n/2 \\ \frac{1}{2} & \text{if} \quad m = n/2. \end{cases}$$

10.29. Proof by induction on the number of variables. Assume that all the variables of f are actual and that $h_f(p)$ is neither a constant nor p. Let us expand f by its last variable:

$$f(x_1, \dots, x_{n-1}, x_n) = \omega(x_1, \dots, x_{n-1})x_n$$
$$\vee\ \psi(x_1, \dots, x_{n-1})\overline{x}_n.$$

Recall that

$$\omega(x_1, \dots, x_{n-1}) = f(x_1, \dots, x_{n-1}, 1),$$

$$\psi(x_1, \dots, x_{n-1}) = f(x_1, \dots, x_{n-1}, 0).$$

It is then clear that

$$h_f = h_\omega p + h_\psi(1 - p).$$

Since ω and ψ are monotone functions, then by the induction hypothesis we may assume that the inequality is true for h_ω and h_ψ, and that it is necessary to prove it for h_f. Rewriting the inequality for h_f, using the inequalities for h_ω and h_ψ, and dividing them by the common factor, we arrive at an inequality whose validity may be easily established directly.

10.30. Use the inductive procedure applied in solving the preceding problem.

10.31. At its point of intersection with the line $h(p) = p$, the derivative $h'_f(p) > 1$. The inequalities $h'_f(p_1) > 1$ and $h'_f(p_2) > 1$ cannot hold simultaneously at two neighboring points of intersection.

10.32. Recall that ε_i are the upper bounds of the error probabilities of the elements; therefore, in particular, there must exist an arbitrarily reliable realization for every function also whenever the network elements operate correctly.

10.33. Express the error probability of the network in terms of and the error probability δ of the network under the assumption that the output element operates correctly.

10.34. Suppose that we are given three copies of networks that realize some function with error probability ε. Connect their outputs to the inputs of the mixer m. Then we have obtained a network that realizes the same function, but more reliably.

10.35. Use Problems 10.29 and 10.31.

10.36. A finite system of functional elements $\Phi = \{f_1, \dots, f_k\}$ containing absolutely reliable constants is h-complete if and only if

(1) the system of functions realized by these elements is complete;

(2) among the absolutely reliable elements occurring in Φ may be found elements that realize;

(a) a nonlinear function ($\omega \notin \mathbf{L}$);

(b) the function $\omega_2 \notin \mathbf{D}^{01}$;

(c) the function $\omega_3 \notin \mathbf{K}^{01}$.

To prove sufficiency, show that the mixer $m(x,y,z)$ may be obtained from ω_1, ω_2, ω_3 and the constants, after which the results of Problem 10.34 may be applied.

The necessity of Condition 1 has already been proved (Problem 10.32). To prove that the remaining conditions are necessary, consider a function that cannot be realized by a network of absolutely reliable elements, and a sufficiently reliable network S that realizes it. In S, find the maximal sub-network P containing the output element and consisting of absolutely reliable elements, what may be called a *reliable output sub-network.* Prove that if P realizes a linear function, the entire network S cannot realize a nonlinear function sufficiently reliably. Bear in mind that the error probability of S cannot be less than the probability that there is an error at one of the inputs of P, but not at the other. Consider the other two cases analogously.

10.37. (a) The solution may be obtained by analogy with the solution of Problem 7.26, using the result of Problem 7.23.

(b) Use part (a).

10.38. The $\mathbf{S}_{01}$-pre-complete classes are $\mathbf{L}_{01}$, the class of linear functions that preserve the additive and multiplicative identities and **MS**, the class of monotone self-dual functions. The inclusion $\mathbf{MS} \subset \mathbf{S}_{01}$ follows from the fact that the functions in **MS** are monotone and not constants, and therefore (Problem 5.30 are in $\mathbf{P}_{01}$; the inclusion $\mathbf{L}_{01} \subset \mathbf{S}_{01}$ was basically found in the hint for Problem 6.14 (cf. also Problem 4.6). Use the solution of Problem 7.27.

10.40. It is best to consider the case in which one of the functions in Φ is not self-dual apart from the case in which all the functions are self-dual. In the first case, use the remark appended to the solution of Problem 7.15, and in the second case use Problems 10.38 and 10.39.

10.42. Use the limiting relation (10.42), which follows from the Chebyshev inequality.

10.43. By Problem 10.41, R is either universally reliable or is in $\mathbf{F}^{(3)}$ or $\mathbf{G}^{(3)}$ (but not both). Suppose that $R \subset \mathbf{F}^{(3)}$. Then by contrast with $\mathbf{F}^{(3)}$, R generates either $\mathbf{F}^{(\mu)}$, $\mathbf{F}_0^{(\mu)}$, or $\mathbf{MF}_0^{(\mu)}$; $\mu < \mu_1$ (by Condition 3b, since $R_1 = R$). In particular, all the functions in $\mathbf{MF}_0^{(\mu_1)}$ may be obtained. The class $\mathbf{MF}_0^{(\mu_1)}$ contains, in particular, the functions $S^{k+1}_{\mu_1 k+1}$.

In the dual case ($R \subset \mathbf{G}^{(3)}$), the functions $S^{\mu_1 k-k+2}_{\mu_1 k+1}$ (which are dual to $S^{k+1}_{\mu_1 k+1}$) must be considered.

Solutions

10.1. (a) $P(x \vee \bar{x}) = P(x) + P(\bar{x}) = 1$.
(b) $P(x) = 1 - P(\bar{x}) \leqslant 1$, since $P(\bar{x}) \geqslant 0$.
(c) $P(0) = P(\bar{1}) = 1 - P(1) = 0$.
(d) We have $x \to y = \bar{x} \vee y$. Thus $\bar{x} \vee y = 1$. Then

$$x = x(\bar{x} \vee y) = x\bar{x} \vee xy = xy;$$

$$y = y(x \vee \bar{x}) = xy \vee y\bar{x} = x \vee \bar{x}y.$$

(Truth tables could also be constructed.)

The events x and $\bar{x}y$ are disjoint, i.e., $P(y) = P(x) + P(\bar{x}y)$, that is (Axiom 1), $P(y) \geqslant P(x)$.

(e) $x \vee y = x \vee \bar{x}y$ (cf. (2.22)), i.e., $P(x \vee y) = P(x) + P(\bar{x}y)$ (x and $\bar{x}y$ are disjoint). Further, $\bar{x}y \to y = 1$, i.e., by (d), $P(y) \geqslant P(\bar{x}y)$; that is, $P(x) + P(y) \geqslant P(x \vee y)$.

(f) Let us find $P(\bar{x}y)$. We have $\bar{x}y \vee xy = y$, and $\bar{x}y$ and xy are disjoint; i.e., $P(\bar{x}y) + P(xy) = P(y)$. We obtain $P(\bar{x}y) = P(y) - P(xy)$ and $P(x \vee y) = P(x) + P(\bar{x}y) = P(x) + P(y) - P(xy)$.

(i) $P(x) = P(xy \vee x\bar{y}) = P(xy) + P(x\bar{y})$; $P(x) \geqslant P(xy)$.

(j) $P(xy) = P(x) + P(y) - P)(x \vee y)$, hence by (b), $P(xy) \geqslant P(x) + P(y) - 1$.

If $P(x \vee y) = 1$, in particular if $x \vee y = 1$, the case of equality ($P(xy) = P(x) + P(y) - 1$) holds.

10.3. $P_y(x) = \dfrac{P(x)P_x(y)}{P(y)}$.

10.4. (a)

$$P(x) = P(a_1x) + P(a_2x) + \dots + P(a_nx) = P(a_1)P_{a_1}(x) + P(a_2)P_{a_2}(x) + \dots + P(a_n)P_{a_n}(x).$$

(b)

$$P_x(a_i) = \frac{P(xa_i)}{P(x)} = \frac{P_{a_i}(x)P(a_i)}{\sum_{j=1}^{n} P(a_j)P_{a_j}(x)}.$$

We have used the formula for $P(x)$ derived in part (a).

10.5. $P(x\bar{y}) = P(x) - P(xy) = P(x) - P(x)P(y) = P(x)(1 - P(y)) = P(x)P(\bar{y})$.

10.6. The assertion for the negation stated in the hint is self-evident (formula (a), Problem 10.1). Further, it suffices to verify the conjunction. Suppose that $P(x)$ and $P(y)$ are equal to 0 or 1; let us compute $P(xy)$. If $P(x)$ or $P(y)$ is equal to 0, then by part (i) of Problem 10.1, $P(xy) \leqslant 0 \Rightarrow P(xy) = 0$. It remains for us to check the case $P(x) = P(y) = 1$; but then by part (j), $P(xy) \geqslant P(x) + P(y) - 1 = 1$, i.e., $P(xy) = 1$.

Thus $P(xy)$ is uniquely defined, and corresponds to the truth table for the conjunction. That is, $P(f(a_1, \dots, a_n)) = f(p_1, \dots, p_n)$, whence $p_i = 0$ or 1 arbitrarily.

Remark. From the foregoing, it follows that the events x and y in which P is equal to 0 or 1 are disjoint.

10.7. First note that

$$P(a_1a_2a_3) = p_{111} = \tfrac{1}{4} \neq P(a_1)P(a_2)P(a_3) = \tfrac{1}{8}\ .$$

The events a_1, a_2, and a_3 are pairwise independent, since

$$P(a_1a_2) = P(a_1a_3) = P(a_2a_3) = p_{111} = \tfrac{1}{4}\ .$$

10.8. We have

$$\begin{aligned} p_{\sigma_1\ldots\sigma_n} &= P(a_1^{\sigma_1}\ldots a_n^{\sigma_n}) \\ &= p_1^{\sigma_1}(1-p_1)^{1-\sigma_1} p_2^{\sigma_2}(1-p_2)^{1-\sigma_2} \\ &\qquad \ldots\ p_n^{\sigma_n}(1-p_n)^{1-\sigma_n}. \end{aligned}$$

Here we have started from the fact that

$$p_i^{\sigma_i}(1-p_i)^{1-\sigma_i} = \begin{cases} p_i & \text{if} \quad \sigma_i = 1 \\ 1-p_i & \text{if} \quad \sigma_i = 0. \end{cases}$$

We have also used Problem 10.5.

10.9. $a = e_2 \vee e_4 \vee e_6$, $P(a) = \frac{1}{2}$; $b = \bar{a}$, $P(b) = \frac{1}{2}$; $c = e_3 \vee e_6$, $P(c) = \frac{1}{3}$; $d = e_1 \vee e_2 \vee e_3 \vee e_4$, $P(d) = \frac{2}{3}$.

The events a and b are disjoint, and since they have non-zero probabilities, they are dependent.

The event ac asserts that a number divisible by 6 appeared, i.e., in this case $ac = e_6$ and $P(ac) = 1/6 = P(a)P(c)$. Thus, these events are independent.

By Problem 10.5, b and c are then also independent (which can be easily verified directly). The events a and d, thus b and d as well, are independent. The events c and d are dependent.

From the foregoing it follows that no three of these events are independent.

10.10. The elementary events will be those events e_{ij} in which the first die has the number i and the second die the number j, $1 \leq i, j \leq 6$. A total of 36 elementary events are obtained. If the dice are tossed independently, it is natural to let all combinations of the numbers (i,j)

be equally probable: $P(e_{ij}) = 1/36$. This obvious fact may be obtained formally if we recall that the probability that a particular number is on one of the dice is equal to 1/6, and if we then compute the probability e_{ij} as the conjunction of two independent events with probabilities 1/6. In each case, it now remains for us to count the number of pairs e_{ij} that possess this property.

10.7'. Suppose that with each number of a face that is a base we may associate one of the elementary events e_1, e_2, e_3 or e_4; $P(e_i) = 1/4$. We have $a_1 = e_1 \vee e_4$, $a_2 = e_2 \vee e_4$, $a_3 = e_3 \vee e_4$; $P(a_i) = \frac{1}{2}$.

Further, $a_1a_2a_3 = e_4$; $P(a_1a_2a_3) = \frac{1}{4} \neq \frac{1}{8}$; $a_1a_2 = a_1a_3 = a_2a_3 = e_4$, i.e., $P(a_1a_2) = P(a_1a_3) = P(a_2a_3) = P(e_4) = \frac{1}{4} = (\frac{1}{2})^2$.

It is easy to verify that this example is no different from the example given in Problem 10.7.

10.11. Let us number the sides of the disc with the digits from 1 to 6. Our assumption that the discs are first shuffled is equivalent to assuming that all possible outcomes are equally probable, i.e., with probability 1/6 the side of the disc with the number i will appear. These events $e_1, \ldots, e_6$ are considered the elementary events. Suppose that a is the event in which the side colored red is shown; in the events b and c, the opposite sides are colored red and blue, respectively. Then $P_a(b) = 2/3$ and $P_a(c) = 1/3$. Thus, it is more probable that the opposite side is colored with the same color. Based on this result, in guessing the color of the opposite side it is natural to think of the color we see. On the average, in two-thirds of the cases we must make a guess if we are assuming that all the elementary events e_i in fact occur the same number of times. Incidentally, until now we have had no grounds for claiming that a precise meaning could be assigned to the phrase, "on the average," on the basis of our previous results. In this case, purely combinatorial considerations favor the above hypothesis. We advise the reader to experiment.

10.12. *First Method.* There are 7·6·5 triplets of distinct digits from 1 to 7. All the sequences may be considered equally probable in a natural way. The word "beg" is equal to $1/7 \cdot 6 \cdot 5$, while the word "rag" corresponds to the two sequences (5,1,2) and (5,7,2), i.e., its probability is equal to $1/7 \cdot 3 \cdot 5$.

Second Method. Suppose that in event a the letter "b" is selected first, in event b the second letter is "e", and in event c the third letter is "g". We wish to find $P(abc) = P(ab)P_{ab}(c) = P(a)P_a(b)P_{ab}(c)$. It is clear that $P(a) = 1/7$. (An elementary event is one in which some definite letter is selected; with all the letters other than "a" we associate the probability "1/7", while with "a", the probability 2/7.)

Analogously, to compute $P_a(b)$ the events in which some definite letter other than "b" is selected are made the elementary events. In this case, $P_a(b) = 1/6$. Finally, if the letters "b" and "e" are eliminated, the probability of selecting "g" $P_{ab}(c) = 1/5$. Thus, $P(abc) = 1/7 \cdot 1/6 \cdot 1/5$.

Analogous reasoning holds when considering the word "rag," except that here $P_a(b) = 1/3$, since the letter "a" is selected on the second step. As a result, in this case $P(abc) = 1/7 \cdot 1/3 \cdot 1/5$.

The second method may prove more time-consuming, though we present it here so as to demonstrate by a very simple example a method of selecting a set of elementary events when computing conditional probabilities.

10.13. In the sum for $M\xi$, we select for every i all terms corresponding to j such that $\tilde{a}_j \to a_i = 1$. For these terms, $\xi(\tilde{a}_j) = \xi(a_i)$. Extracting the term $\xi(a_i)$, we obtain the sum $P(\tilde{a}_j)$, which is equal to $P(a_i)$, since the event a_i is equal to the disjunction of the $\tilde{a}_j$, whereas the $\tilde{a}_j$ are disjoint.

10.14. We compute (using Problem 10.13) $M\xi$ and $M\eta$ for the system of events $\{a_ib_j\}$. Adding, we obtain $M(\xi + \eta)$ for this system of events.

10.15. The mathematical expectation of a constant may be computed on the basis of an arbitrary complete system of disjoint events.

10.17. The elementary events correspond to the pairs (i,j) $(1 \leq i, j \leq 6)$.

1. The random variable ξ takes three values: 0 if i and j are odd; 2 if i and j are even; and 1 if one of them is even

and the other odd. It can be easily verified that the first possibility holds in nine elementary events, the second in 18 elementary events, and the third in nine elementary events. Therefore, ξ takes the value 0 with probability 1/4, the value 1 with probability 1/2, and the value 2 with probability 1/4. Hence, $M\xi = 1 \cdot 1/2 + 2 \cdot 1/4 = 1$.

We could have proceeded differently. Note that $\xi = \xi_1 + \xi_2$, where ξ_1 is a random variable equal to 1 if there is an even number on the first die, and 0 otherwise; ξ_2 is associated with the second die. Obviously, $M\xi = M\xi_1 + M\xi_2 = 1$.

2. The random variable η takes values from 2 to 12. For each value of η, it would be possible to determine the number of elementary events in which this value occurs. However, it is simpler to note that $\eta = \eta_1 + \eta_2$, where the random variable η_1 is equal to the number on the first die and η_2 is equal to the number on the second die. By Problem 10.16, $M\eta_1 = M\eta_2 = 1/2$, that is, $M\eta = M\eta_1 + M\eta_2 = 7$.

10.18. A system of elementary events E may be interpreted as a set of permutations $(\alpha_1, \dots, \alpha_n)$ of a sequence of numbers $(1, \dots, n)$, where α_i is the number of the counter in the i-th box. The probability of each elementary event is equal to $1/n!$. It is not a simple matter to find the number of elementary events in which ξ takes a fixed value in this case. Instead, as we have done previously, we represent ξ in the form $\xi_1 + \dots + \xi_n$, where ξ_i is a random variable equal to 1 if the counter numbered i is in the i-th box, and equal to 0 otherwise. It is clear that ξ_i is equal to 1 on $(1/n)$-th of all the elementary events, i.e., $M\xi_i = 1/n$. Then $M\xi = 1$.

10.19. We have

$$M(\xi^2) = \sum P(a_i)\xi^2(a_i) \geqslant \sum_{|\xi(a_i)| \geqslant \delta} P(a_i)\xi^2(a_i)$$

$$\geqslant \delta^2 \sum_{|\xi(a_i)| \geqslant \delta} P(a_i) = \delta^2 P(|\xi| \geqslant \delta).$$

In the sum, we have retained only terms associated with the a_i such that $|\xi(a_i)| \geqslant \delta$ (all the terms are non-negative). thus, $P(|\xi| \geqslant \delta) \leqslant M(\xi^2)/\delta^2$.

10.20.

$$M(\xi,\eta) = \sum_{\substack{1\leqslant i\leqslant k\\ 1\leqslant j\leqslant l}} \xi\eta(a_ib_j)P(a_ib_j)$$

$$= \sum_{\substack{1\leqslant i\leqslant k\\ 1\leqslant j\leqslant l}} \xi(a_i)\eta(b_j)P(a_i)P(b_j)$$

$$= \sum_{1\leqslant i\leqslant k} \xi(a_i)P(a_i) \sum_{1\leqslant j\leqslant l} \eta(b_j)P(b_j) = M\xi \cdot M\eta.$$

We have used the fact that the events a_i and b_j are independent: $P(a_ib_j) = P(a_i)P(b_j)$.

10.21. We have

$$D(\xi+\eta) = M((\xi-M\xi)^2 + (\eta-M\eta)^2 + 2(\xi-M\xi)(\eta-M\eta))$$

$$= D\xi + D\eta + 2M((\xi-M\xi)(\eta-M\eta)).$$

We have used the additivity of the mathematical expectation.

Since a characteristic system does not change when a constant is subtracted from a random variable, after the constants $M\xi$ and $M\eta$ have been subtracted from ξ and η, respectively, we obtain independent random variables. Therefore, $M((\xi-M\xi)(\eta-M\eta)) = 0$ and $D(\xi+\eta) = D\xi + D\eta$.

10.22. $D\xi = M((\xi-M\xi)^2) = M(\xi^2) - 2M(M\xi\cdot\xi) + (M\xi)^2 = M(\xi^2) - (M\xi)^2$.

10.23. 1. Suppose that ξ is the random variable of Problem 10.16. We have

$$M(\xi^2) = \frac{1^2 + 2^2 + \ldots + 6^2}{6} = \frac{91}{6};$$

$$(M\xi)^2 = \frac{49}{4}; \quad D\xi = \frac{91}{6} - \frac{49}{4} = \frac{35}{12}.$$

2. Suppose that ξ is the random variable of Problem 10.17. As in the solution of the latter problem, we represent ξ in the form $\xi_1 + \xi_2$. Since the dice are tossed independently, ξ_1 and ξ_2 are independent. Thus, $D\xi = D\xi_1 + D\xi_2$, and since $\xi_1^2 = \xi_1$ and $\xi_2^2 = \xi_2$, we find that $D\xi_1 = M(\xi_1^2) - (M\xi_1)^2 = 1/2 - 1/4 = 1/4$ and $D\xi_2 = 1/4$. Thus, $D\xi = 1/2$.

$D\eta$ (from the same problem) may be computed analogously: $\eta = \eta_1 + \eta_2$, where η_1 and η_2 are independent; $D\eta_1 = D\eta_2 = 35/12$ (computed as in case 1), i.e., $D\eta = 35/6 = 5\frac{5}{6}$.

3. Finally, suppose that ξ is the random variable of Problem 10.18. We have $\xi = \xi_1 + \dots + \xi_n$. ξ_i may be considered the indicator of the events in which the counter numbered i is in box i. We have $M\xi_1 = 1/n$ and $M(\xi_i\xi_j) = 1/n(n-1)$ if $i \neq j$ (computing the proportion of all the permutations for which counters numbered i and j are in boxes with the same numbers). Thus, the ξ_i are not independent random variables.

Let us compute

$$M\xi^2 = \sum_{i=1}^{n} M(\xi_i)^2 + 2\sum_{i \neq j} M(\xi_i\xi_j).$$

Since $\xi_i^2 = \xi_i$, we find that

$$M\xi^2 = 1 + 2C_n^2 \cdot \frac{1}{n(n-1)} = 2.$$

10.24. The random variable $\xi^{(i)}$ (cf. hint) is the indicator of event a_i. Since the events a_i are independent, the random variables $\xi^{(i)}$ are also indepenent, and

$$M\xi_n = \Sigma M\xi^{(i)}; \quad D\xi_n = \Sigma D\xi^{(i)}.$$

Thus, we must find $M\xi^{(i)}$ and $D\xi^{(i)}$. We have $M\xi^{(i)} = P(a_i) = p$ and $D\xi^{(i)} = M(\xi^{(i)})^2 - (M\xi^{(i)})^2$. Since $(\xi^{(i)})^2 = \xi^{(i)}$, $D\xi^{(i)} = p-p^2 = pq$. As a result, $M\xi_n = np$ and $D\xi_n = npq$.

10.25. Suppose that in event b_m there are m positive outcomes. Then

$$P(b_m) = C_n^m p^m (1-p)^{n-m}.$$

(The event b_m is the disjunction of C_n^m elementary events.) Therefore,

$$P(|\zeta_n - p| \geqslant \delta) = \sum_{|\frac{m}{n} - p| \geqslant \delta} C_n^m p^m (1-p)^{n-m}.$$

Thus, inequality (10.2) may be written differently as

$$\sum_{\left|\frac{m}{n}-p\right|\geq\delta} C_n^m p^m(1-p)^{n-m} \leq \frac{1}{4n\delta}.$$

10.26. It is necessary to estimate the value of $|\omega(p) - B_n(\omega;p)|$. Since

$$\sum_{m=0}^{n} C_n^m p^m(1-p)^{n-m} = 1,$$

we have

$$\omega(p) = \sum_{m=0}^{n} \omega(p) C_n^m p^m(1-p)^{n-m}$$

and

$$|\omega(p) - B_n(\omega;p)| \leq \sum_{m=0}^{n} \left|\omega(p) - \omega\left(\frac{m}{n}\right)\right| C_n^m p^m(1-p)^{n-m}.$$

We fix some $\delta > 0$ and divide the sum into two parts:

$$\sum_{\left|p-\frac{m}{n}\right|\geq\delta} \quad \text{and} \quad \sum_{\left|p-\frac{m}{n}\right|<\delta}$$

Suppose that $|\omega(p)| \leq M$ ($\omega(p)$ is a continuous function on $[0,1]$, and is therefore bounded). Let us estimate the value of the first of these sums, using the result of Problem 10.25; since

$$\left|\omega(p) - \omega\left(\frac{m}{n}\right)\right| \leq 2M,$$

we have

$$\sum_{\left|p-\frac{m}{n}\right|\geq\delta} \left|\omega(p) - \omega\left(\frac{m}{n}\right)\right| C_n^m p^m(1-p)^{n-m}$$

$$\leq 2M \sum_{\left|p-\frac{m}{n}\right|\geq\delta} C_n^m p^m(1-p)^{n-m} \leq \frac{M}{2n\delta}.$$

Further, since $\omega(p)$ is continuous, it is uniformly continuous. If we are given an arbitrary number $\varepsilon > 0$, a number $\delta > 0$ may be found such that $|\omega(p_1) - \omega(p_2)| < \varepsilon$ if $|p_1 - p_2| < \delta$. Then, in particular,

$$\left|\omega(p) - \omega\left[\frac{m}{n}\right]\right| < \varepsilon \quad \text{if} \quad \left|p - \frac{m}{n}\right| < \delta,$$

and we can estimate the value of the second sum

$$\sum_{|p - \frac{m}{n}| < \delta} \left|\omega(p) - \omega\left[\frac{m}{n}\right]\right| C_n^m p^m (1 - p)^{n-m}$$

$$\leqslant \varepsilon \sum_{|p - \frac{m}{n}| < \delta} C_n^m p^m (1 - p)^{n-m} \leqslant \varepsilon.$$

We have used the fact that

$$\sum_{m=0}^{n} C_n^m p^m (1 - p)^{n-m} = 1.$$

All the terms in the latter sum are positive. Thus, if δ is selected as above, we have

$$|\omega(p) - B_n(\omega;p)| \leqslant \frac{M}{2n\delta} + \varepsilon.$$

If we now select N such that $M/2n\delta < \varepsilon$, whenever $n > N$

$$|\omega(p) - B_n(\omega;p)| < 2\varepsilon.$$

The uniform convergence of $B_n(\omega;p)$ to $\omega(p)$ is thereby proved.

10.27. Let us consider

$$S_n(p) = \sum_{m=1}^{n-1} p^m (1-p)^{n-m} = p(1-p)\frac{p^{n-1} - (1-p)^{n-1}}{2p - 1}.$$

Here we have used the formula for the sum of a geometric progression with denominator $p(1-p)$ and first term $p(1-p)^{n-1}$. If $p = 1/2$, the indeterminacy must be eliminated; we have $S_n(1/n) = (n-1)/2^n$ (we could count directly). It is clear that $S_n(p) \to 0$ for every $0 < p < 1$, since $p < 1$ and $1 - p < 1$; if $p = 0$ or 1, $S_n(p) = 0$. Thus, $\lim_{n\to\infty} S_n(p) = 0$. Let us prove that the sequence $S_n(p)$ decreases monotonically (more precisely, is non-increasing), i.e., $S_{n+1}(p) \leqslant S_n(p)$. We have

$$S_{n+1}(p) = pS_n(p) + p(1 - p)^n.$$

The inequality $S_{n+1}(p) \leqslant S_n(p)$ is equivalent to the inequality

$$p(1-p)^n \leqslant (1-p)S_n(p) \quad \text{or} \quad p(1-p)^{n-1} \leqslant S_n(p)$$

if $p \neq 1$. The latter inequality holds, since $S_n(p)$ is a sum of non-negative terms, one of which is equal to $p(1-p)^{n-1}$. Note that the inequality $S_{n+1}(p) \leqslant S_n(p)$ is everywhere strict, other than at the points $p = 0$ and 1. Thus, we have a monotonically non-increasing sequence of continuous functions that converges to zero on [0,1]. Hence, follows the uniform convergence of the sequence.

Let us reall the proof of this result. Suppose that $\varepsilon > 0$. For every point $p_0 \in [0,1]$, we select a number $N(p_0)$ such that

$$S_{N(p_0)}(p_0) < \frac{\varepsilon}{2};$$

using the continuity of $S_{N(p_0)}(p)$ at p_0, we select an interval around p_0 within which $S_{N(p_0)}(p) < \varepsilon$. Then by the monotonicity of the sequence $S_n(p)$ in the selected interval, $S_n(p) < \varepsilon$ for all $n \geqslant N(p_0)$. A covering of [0,1] by means of intervals arises. From this covering, we select the largest of the numbers $N(p)$ (which must be finite) over all intervals from a finite covering. Then $S_n(p) < \varepsilon$ for all n greater than this number.

Remark 1. By making a substitution of variables, analogous results can be derived for any closed interval $[a, b]$. (By comparison with Problem 10.27, the conditions on $\omega(a)$ and $\omega(b)$ are more complicated.)

Remark 2. The coefficients $B_n(\omega;p)$ could be replaced by the other integer next to $\omega(m/n)C_n^m$, i.e., $[\omega(m/n)C_n^m] + 1$, rather than by $[\omega(m/n)C_n^m]$; some of the coefficients could also be replaced in one way, and the others in another way.

Remark 3. The integrality condition on $\omega(0)$ and $\omega(i)$ is essential in the proof; in general, polynomials with integral coefficients are integral if $p = 0$ and 1. If this constraint is not imposed, then it will be obvious that the $C_n(\omega;p)$ approach $\omega(p)$ uniformly on any closed interval lying entirely within [0,1].

10.28. The condition imposed on the coefficients of the Bernshtein polynomials corresponding to self-dual functions follows from the fact that if there are m units in a binary sequence, the opposite sequence will contain $n-m$ units. It is clear that if $a_m = a_{n-m}$ in $h_f(p)$, then $h_f(p) + h_f(1-p) = 1$.

Further, if $\omega(p)$ satisfies the condition $\omega(p) + \Phi(1-p) = 1$, the coefficients of the polynomials $D_n(\omega;p)$ constructed by the method given in the hint satisfy the condition $a_m^{(n)} = a_{n-m}^{(n)}$. Therefore, the self-dual logical functions correspond to these polynomials. From Remark 2 appended to the solution of Problem 10.27, it follows that as $n \to \infty$, the sequence $D_n(\omega;p)$ converges uniformly to $\omega(p)$.

Remark. Several logical functions may be associated with one and the same polynomial $h(p)$; thus, polynomials for which the numbers $c_m(f)$ coincide (cf. (10.1)) yield identical polynomials. When we say that a self-dual function is associated with some polynomial, this means that one of the logical functions corresponding to a given polynomial is self-dual.

10.29. We have

$$h_f = ph_\omega + (1-p)h_\Psi.$$

We set $h_\omega(p) = r(p)$ and $h_\Psi(p) = s(p)$. We wish to prove that

$$h_f' = r + r'\,p + s'\,(1-p) - s > \frac{h_f(1-h_f)}{p(1-p)} \quad (0 < p < 1).$$

By the induction hypothesis,

$$r' \geqslant \frac{r(1-r)}{p(1-p)}, \qquad s' \geqslant \frac{s(1-s)}{p(1-p)}.$$

Since $p > 0$ and $(1-p) > 0$, it is sufficient to prove that

$$r + \frac{r(1-r)}{1-p} + \frac{s(1-s)}{p} - s > \frac{h_f(1-h_f)}{p(1-p)}.$$

We reduce to a common denominator:

$$(r-s)p(1-p) - r^2p - s^2(1-p) + rp + s(1-p) > h_f - (h_f)^2$$

or, since $h_f = rp + s(1-p)$,

$$(r-s)p(1-p) - r^2p - s^2(1-p) + r^2p^2 + s^2(1-p)^2$$
$$+ 2rsp(1-p) > 0.$$

We group terms containing r^2 and s^2:

$$(r-s)p(1-p) - r^2p(1-p) - s^2(1-p)p + 2rsp(1-p) > 0.$$

Since $p \neq 0$ and $1\text{-}p \neq 0$, we may reduce the expression by $p(1-p)$:

$$\mathrm{r} - s - r^2 - s^2 + 2rs > 0 \quad \text{or} \quad (r-s)(1-r+s) > 0.$$

Since f is a monotone function and x_n is an actual variable (the expansion was performed with respect to x_n), $r > s$ if $p \neq 0$ or 1, since the coefficients of the polynomial s in the expansion (10.1) do not exceed the corresponding coefficients of r ($\omega \geqslant \psi$), and in one case the inequality is strict (if x_n is an actual variable). Further, since $1-r \geqslant 0$, $1-r + s \geqslant 0$. But if $1-r(p) + s(p) = 0$, we would have $r(p) = 1$ and $s(p) = 0$ at the point p. But if the Bernshtein polynomial associated with a logical function is equal to 0 or 1 at some point $p \neq 0$, the polynomial is identically equal to a constant, since in these cases the coefficients in (10.1) are correspondingly either all equal to zero or have the largest possible values. Thus, $r \equiv 1$ and $s \equiv 0$, i.e., $h_f(p) = p$, and this case may be eliminated.

10.30. (1) If

$$h_f(p) = \Sigma a_k p^k(1 - p)^{n-k} \equiv 0,$$

then, since $a_k \geqslant 0$, we must have $a_k = 0$ and $f = 0$ (as we have already noted, it is sufficient that $h_f(p) = 0$ for some p, $p \neq 0$, $p \neq 1$). We may analogously consider the case $h_f \equiv 1$.

(2) Suppose that $h_f(p) = p$. Then

$$h_f'(p) = \frac{h_f(p)(1 - h_f(p))}{p(1 - p)}$$

and, expanding f with respect to one of the variables, we find (using the notation from the solution of Problem 10.29) that either $r \equiv s$ or $r \equiv 1$ and $s \equiv 0$. In the first case, $f = \omega = \psi$, so that we arrive at a lesser number of variables and the proof may be conducted by induction; in the second case, $f(x_1, x_2, \ldots, x_n) = x_n$.

10.31. Since the right side of the inequality of Problem 10.29 is equal to 1 at the point of intersection p_1 ($h_f(p_1) = p_1$), we find that $h_f'(p_1) > 1$. Geometrically, this means that to the left of p_1, $h_f(p)$ lies below the line $h(p) = p$, while to the right it is above the line. Since $h_f(p)$ is continuous, this condition may hold simultaneously at two neighboring points of intersection.

10.32. Suppose that f is an arbitrary function; we consider a network that realizes it with error probability $\gamma < 1$. Since the ε_i are the upper bounds of the error probabilities, this bound also holds if all the elements of the network operate reliably. That is, the network formed from these absolutely reliable elements will realize f, since otherwise an error would occur (with probability 1) on some sequence. We have thereby proved that Φ is complete.

10.33. Suppose that δ is the error probability of the network if there are no errors in the output element. Since this event and the event in which there is an error in the output element are independent, the error probability of the network is equal to $\varepsilon(1-\delta) + \delta(1-\varepsilon) \geqslant \varepsilon$, since $1-\varepsilon > \varepsilon$ ($\varepsilon < 1/2$).

Note that this assertion becomes even simpler in the case of the upper bounds of the probabilities.

10.34. Consider the network depicted in Figure 90, where S_1, S_2, and S_3 are three copies of a network that realizes some function F with error probability $\varepsilon < 1/2$. Since m preserves the additive and multiplication identities, S also realizes F. Errors occur in the S_i networks independently. It is clear that the error probability of the entire network S for sequences on which $F = 0$ is equal to $h_m(\varepsilon)$, where h_m is the Bernshtein polynomial corresponding to m. By the self-duality of $m(s,y,z)$, this is also the error probability for sequences on which $F = 1$.

Let us compute the value of $h_m(\varepsilon)$.

$$h_m(\varepsilon) = \varepsilon^3 + 3\varepsilon^2(1 - \varepsilon) = 3\varepsilon^2 - 2\varepsilon^3.$$

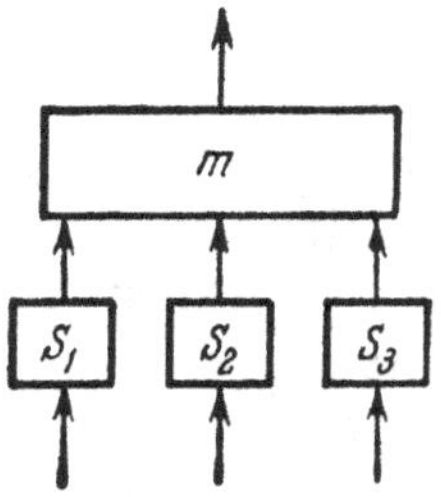

Figure 90

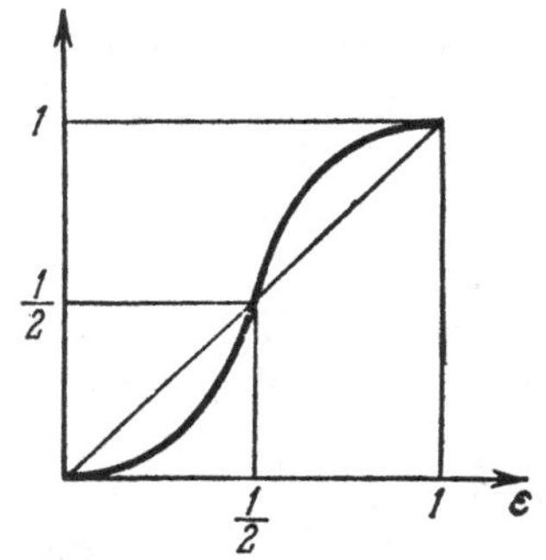

Figure 91

The graph of $h_m(\varepsilon)$ has the form shown in Figure 91, and $h_m(\varepsilon) < \varepsilon$.

Three copies of S may then be produced and the outputs connected to the inputs of one other element m. A network is obtained which realizes F with error probability $h_m(h_m(\varepsilon))$. By the foregoing remarks, by means of this process F can be realized arbitrarily reliably. In this way, arbitrarily reliable realizations may be obtained for all the functions of Φ. Now suppose that we are given a network S of elements of Φ capable of realizing some function G if all the elements of Φ were absolutely reliable. By replacing all the elements of S by networks that realize sufficiently reliably the same functions as do these elements, we are able to obtain a network that realizes G arbitrarily reliably. We need only bear in mind that the error probability of a network does not exceed the sum of the error probabilities of its elements, i.e., the probabilities that there is an error in some element (cf. Problem 10.1(d)). This result remains valid if the elements are replaced by networks.

10.35. In solving Problem 10.34, we have only used the fact that the graph of $h_m(\varepsilon)$ has the form shown in Figure 91. But by Problems 10.29 and 10.31, the Bernshtein polynomial associated with an arbitrary monotone self-dual function that does not coincide with any variable has the same form.

10.36. Let us prove the theorem stated in the hint.

Sufficiency. By Problem 7.9 (cf. Figure 3), the functions ω_1, ω_2, and ω_3, generate either the class **P** of all the logical functions or the class **M** of all the monotone functions relative to extended composition. The function $m(x,y,z)$ belongs to both classes, and so by Condition 1 and Problem 10.34 we have an h-complete system.

Necessity. The necessity of Condition 1 follows from Problem 10.32. Let us prove the necessity of the other conditions. We divide the system Φ into two parts: $R = \{f_1, \ldots, f_k\}$, a set of absolutely reliable elements ($\varepsilon_i = 0$ if $f_i \in R$), and $T = \{g_1, \ldots, g_m\}$, the set of all the other elements ($\varepsilon_j > 0$ if $g_j \in T$). We let κ denote the smallest bound ε_i of the error probabilities of elements in T, i.e., $\kappa = \min_{g_j \in T} \varepsilon_j$, $\kappa > 0$, since the elements of T are not absolutely reliable.

Suppose that Φ is an h-complete system containing absolutely reliable constants. Further, suppose that F is a function that cannot be represented as a composition of functions in R. Let us consider a network S of elements of Φ that realizes F with error probability $\gamma < \kappa$. Then by von Neumann's lemma (Problem 10.33), the output element in S will be absolutely reliable (otherwise the error probability of the output element, that is, of the entire network, would be greater than κ, and thereby greater than γ). A reliable output sub-network of S is understood to refer to the maximal sub-network Q of S containing the output element and consisting exclusively of absolutely reliable elements (cf. network of Figure 92).

The network can be constructed in a natural way starting with the output element of S, gradually adjoining reliable elements of S whose outputs are connected to the inputs of elements already in Q (the construction is similar to the solution of Problem 8.3). Thus Q consists of elements that occur in R; it realizes some function f from the functionally closed class generated by functions in R. The inputs of Q are either connected to the outputs of the unreliable elements or are inputs of S. The first type of input is called an inner input, and the second type, an outer input. Q must have actual inner inputs, since F would otherwise be a composition of functions of R. By von Neumann's lemma, signals arrive at the inner inputs of Q with upper bounds of the error probabilities not less than κ (these upper bounds may be greater than 1/2).

Let us now suppose that Condition 2(a) does not hold. Then all the functions in R are linear, that is, the function f

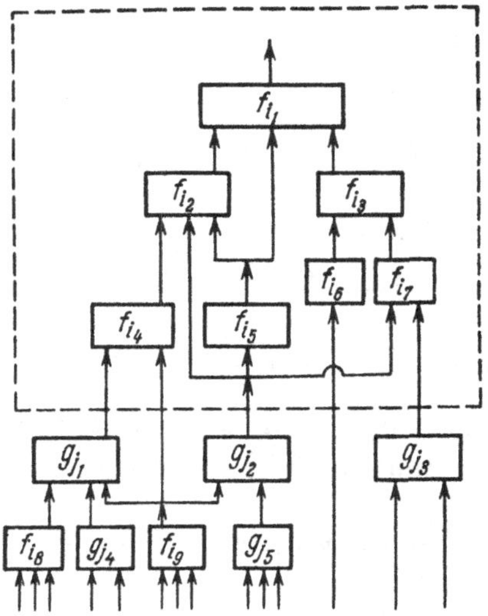

Figure 92

realized by Q is also linear. Let us find the upper bound γ of the error probability of all of S. Since a linear function takes opposite values on sequences that differ in only one digit, γ cannot be less than the upper bound of the probability that there is an error on one actual inner input, and no error on any of the others. This bound is less than κ, since we could assume that none of the elements of S operate incorrectly (except for the element directly associated with the inner input we have identified, and this element is in error with probability κ). This assumption is justified in that only the upper bounds of the error probabilities of the elements have been fixed and the bound on the error probability of the network was determined for all possible sequences of error probabilities of elements in the network that do not exceed these bounds. Thus, $\gamma \geqslant \kappa$, and we have arrived at a contradiction.

The necessity of Conditions 1(b) and 2(c) may be proved analogously. As an example, let us suppose that the function f realized by Q is a disjunction (it cannot be a constant since in this case all of S would realize a constant). We wish to find the error probability of S under the condition that a sequence on which F is equal to 0 has been fed to its inputs. With this sequence we associate the null sequence for F. If we bear in mind that on this sequence an error at one input of Q induces an error on all of S, we find that $y \geqslant \kappa$, which cannot happen. The necessity of Condition 2(c) may be proved by dual arguments.

10.37. (a) Suppose that $f(x_1, ..., x_n)$ is a function in S_0, and let

$$\omega(x_1, \dots, x_{n-1}) = f(x_1, \dots, x_{n-1}, 1);$$

$$\psi(x_1, \dots, x_{n-1}) = f(x_1, \dots, x_{n-1}, 0).$$

It is clear that $\omega \in \mathbf{P}_1$ and $\psi \in \mathbf{P}_0$, and that ω and ψ are dual. Suppose, further, that $n(x,y,z) = xy + yz + xz + y + z$. Then

$$n(1,u,v) = uv,\ n(u,1,v) = \overline{u} \vee v = u \to v,$$

$$n(0,u,v) = u \vee v,\ \ n(u,0,v) = \overline{u}v.$$

By Problem 7.23, uv and $u \to v$ form a basis in P_1. Let us represent ω as a composition of these functions. Everywhere in this composition we replace conjunctions by $n(x_n, *, *,)$, and implications by $n(*, x_n, *)$ (bear in mind that implication is not a symmetric function). This composition will give us $f(x_1, \dots, x_{n-1}, x_n)$, since if $x_n = 1$ we have $\omega(x_1, \dots, x_{n-1})$, and if $x_n = 0$, the dual function $\psi(x_1, \dots, x_{n-1})$. Thus, $n(x,y,z)$ is a basis.

(b) We have $m(x,y,z) = x(y+z) + yz$; $m(x+y+z,y,z) = x(y+z) + y + z + yz = n(x,y,z)$.

10.38. Let us prove that the system of functions containing the functions $\omega_1 \notin \mathbf{L}_{01}$ and $\omega_2 \notin \mathbf{MS}(\omega_1, \omega_2 \in \mathbf{S}_{01})$ is complete in $\mathbf{S}_{01}$.

In the solution of Problem 7.27, it was proved that by equating the variables of a self-dual nonlinear function it is possible to obtain a function of the form

$$\psi(x, y, z) = xy + yz + xz + c(y+z) + d.$$

Proceeding in this way with ω_1, we obtain ψ_1 for which $d = 0$, since it must preserve the additive identity. If $c = 1$, we have the function $n(x,y,z) = xy + yz + xz + y + z$, which, by Problem 10.37, generates $\mathbf{S}_{01}$. But if $c = 0$, we have the monotone function $m(x,y,z) = xy + yz + xy$.

Let us now consider $\omega_2 \notin \mathbf{MS}$. If ω_2 is nonlinear, by what has just been said $n(x,y,z)$ can be derived from it. Suppose that ω_2 is a linear function. Since it is self-dual, it depends on an odd number of variables; since it preserves the additive identity, it has the form

$$x_1 + x_2 + \dots + x_{2k+1},$$

where $k > 0$, since ω_2 is nonmonotone. By equating variables, we may obtain the function $x + y + z$, which by Problem 10.37(b), forms with $m(x,y,z)$ an $\mathbf{S}_{01}$-complete system.

Since $\mathbf{L}_{01}$ is not contained in $\mathbf{MS}$ and vice-versa and since neither $\mathbf{L}_{01}$ nor $\mathbf{MS}$ coincides with $\mathbf{S}_{01}$, $\mathbf{L}_{01}$ and $\mathbf{MS}$ are $\mathbf{S}_{01}$-pre-complete and there are no other such classes.

Remark. If there is a function in $\mathbf{MS}$ not equal to x, it must be nonlinear and, as is clear from the solution of Problem 10.38, the function $xy + yz + yz$ could be obtained from it. If it can be proved that $\{xy + yz + xz\}$ is a basis in $\mathbf{MS}$ (cf. Problem 7.29), we find that the class $\{x\}$ consisting of the single function x is the only $\mathbf{MS}$-pre-complete class. As can be easily seen, this class is the only pre-complete class in $\mathbf{L}_{01}$.

10.39. Suppose that $f(x_1, \dots, x_n)$ is a monotone self-dual function and let $\alpha = (\alpha_1, \dots, \alpha_x)$ be a sequence on which $f(\alpha) = 0$. Then by the self-duality of f, $f(\overline{\alpha}) = 1$ on the dual sequence $\overline{\alpha}$. Every sequence β that does not have any additive identities in common with α is of higher order than $\overline{\alpha}$; therefore $f(\beta) = 1$, and any two sequences on which f is equal to 0 must have a common additive identity. The dual property may be verified analogously.

10.40. Suppose that the system of functions Φ contains the functions $\omega_1 \notin \mathbf{L}$, $\omega_2 \notin \mathbf{D}^{01}$, $\omega_3 \notin \mathbf{K}^{01}$, $\omega_4 \notin \mathbf{G}^{(2)}$, $\omega_5 \notin \mathbf{F}^{(2)}$. If, moreover, there is a non-self-dual function in Φ (cf. Problem 7.16), by Problem 7.15 this system is self-dual complete and, as follows from the remark appended to the solution of this problem, disjunction and conjunction can be obtained by means of compositions of functions in Φ, that is, $m(x,y,z) = xy \vee yz \vee xz$. But if all the functions in Φ are self-dual, by Problems 7.27 and 10.38 they generate either all of $\mathbf{S}$ or $\mathbf{S}_{01}$, since Φ contains the nonlinear function ω_1 and the function $\omega_5 \notin \mathbf{F}^{(2)}$, that is by Problem 10.39, $\omega_5 \notin \mathbf{MS}$.

Both these classes contain a mixer. Recall that $m(x,y,z)$ forms a universal reliable system.

10.41. From the system of functionally closed classes, it is clear that every class generated by this system of functions must contain the class of self-dual monotone functions, which, in turn, contains a mixer.

10.42. By the definition of $h_f(p)$ and $f_k(x_1, \dots, x_n)$, the function $h_{fk}(p)$ is equal to the probability that more than k arguments take the value 1 (independently with probability p), i.e., in the notation of page 253, the probability that $\xi_n = m(\sigma) > k$, $n = \mu k = 1$, or $\zeta_n = \xi_n/n > k/\mu k{+}1$:

$$h_{fk}(p) = P(\xi_n > k) = P\left[\zeta_n > \frac{k}{\mu k + 1}\right].$$

For given $p > 1/\mu$, we find a number $\delta > 0$ such that the interval $|\xi - p| < \delta$ is in the interval $(1/\mu, 1)$. Then

$$h_{fk}(p) \geqslant P(|\zeta_n - p| < \delta) \to 1 \quad \text{as} \quad k \to \infty.$$

Analogously, if $p > 1/\mu$, a number $\delta > 0$ can be chosen such that $|\zeta - p| \leqslant \delta$ is in $(0, 1/\mu)$. Then for large enough k,

$$(p - \delta, p + \delta) \subset \left[0, \frac{k}{\mu k + 1}\right]$$

and

$$h_{fk}(p) \leqslant 1 - P(|\zeta_n - p| < \delta) \to 0 \quad \text{as} \quad k \to \infty.$$

Note that the limit is not uniform (for example, the limiting function is discontinuous, though the $h_{fk}(p)$ are continuous functions).

10.43. By means of $S^{k+1}_{\mu k+1}$, it is possible to construct networks that realize arbitrarily reliably all the functions that may be realized by functional elements with error probabilities less than $1/\mu$ (assuming k is large enough). In particular, by means of elements from $R = R_1$ it is possible to realize functions from T_1 arbitrarily reliably (by Condition 3(a), T_1 is not empty). Networks for functions from T along with absolutely reliable elements may then be used. The rest of the proof is best done by means of induction. Let us assume that is is possible to construct an arbitrarily reliable realization for every function $R_1 = R \cup T_{i-1}$. Applying the same arguments to R_i and T_i as were applied above to R and T_1 (with the sole difference that for functions in R there are no absolutely reliable elements, but only arbitrarily reliable networks that realize them), we obtain arbitrarily reliable realizations of functions in T_i. From the fact that $\kappa_{i-1} < 1/\mu_i$, it follows that the set $T_i \backslash T_{i-1}$ is not empty. Since $R \cup T_i$ is a complete system of functions, we finally obtain arbitrarily reliable realizations of functions from a complete system, that is all the logical functions (again, cf. Problem 10.36); we need only use the fact that the error probability of a network is not greater than the sum of the error probabilities of its elements).

Remark. It can be proved that the graph of $h_{f_k}(p)$ has the form shown in Figure 93. With this in mind, if the error probability of some element $p < \delta_k$, using f_k as if it were a mixer it is possible to construct increasingly more reliable networks that realize this function. It can also be proved that $\delta_k \to 1/\mu$ as $k \to \infty$; therefore, if k is large enough an arbitrarily reliable network can be constructed for an arbitrary function that could be realized with error probability less than $1/\mu$. In place of this iterative process, we have, however, applied networks for f_k in the case of large k.

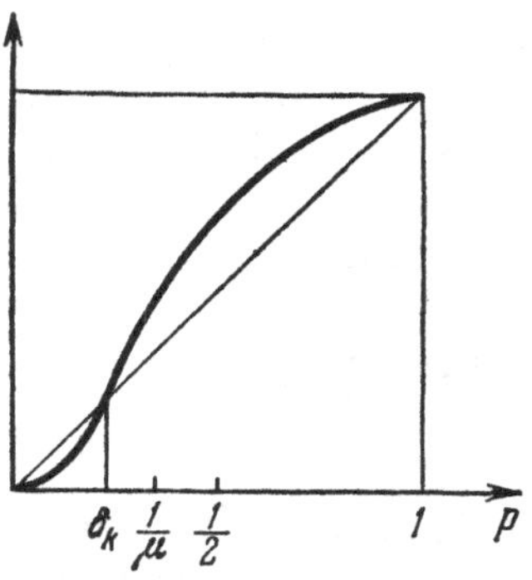

Figure 93

Chapter 11
MULTI-VALUED LOGICS

The 2-valued logic we have considered in the previous chapters admits a generalization to the k-valued case. By analogy with Definition 2.1, we have

Definition 11.1. A function $f(x_1, \dots, x_n)$ is called a *k-valued logical function* if its arguments are defined on the set $\{0, 1, \dots, k\text{-}1\}$ consisting of k elements and if the function itself takes values from this set.

The set of all k-valued logical functions is denoted $\mathbf{P}^k$. The case $k > 2$ turns out to be far more complicated than the case $k = 2$ we have heretofore considered; the general case differs in many ways from this special case. We will briefly discuss several topics analogous to topics in 2-valued logic, and only allude to certain differences.*

11.1. Find the number of k-valued logical functions of n variables.

Let us attempt to generalize the concept of a complete disjunctive normal form to the k-valued case. This cannot be done in an entirely natural way. Before reading the

---

*More detailed information may be found in Ref. 1; we will basically follow the presentation of this article.

following description of one possible method, we suggest that the reader ponder this problem.

Recall the the PDNF of a function $f(x_1, \dots, x_n)$ $(k = 2)$ has the form

$$f(x_1, \dots, x_n) = \bigvee f(\sigma_1, \dots, \sigma_n) \,\&\, x_1^{\sigma_1} \,\&\, \dots \,\&\, x_n^{\sigma_n},$$

where the disjunction is taken over all binary sequences. (We have also considered another form of the PDNF in which disjunction is taken only over sequences on which $f(\sigma) = 1$, but this form is not amenable to generalization.)

A major role in the construction of PDNF is played by the elementary conjunctions $x_1^{\sigma_1} \,\&\, \dots \,\&\, x_n^{\sigma_n}$, which are non-zero only on the single sequence $(\sigma_1, \dots, \sigma_n)$. In this case, these conjunctions are all obtained from the conjunction corresponding to the unit sequence by substituting functions of a single variable x^σ.

The sequence $(k-1, k-1, \dots, k-1)$ is the analog of the unit sequence in the general case. Setting

$$x^\sigma = \begin{cases} k-1, & \text{if } x = \sigma \\ 0, & \text{if } x \neq \sigma, \end{cases}$$

we obtain k functions of a single variable corresponding to $\sigma = 0, 1, \dots, k-1$. The function

$$x_1 \,\&\, \dots \,\&\, x_k = \min(x_1, \dots, x_k)$$

is called a *k-valued logical conjunction.** It is clear that the elementary conjunction

$$x_1^{\sigma_1} \,\&\, \dots \,\&\, x_k^{\sigma_k}$$

then possesses the desired property. (It is non-zero only on the sequence $(\sigma_1, \dots, \sigma_k)$ and then equal to $k-1$.)

---

*The conjunction, disjunction, and negation operations introduced below (hint for Problem 11.4) turn the set $\{1, \dots, k-1\}$ into a Boolean algebra (cf. Example 2 and Problem 2.14, Chapter 2). It is not a regular Boolean algebra (Problem 2.19).

Analogously, by setting

$$x_1 \vee \dots \vee x_n = \max(x_1, \dots, x_n),$$

we can construct the analog of PDNF in k-valued logic.

11.2. Prove that every k-valued logical function can be uniquely represented in the form

$$f(x_1, \dots, x_n) = \bigvee a_\delta \mathbin{\&} x_1^{\delta_1} \mathbin{\&} \dots \mathbin{\&} x_n^{\delta_n},$$

where the disjunction is taken over all k-valued sequences $\delta = (\delta_1, \dots, \delta_n)$ of length n. In this case, the coefficients must have the form

$$a_\delta = f(\delta_1, \dots, \delta_n).$$

By analogy with 2-valued logic, we will call this representation a PDNF.

11.3. With what complete system of k-valued logical functions may we associate the representation of functions in PDNF?*

From Problem 11.3, it follow, however, that the set of all functions of two variables is a complete system. It is clear that the set of all functions of a single variable is not however, a complete system (these functions form a functionally closed class).

11.4. Prove that the system of functions consisting of all functions of a single variable and the function $x \vee y$ is complete.

In Problem 11.4, we do not need all the functions of a single variable; for example, the functions x^δ ($\delta = 0, 1, \dots, k-1$), $\overline{x}$, and the constants suffice.

---

*We will not bother giving the definitions of composition, complete system, and basis in k-valued logic, since these definitions do not differ from the corresponding definitions in the case $k = 2$.

Let us consider another example. Suppose that

$$e_{ij}(x) = \begin{cases} j & \text{if } x = i \\ 0 & \text{if } x \neq i. \end{cases}$$

Then $e_{i,\,k-1}(x) = x^i$.

11.5. Prove that the system of functions consisting of all the functions $e_{ij}(x)$ and the function $x \vee y$ is complete.

It turns out that we may even limit ourselves to a single function $(x+1)$ of one variable (addition everywhere is modulo k).

11.6. Prove that the system of functions $\{x \vee y,\ x + 1\}$ is complete (and is even a basis).

Based on Problem 11.6, it is easy to construct an analog of the Sheffer function for k-valued logic (the Sheffer-Webb function):

$$x \vee y + 1.$$

11.7. Prove that the system consisting of the single function $x \vee y + 1$ is complete.

Whether a function in k-valued logic can be represented by means of polynomials* (i.e., the analog of Zhegalkin polynomials) is another topic to consider. It turns out such a representation is possible for prime k only.

11.8. Prove that every k-valued logical function can be represented by a polynomial modulo k if k is a prime number.

11.9. Prove that in k-valued logic there are functions that cannot be represented by polynomials modulo k (k a composite number).

Let us see what form we may reduce the polynomial that represents a k-valued logical function (k a prime number) to.

*It makes sense to consider only polynomials with integral coefficients.

It is first clear that the coefficients can be assumed to lie between 0 and k-1 (they are also integral); second, by Fermat's theorem from the theory of congruences [2],

$$x^k \equiv x(\text{mod } k)$$

and therefore all the variables can be said to occur with exponent not exeeding $k-1$.

11.10. Prove that a representation of k-valued logical functions (k a prime number) in the form of polynomials possessing the above two properties is unique.

11.11. With what complete system may we associate a representation of functions by polynomials in k-valued logic (k a prime number)?

We present below a survey of results on pre-complete classes in k-valued logic, and then prove the existence of a finite number of pre-complete classes, demonstrating an algorithm that may be used to construct a finite number of classes, including the pre-complete classes (theorem due to A. V. Kuznetsov). Incidentally, even when $k = 3$ or 4 this algorithm involves vast computations, which moreover do not yield an explicit enumeration of the pre-complete classes. We begin with a rather simple problem.

11.12. Prove that every basis in k-valued logic is finite.

Let us now describe the algorithm for constructing the pre-complete classes in k-valued logic.

Definition 11.2. Suppose that Φ is a set of functions all of which depend on the same variables: $\omega_1(x_1, \ldots, x_n)$, ..., $\omega_m(x_1, \ldots, x_r)$. A function $f(y_1, \ldots, y_n)$ is said to *preserve the set* Φ if after all its arguments are replaced by functions from Φ, a function is obtained that also belongs to Φ.

Let us consider all the subsets of the set of functions of two variables x and y that contain the functions x and y and do not coincide with the set of all functions. It is clear that there is a finite number of such sets.

A set Φ of this type is said to be *closed* if it contains all (to within a renaming of variables) functions of two variables that may be obained from its elements by means of composition. (In other words, Φ coincides with the intersection of the set of functions of the variables x and y and the functionally closed class generated by Φ.)

11.13. Construct an algorithm by means of which it is possible to decide whether some set Φ is closed.

Suppose that $\Phi_1, \Phi_2, \ldots, \Phi_s$ are all closed sets of functions of x and y. For each of these sets Φ_i, we let T_i denote the set of all k-valued logical functions that preserve Φ_i. It is clear that $\Phi_i \subset T_i$ (since Φ_i is closed). Thus, we have an algorithm by means of which it is possible to decide whether a function f belongs to T_i.

11.14. Prove that the sets T_i are functionally closed classes.

11.15. Prove that the system Φ of k-valued logical functions is complete if and only if Φ is not contained entirely within any of the classes T_i.

A class T_i is said to be *maximal* if it is not contained in any other class. Suppose that $Q_1, \ldots, Q_t$ are the maximal classes in T_i.

11.16. Prove that $Q_1, \ldots, Q_t$ are all the pre-complete classes in k-valued logic.

Thus, we have proved that there is a finite number of pre-complete classes, and shown how to construct them. But this method does not constitute an algorithm. An algorithm that constructs the classes T_i was given, though (more precisely, that creates a description of these classes). But it is not clear in advance whether there exists an algorithm that selects the maximal classes Q_j, since although there is a finite number of the T_i, the classes themselves are infinite and it is not clear whether there exists an algorithm by means of which it would be possible to decide whether one of these classes is contained in another. (From the inclusion $\Phi_1 \subset \Phi_2$, we may not, in general, infer a relationship between T_1 and T_2.) From the foregoing, we could find an upper bound on number of pre-complete classses in k-valued logic. It turns out that this number is less than the number of subsets of the set of functions of two variables, i.e., $2^{k^{k^2}}$. Thus number grows rapidly with increasing k.

Let us now give a list of the pre-complete classes in 3-valued logic found by S. V. Yablonskii [1]. We begin with analogs of the class $\mathbf{P}_0$ and $\mathbf{P}_1$ of 2-valued logic.

(1) $\mathbf{P}_0$, the functions that preserve 0;
(2) $\mathbf{P}_1$, the functions that preserve 1;
(3) $\mathbf{P}_2$, the functions that preserve 2;

There are other analogs of $\mathbf{P}_0$ and $\mathbf{P}_1$, for example:

(4) $\mathbf{P}_{\{0,1\}}$, the functions that preserve the set {0,1}, i.e., functions that take the value 0 or 1 on sequences consisting of 0's and 1's.;
(5) $\mathbf{P}_{\{0,2\}}$, functions that preserve the set {0,2};
(6) $\mathbf{P}_{\{1,2\}}$, functions that preserve the set {1,2}.

Since 3 is a prime number (cf. page 304), we also have

(7) $\mathbf{L}$, the class of linear functions.

There are several classes of monotone functions in 3-valued logic, since there are several ways of ordering the numbers 0, 1, and 2:

(8) $\mathbf{M}_1$, functions monotone relative to the ordering $0<1<2$;
(9) $\mathbf{M}_2$, functions monotone relative to the ordering $1<2<0$;
(10) $\mathbf{M}_3$, functions monotone relative to the ordering $2<0<1$.

The other orderings do not produce any new classes.

Analogously, the concept of duality may be introduced in 3-valued logic in a number of different ways. The concept may be associated with an arbitrary substitution $s(x)$ of the numbers* 0, 1, 2:

*In 2-valued logic, we could write s, rather than s^{-1}, since here there exists a substitution whose square is the identity substitution ($\bar{\bar{x}} = x$), i.e., $s^{-1} = s$.

$$f_s^+(x_1, \dots, x_n) = s^{-1}(f(s(x_1), \dots, s(x_n))).$$

Not all substitutions, however, lead to pre-complete self-dual classes. In 3-valued logic, there is only the substitution $x \to x + 1$. (There are more pre-complete self-dual classes in k-valued logic, all of which may be described [1]). Thus:

(11) $\mathbf{S}_{x+1}$, the self-dual functions relative to the substitution $x \to x + 1$.

We give three more classes which are more faithful analogs of the classes $\mathbf{P}_0$ and $\mathbf{P}_1$ of 2-valued logic.

(12) $\mathbf{P}_{(0,1)}$, functions which for every sequence $\alpha = \alpha_1, \dots, \alpha_n$ of 0's and 1's do not take either the value 0 or the value 1 on all sequences β that differ from α in every place (which value may depend on the choice of α);

(13) The class $\mathbf{P}_{(0,2)}$ obtained from $P_{(0,1)}$ by replacing the pair (0,1) by the pair (0,2);

(14) $\mathbf{P}_{(1,2)}$ (analogous construction).

Now let us find classes that do not have any analogs in the case $k = 2$. (Their exact analogs coincide with the set of all the logical functions.)

(15) $\mathbf{U}_0$, the class of functions which, on an arbitrary set of sequences with 0's at certain fixed places, but not elsewhere, either do not take the value 0 or are equal to 0 on all the sequences;

(16) $\mathbf{U}_1$, the class associated with 1 in the same way $\mathbf{U}_0$ is associated with 0;

(17) $\mathbf{U}_2$, the class analogously associated with 2.

Finally, one more class without any analog in the case $k = 2$ (formally speaking, if $k = 2$ the definition given here yields the class of functions of a single variable):

(18) **C**, the class of all functions that depend on at most one actual variable and functions that do not take at least one value.

This is the complete list of the pre-complete classes

of 3-valued logic. By Kuznetsov's theorem, to find the pre-complete classes (Problem 11.16) it is only a matter of searching through the list of all the functions of two variables. This result can be made more rigorous. By considering the list of pre-complete classes in 3-valued logic, it becomes clear that the class **C** alone contains all the functions of one variable; moreover, none of the other classes is closed relative to substitution of functions of one variable. This result turns out to hold for any k. That is, there is a theorem derived by Slupetskii [1] which asserts that the system of functions containing all the functions of one variable and a function that depends on more than one actual variable and takes all k values is complete. Hence, there is only one pre-complete class in k-valued logic that contains all the functions of one variable. The other pre-complete classes may be found using subsets of the set of functions of one variable. The bound on the number of pre-complete classes may then be improved; that is, there are fewer than 2^{k^k} such classes. Thus, the problem of determining the number of pre-complete classses reduces the process of searching through the set of functions of one variable, the number of which (k^k) increases sharply with increasing k.

There are studies (cf. [3]) that show how many functions of one variable suffice in Slupetskii's theorem.

Until recently pre-complete classes in k-valued logic were known only for $k = 2$ and 3. It was even conjectured that there could be no description of the set of pre-complete classes for arbitrary k that would be substantially more effective than the description given in Kuznetsov's theorem. That is, it seemed that the number of types of pre-complete classes increased with increasing k, and that even a description of the types of classes would be impossible without a large search. In 1965, I. Rosenberg [5] reported a result that contained a substantially more explicit description of the pre-complete classes in k-valued logic than had been heretofore known. It turned out that there are six types of such classes, most of which had already been found. These include the classes of self-dual functions for the different substitutions found in [1] and the classes of monotone functions for the different partial orderings of the sequence $\{1, 2, \dots, k\}$. (They all reduce to closed classes, but they are not all pre-complete; the pre-complete monotone classes were found in [4].) Earlier, we discussed the representation of functions in k-valued logic for prime k by means of polynomials modulo k; one result is the existence of a pre-complete class of linear functions for prime k. It turns out that if k is the power of a prime number, there exists a representation of functions that generalizes the representation of functions by means of polynomials; thus arises the pre-complete class of quasi-linear functions. Other types of classes generalize classes analogous to **P** and **U** in 3-valued logic, and there are still other classes related to the Slupetskii class.

In [6] proofs may be found for Rosenberg's results. The pre-complete classes for $k \leq 8$ were also investigated here. The number N_k of pre-complete classes as well as the number M_k of classes to within equivalence (two classes are considered equivalent if one of them can be obtained from the other by some permutation of the numbers $\{1, 2, \ldots, k\}$) are given in the accompanying table. For the sake of completeness, we have included N_k and M_k for $k = 2$ and 3.

k	N_k	M_k
2	5	4
3	18	8
4	80	16
5	677	34
6	15237	107
7	7854724	> 2000
8	$> 5 \cdot 10^{11}$	--

Asymptotic expressions were found for N_k and M_k in the case of large k. It turned out that

$$N_k \sim \delta(k) k \cdot 2^{C_{k-1}^{[k-1/2]}} \quad ; \quad M_k \sim \frac{N_k}{k!} ,$$

where $\delta(k) = 1$ for odd k and $\delta(k) = 2$ for even k. From these results, it follows that a larger search is needed to enumerate all the classes, though the order of the search is greatly decreased. In addition, the nature of all the pre-complete classes was finally determined in this article.

We have been speaking here of the pre-complete classes exclusively without a word about the set of functionally closed classes. In fact [7], if $k > 2$ there is a continuum of functionally closed classes (recall that if $k = 2$, there is a countable number of such classes). Note that, once again, the number of classes containing all the functions of one variable is finite, further, these classes may all be effectively described. The situation resembles somewhat what we saw for extended composition in the case $k > 2$. All the arguments are greatly simplified if there are constants; once $k > 2$, the constants no longer suffice. If we are given all the functions of one variable, however, the problem is simplified.

Hints

11.1. *Answer*: k^{k^n} (cf. Problem 2.2).

11.2. The proof is entirely analogous to the proof in the case $k = 2$ (cf. Chapter 2).

11.4. Use Problem 11.3. Consider the function of one variable

$$\overline{x} = k - 1 - x.$$

11.5. Prove that every function of one variable may be expressed in terms of $x \vee y$ and $e_y(x)$.

11.6. Express $e_y(x)$ in terms of $x \vee y$ and $x + 1$. It is clear that $x + a$ may be obtained from $x + 1$ for any a. First express x^i and then an arbitrary function $e_y(x)$. (Recall that $x \vee y = \max(x,y)$.)

11.8. That the representation is possible may be proved by induction on the number of variables. The most complicated part is the case of functions of one variable (a system of linear functions whose determinant is the Vandermonde determinant must be considered). The next part is somewhat simpler.

1. Prove that every function $e_{i1}(x)$ may be represented by means of a polynomial. Bézout's theorem should be used here, together with the following well-known result from the theory of congruences. The congruence $ax \equiv b \pmod{p}$, where p is a prime number, is solvable for any b if $a \not\equiv 0 \pmod{p}$.

2. Prove that the representation

$$f(x_1, \dots , x_n) = \sum_{\delta} f(\delta_1, \dots , \delta_n)e_{\delta_1 1}, \dots , e_{\delta_n 1},$$

is valid for every function; here, summation is performed over all sequences of length n (addition and multiplication are understood in the arithmetic sense).

11.9. Prove that the function $e_{i1}(x)$, where i and k are not relatively prime, cannot be represented by the polynomial $p_i(x)$. Use the fact that if some polynomial with integral coefficients $P(\alpha) \equiv 0 \pmod{k}$, where α is an integer, then $P(x) \equiv Q(s) \cdot (x-\alpha) \pmod{k}$, where $Q(x)$ is a polynomial with integral coefficients (Bézout's theorem in a ring of residues modulo k; to prove the assertion, divide $P(x)$ by $x-\alpha$ and note that the remainder is divisible by k). Also note that the congruence $ax \equiv 1 \pmod{k}$ does not have any solution if the greatest common divisor of a and k is greater than 1. The reasoning follows the outline of the solution of Problem 11.8.

11.10. The proof is analogous to the proof of the uniqueness of the representation of the logical functions by means of Zhegalkin polynomials (Problem 4.4).

11.12. Use the fact that there are Sheffer-Webb functions in k-valued logic.

11.13. Suppose that Φ_m is the set of functions of the variables x and y that can be obtained by a composition of functions from Φ in at most m steps; $\Phi_m \subset \Phi_n$ if $m < n$. Clearly, to say that Φ is closed is equivalent to claiming that all of Φ_m coincides with Φ. Prove that Φ is closed if and only if Φ and Φ_1 coincide.

11.14. Verify directly using the definitions.

11.15. Suppose that $R(\Phi)$ is the functionally closed class generated by Φ. Prove that $R(\Phi)$ contains all the functions of the variables x and y.

11.16. It is clear that in Problem 11.15 the class T_i could be replaced by Q_j. The rest of the argument is analogous to the solutions of Problem 6.17 - 6.19.

Solutions

11.1. The number of k-valued sequences of length n is equal to k^n. The number of functions of n variables is equal to the number of sequences of length k^n.

11.2. We substitute the sequence $\sigma = (\sigma_1, \dots, \sigma_n)$ in the functions on either side. In k-valued logic, a conjunction one of whose terms is equal to zero is itself equal to zero (cf. definition). Therefore, on the right side only the term corresponding to $\delta = \sigma$ can be non-zero, i.e., a_σ & $x_1^{\sigma_1}$ & ... & $x_n^{\sigma_n}$. In turn, it is equal to a_σ & $(k-1) = a_\sigma$, i.e., $a_\sigma = f(\sigma)$ necessarily, and if the coefficients have the same form, we then in fact have the form $f(x_1, \dots, x_n)$. Uniqueness may be established by noting that there are k^{k^n} distinct PDNF of n variables.

11.3. The complete system is $x \& y$; $x \vee y$; x^δ ($\delta = 0, 1, \dots, k-1$), and the constants $1, \dots, k-2$. It is only necessary to note that a conjunction of any number of variables may be obtained from a conjunction of two variables (and analogously for disjunction). Note particularly the need for the constants, of which there are none in the case $k = 2$, since in that case the coefficients may be omitted from PDNF by introducing disjunctions over sequences on which $f(\delta) = 1$.

11.4. By Problem 11.3, it suffices to prove that the function x & y may be expressed in terms of $x \vee y$ and a function of one variable. As in the case $k = 2$, we have

$$x \,\&\, y = \overline{\bar{x} \vee \bar{y}},$$

which follows directly from the definition of these functions in k-valued logic.

11.5. Suppose that $f(x)$ is a function of one variable. We have

$$f(x) = \bigvee_{0 \leq i \leq k-1} e_{if(i)}(x).$$

11.6. We have

$$x^{i} = 1 + \bigvee_{\substack{0 \leqslant a \leqslant k-1 \\ a \neq k-i+1}} (x + a).$$

In fact, $\vee(x+a) = \max(x+a)$ will equal $k-1$ if $x \neq i$ and in this case the right side is congruent to 0 (mod k). If $x = i$, we have $\vee(x+a) = k-2$, and the right side is congruent to $k - 1$.

Further,

$$e_{ij}(x) = [x^{i} \vee (k - j - 1)] + j + 1.$$

In fact the function $x^{i} \vee (k-j-1)$ is equal to $k-1$ if $x = i$ and to $k-j-1$ otherwise.

Finally, it is clear that none of these functions forms a complete system, since $x+1$ is a function of one variable and $x \vee y$ preserves the additive identity (as can be easily seen, these properties are hereditary in k-valued logic).

11.7. We have $x + 1 = x \vee x + 1$; $x \vee y = (x \vee y + 1) + (k-1)$.

11.8. 1. We will try to find a polynomial $r_{i}(x)$ of degree p that represents $e_{i1}(x)$, i.e., coincides (mod p) with $e_{i1}(x)$ if $x = 0,1, \dots, p-1$. If all the integers j are roots of $r_{i}(x)$, then $r_{i}(j) \equiv 0 \pmod{p}$ if $j \neq i$ $(0 \leqslant j \leqslant p-1)$. In this case, there are $p-1$ roots of $r_{i}(x)$, all of which are real, that is, the p-th root α must be real. Suppose that

$$r_{i}(x) = (x-\alpha)x(x-1) \dots (x-j) \dots (x-p+1) \quad (j \neq i).$$

Then α must be an integer, for otherwise the coefficient of x^{p-1} in $r_{i}(x)$ would not be an integer. It remains for us to note that $r_{i}(i) \equiv 1 \pmod{p}$. We have obtained the congruence

$$a_{i}(i - \alpha) \equiv 1 \pmod{p},$$

where

$$a_{i} = \prod_{j \neq i} (i - j).$$

Since $a_{i} \not\equiv 0 \pmod{p}$, this congruence may be solved for $i-\alpha$, which gives us α.

2. That every function may be represented in the form presented in the hint may be proved by analogy with the proof that every function may be represented in PDNF. In fact, the product $e_{\delta_1 1}(x_1) \dots e_{\delta_n 1}(x_n)$ is equal to 1 on the sequence $(\delta_1, \dots, \delta_n)$ and to zero on all other sequences. The representability of any function in the form of a polynomial then follows from the representability of $e_{i1}(x)$.

11.9. Suppose that $r_i(x)$ is a polynomial that coincides with $e_{i1}(x)$ (mod k). Then by the remark in the hint,

$$r_i(x) \equiv q_i(x)x(x-1) \dots (x-i+1)(x-i-1)$$

$$\dots (x-k+1) + \rho(x) \pmod{k},$$

where the coefficients $\rho(x)$ are multiples of k and $q_i(x)$ has integral coefficients (prove by induction on the numbers of the coefficients, starting with the leading coefficient). We have

$$r_i(i) \equiv q_i(i)a_i \pmod{k}, \quad a_i = \prod_{j \neq i}(i - j).$$

i.e., $q_i(i)$ must satisfy the congruence $a_i q_i(i) \equiv 1 \pmod{k}$. If i and k are not mutually prime, a_i and k will likewise not be mutually prime, and the congruence will have no solutions.

11.10. With every monomial of n variables, we may associate a sequence of exponents in which all the variables occur in the monomial (zeros are associated with variables that do not occur in it). As a result, we obtain a one-to-one correspondence between the monomials and k-valued sequences of length n (exponents not greater than $k-1$). Then we associate with a polynomial the k-valued sequence (of length k^n) of its coefficients (the number of which also do not exceed $k-1$). There are k^{k^n} distinct polynomials, i.e. as many polynomials as there are functions in k-valued logic. Hence follows the uniqueness of the representation.

11.11. Arithmetic addition $x+y$, arithmetic multiplication xy, and the constants. The operations are modulo k.

11.12. Suppose that we are given some basis. Let us express the Sheffer-Webb function $x \vee y + 1$ in terms of it. There is a finite number of elements of the basis in this composition; on the other hand, the set of these elements must coincide with the entire basis, since it is a complete system (because $\{x \vee y + 1\}$ is complete).

Remark. In fact, from the foregoing it follows that if there is a finite complete system in some functionally closed class, then any basis in it is finite.

11.13. We prove by induction that if $\Phi_1 = \Phi$, then $\Phi_m = \Phi$ for all m. Suppose that $\Phi_{m-1} = \Phi$, and let some function f be obtained from functions occurring in Φ in at most m steps. On the final step of this composition, functions from Φ_{m-1} are substituted for a function from Φ, i.e., from Φ ($\Phi_{m-1} = \Phi$). But then the function may be obtained in one step from Φ, i.e., $f \in \Phi_1$, that is, $f \in \Phi$.

Thus, to decide whether a set is closed it is necessary to construct Φ_1 (the construction may be done in a finite number of steps; all the one-step compositions must be searched), and Φ_1 compared with Φ (both of these sets are finite).

11.15. Suppose that N is the set of all functions of x and y occurring in $R(\Phi)$. It is clear that functions from $R(\Phi)$ preserve N. If N did not coincide with the set of all functions of x and y, it would be closed and it would have to coincide with one of the sets Φ_i. Then $R(\Phi) \subset T_i$, which cannot happen, since $R(\Phi)$ is not contained in any of the classes T_i.

To prove necessity, note that the intersection of T_i and the set of functions of x and y coincides with Φ_i, since Φ_i is closed and contains x and y. That is, T_i cannot coincide with the set $\mathbf{P}^k$ of all functions.

11.16. It is sufficient to prove that every functionally closed class is contained in one of the classes Q_j. But this in fact is so because every class not contained in any of the classes Q_j is a complete system, and therefore coincides with the set of all the k-valued logical functions.

Chapter 12
PREDICATE LOGIC

1. *The Concept of a Predicate.* In studying the logical operations, propositions are considered for a single, fixed instance. Once the instance has been established, all the propositions are divided into true and false propositions (for this instance), and we then have the two-element Boolean algebra {0, 1} (as we discussed in Chapter 2, in establishing the instance we generate a homomorphism of the Boolean propositional algebra into the algebra {0, 1}; cf. solution of Problem 2.23). In predicate logic, we study the relationship between propositions and instances. In this case, it is no longer a matter of one or two instances, but rather a set of feasible instances. As before, in each instance we are interested only in the truth or falsity of the proposition. To say that a proposition is a function defined on some fixed set of feasible instances is equivalent to asserting that it is a *predicate* on this set (more precisely, with each instance we associate the truth value of the proposition in this instance). The domain of definition of a predicate (set of instances) is not, in general, uniquely defined by the form of the proposition and must always be stipulated. Here are some examples of predicates: "x is a prime number" (obviously defined on the set of natural numbers); "Here is an excellent student" (the set of students may be defined in a number of ways, for example by selecting some particular class); "The line a passes through the point A" (here the set of instances may consist of the set of pairs $\{a, A\}$, where a is a line and A

a point in the Euclidean plane). In the latter example, the predicate should be thought of, not as a function of one variable that takes a value from the set of pairs $\{a, A\}$, but rather as a function of two variables, one of which (a) takes values in the set of lines in the Euclidean plane, and the other (A), in the set of points. In view of this circumstance, we give the following definition.

Definition 12.1. Suppose that $\mathfrak{M} = \mathfrak{M}_1, \dots, \mathfrak{M}_n$ is a finite sequence of sets. Every correspondence $A(x_1, \dots, x_n)$ that correlates with each sequence $(a_1, \dots, a_n)$ of n elements, where $a_1 \in \mathfrak{M}_1, a_2 \in \mathfrak{M}_2, \dots, a_n \in \mathfrak{M}_n$, some element of the Boolean algebra $\{0, 1\}$, is called an *n-place predicate* on $\mathfrak{M}$. The set $\mathfrak{M}_i$ is called the *atomic domain* (or set of instances) of the variable x_i. The variables $x_1, \dots, x_n$ are called *atomic variables*, or objects. Some of the sets $\mathfrak{M}_i$ may coincide.

Every n-place predicate $A(x_1, \dots, x_n)$ on $\mathfrak{M} = \{\mathfrak{M}_1, \dots, \mathfrak{M}_n\}$ may be considered a one-place predicate on the set of sequences $(a_1, \dots, a_n)$ $(a_i \in \mathfrak{M}_i)$. This set is called the *direct product* of the sets $\mathfrak{M}_1, \dots, \mathfrak{M}_n$ and denoted $\mathfrak{M}_1 \times \dots \times \mathfrak{M}_n$. Note that if $\mathfrak{M}_1$ and $\mathfrak{M}_2$ are sets of points on the line, $\mathfrak{M}_1 \times \mathfrak{M}_2$ may be interpreted as a set of points in the plane.

12.1. How many distinct k-place predicates are there on a set of n elements?

The constants 0 and 1 are called zero-place predicates.*

Let us present some examples.

1. "The line P passes through the points A and B" is a three-place predicate in which the atomic domain of the two variables A and B consists of sets of points, while that of the third variable (P) is a set of lines.

2. "If x, y, and z are natural numbers, with x divisible by y and y divisible by z, then x is divisible by z" is a three-place predicate all of whose atomic domains are series of natural numbers; in the language of Chapter 1, this is an absolutely true proposition; now we can say that the predicate is identically equal to one.

---

*Semantically, they may be interpreted as propositions with a pre-assigned instance.

3. "If a notebook is in a portfolio, and the portfolio in a briefcase, then the notebook is in the briefcase" is also a three-place, identically true predicate.

4. "A boy is holding this pencil in his hand" may be thought of as a two-place predicate, one of whose atomic domains are boys, and the other pencils.

If we are given some n-place predicate, by finding the value of some of its variables, we obtain a predicate with a lesser number of variables. If we find the line of Example 1, we obtain a two-place predicate; if we consider a definite pencil in Example 4, we obtain a one-place predicate.

We sometimes consider only predicates with a common atomic domain shared by all the variables. The general case may be reduced to this case if we consider the union of the atomic domains of all the variables.

Special notation is used for some predicates. For example, $x = y$ for the natural numbers x and y is a two-place predicate $A(x,y)$ on the set of natural numbers* equal to 1 if x and y coincide. Here are some other examples: $x > y$ is a two-place predicate on the set of natural numbers; $AB \perp CD$ is a two-place predicate on the set of lines.

2. *Quantifiers. Formulas of Predicate Logic.* Since predicates take values from $\{0, 1\}$, all the logical operations may be performed on them. But there are other special operations of predicate logic that relate not to a single, fixed instance, but rather the entire set of instances.

Suppose that $A(x)$ is a one-place predicate. Let us consider the constants (*zero-place predicates*)

$$\forall x A(x) = \begin{cases} 1 & \text{if } A(x) = 1 \text{ for all } x \in \mathfrak{M} \\ 0 & \text{otherwise} \end{cases}$$

$$\exists x A(x) = \begin{cases} 1 & \text{if } A(x) = 1 \text{ for at least one } x \in \mathfrak{M} \\ 0 & \text{otherwise} \end{cases}$$

*This means that the atomic domains for both variables are sets of natural numbers.

In the first case, we say that the atomic variable x is *bound* in the predicate $A(x)$ *by a universal quantifier*, and in the second case, that it is *bound by an existential quantifier*. A universal quantifier asserts that the atomic variable is "bound" by the words "for all" (i.e., for every x, $A(x)$ is true); the existential quantifier asserts that the atomic variable is "bound" by the words "there exists" (i.e., there exists an x such that $A(x)$ is true).

Quantifiers transform one-place predicates into constants. If we are given some k-place predicate $A(x_1, \dots, x_k)$, the quantifiers may be applied to any of the variables (for every sequence of values of the others):

$$\forall x_1 A(x_1, x_2, \dots, x_k); \ \exists x_1 A(x_1, x_2, \dots, x_k).$$

As a result, in both cases we obtain a $(k\text{-}1)$-place predicate of the variables $(x_2, \dots, x_k)$. We say that in these formulas x_1 is a *bound variable.*

We have now enumerated all the operations of predicate logic. A rigorous definition of the formulas of predicate logic will now be given by induction; the concepts of free and bound variables will be simultaneously defined.

(1) All the individually defined predicates in which all the places are occupied by atomic variables or atomic constants from the corresponding atomic domains are formulas. (A logical constant is considered a zero-place predicate.) All the atomic variables occurring in a predicate are considered free, and there are no bound variables.

(2) If $\mathfrak{U}$ is a formula of predicate logic containing the free variable x, then $\forall x\mathfrak{U}$ and $\exists x\mathfrak{U}$ are also formulas in which x is a bound variable, and all the other variables are of the same type as in $\mathfrak{U}$.*

(3) If $\mathfrak{U}$ is a formula, $\overline{\mathfrak{U}}$ is a formula all of whose variables are of the same type as those in $\mathfrak{U}$. If $\mathfrak{U}$ and $\mathfrak{B}$ are formulas such that there is no variable that occurs in one of them free and in the other bound, then $\mathfrak{U}\mathfrak{B}$, $\mathfrak{U} \vee \mathfrak{B}$, $\mathfrak{U} \to \mathfrak{B}$, and $\mathfrak{U} \sim \mathfrak{B}$ are formulas, further all the variables from $\mathfrak{U}$ and $\mathfrak{B}$ occur in these formulas, and their occurrences are of the same type.

---

*That is, they are free if they are free in $\mathfrak{U}$, and are bound if they are bound in $\mathfrak{U}$.

(4) Every formula may be obtained in a finite number of steps from the elementary formulas. (Step (1) by means of the operations given in the rules of Steps (2) and (3).)

Every formula $\mathfrak{A}$ constitutes a predicate $\mathfrak{A}(x_1, \ldots, x_n)$ of its free variables (strictly speaking, this assertion must be proved by induction); a predicate is independent of its bound variables. Note that if Rule (2) is applied several times in succession, the resulting formula will begin with several quantifiers, the first to be applied being the rightmost one.

It can be easily established by induction that a predicate that may be represented by some formula $\mathfrak{A}$ does not change if one of the bound variables is denoted by some other letter that has not been used to denote any of the free variables. From this simple remark follow two important results. First, the constraint in Step (3) on formulas connected by a logical operation does not lead to a constraint on the class of predicates we may represent (we could first rename the bound variables). Second, the bound variables could be renamed in such a way that all the quantifiers are applied to variables denoted by distinct letters. We will henceforth assume that the formulas have this form, so that there is no need to parenthesize the part of a formula to which some quantifier applies.

There are two points to consider in discussing an equivalence between formulas of predicate logic. First, suppose that the atomic domains $\mathfrak{M} = \{\mathfrak{M}_1, \ldots, \mathfrak{M}_n\}$ have been assigned to all the free variables occurring in the formulas $\mathfrak{A}$ and $\mathfrak{B}$, and likewise that the atomic constants in $\mathfrak{A}$ and $\mathfrak{B}$ have been assigned. The formulas $\mathfrak{A}$ and $\mathfrak{B}$ are said to be (*logically*) *equivalent on the system of domains M* if the predicates $\mathfrak{A}(x_1, \ldots, x_r)$ and $\mathfrak{B}(y_1, \ldots, y_m)$ represented by them are equal for all substitutions of the predicates occurring in them. (This substitution must be well-formed, i.e., it must take account of the number of places in the predicates, and predicates denoted by identical letters must be replaced by identical predicates.) Two predicates A and B are said to be *equal* if their values coincide for all values of the variables occurring in them. (If the sequences of variables in A and B do not coincide, it may be assumed that there are actual variables on which A and B do not depend.)

Two formulas $\mathfrak{A}$ and $\mathfrak{B}$ that do not contain the symbols of the individual objects are said to be *absolutely equivalent* if they are equivalent on any sequence of atomic domains.

A designated number of predicates, called *individual predicates*, is sometimes defined on $\mathfrak{M}$; a set of individual predicates is called a *signature.* In considering an equivalence of formulas that contain individual predicates, the latter may be replaced by the correspondingly designated predicates.

Let us illustrate the difference between the definitions of equivalence and equality by means of a very simple example.

12.2. Prove that the formula $A(x) \vee A(y)$ is equivalent to 1 on any domain $\mathfrak{M}$ (more precisely, on $\mathfrak{M} \times \mathfrak{M}$) consisting of a single element, but is not absolutely equivalent to 1.

3. *Examples of Predicates.* We will now present several examples to accustom the reader to the language of predicate logic.

12.3. Introduce one-place predicates on suitable domains and use the predicates to write out the following propositions in the form of formulas of predicate logic.

(a) Every natural number divisible by 12 is divisible by 2, 4, and 6.

(b) The inhabitants of Switzerland speak either French, Italian, or German.

(c) A function continuous on the closed interval [0, 1] preserves the sign or is zero.

The assertion in Problem 12.3 is somewhat vague, since every predicate may be considered one-place. We could consider the problem of finding maximally elementary one-place predicates so as to obtain a maximally meaningful formula (we will not bother refining the meaning of these requirements).

12.4. In the following examples, perform the same operation as in the above problem without limiting yourself to one-place predicates.

(a) If α is a root of a polynomial of one variable with real coefficients, then $\overline{\alpha}$ is also a root of this polynomial.

(b) There is at least one point between any two points on a line (which does not coincide with either of these points).

(c) A unique line passes through two distinct points.

(d) Every student completed at least one laboratory assignment.

(e) If a product of natural numbers is divisible by a prime number, at least one of its factors is divisible by this prime number.

(f) A unique plane passes through three points not all lying on the same line.

4. *Predicates on Finite Domains. The Logic of One-Place Predicates.* Let us list certain properties of the formulas of predicate logic related to a fixed set. For the sake of simplicity, we will assume that some fixed set $\mathfrak{M}$ is the atomic domain of all the atomic variables.

Let us first suppose that the domain $\mathfrak{M}$ is finite and that a_1, a_2, ... , a_n are its elements. Then on $\mathfrak{M}_1$ every formula is equivalent to a formula without any quantifiers. This result follows from two directly verifiable equivalences:

$$\forall x A(x) = A(a_1) \ \& \ \dots \ \& \ A(a_n);$$

$$\exists x A(x) = A(a_1) \lor \dots \lor A(a_n).$$

By means of these relations, the proof may be performed by induction. In the case of an infinite field, the quantifiers may be considered analogs of the conjunction or disjunction of an infinite number of terms* (values of the predicate for all the elements of the field). This analogy is useful to keep in mind in establishing various properties of the quantifiers.

12.5. Suppose that $\mathfrak{M}$ is a formula of predicate logic on a fixed domain $\mathfrak{M}$ (of arbitrary cardinality) containing only individual one-place predicates; prove that there then exists a formula $\tilde{\mathfrak{U}}$ equivalent to $\mathfrak{U}$ and containing the same predicates, but without any quantifiers.

Important results demonstrating the limitations of the language of the logic of one-place predicates are apparent from Problem 12.5.

*The symbol $\bigwedge x$ is sometimes used to denote the universal quantifier, and the symbol $\bigvee x$ the existential quantifier (cf. footnote on page 3).

12.6. Prove that every individual predicate on a finite domain may be represented by a formula containing only one-place predicates.

12.7. Prove that there exist individual predicates that may not be represented by means of a formula containing only one-place predicates on the same atomic domain as the initial predicate. Find necessary and sufficient conditions under which such a representation is possible.

Note that it is essential that the requirement be restricted to the initial domain, since (cf. Definition 12.1) every n-place predicate can always be considered a one-place predicate by introducing a direct product of domains.

5. *Properties of the Quantifiers.* We begin with the distributive properties of the quantifiers.

12.8. Prove the following (absolute) equivalences:

$$\forall x(A(x) \,\&\, B(x)) = \forall x A(x) \,\&\, \forall y B(y);$$

$$\exists x(A(x) \vee B(x)) = \exists x A(x) \vee \exists y B(y).$$

12.9. Prove that the distributive laws for the universal quantifier relative to disjunction and for the existential quantifier relative to conjunction do not, in general, exist, i.e., the following formulas are not absolutely equivalent.

$$\forall x(A(x) \vee B(x)) \quad \text{and} \quad \forall x A(x) \vee \forall y B(y);$$

$$\forall x(A(x) \,\&\, B(x)) \quad \text{and} \quad \exists x A(x) \,\&\, \exists y(B).$$

Let us see what the distributive laws for finite domains look like if we introduce quantifier-free formulas. The laws from Problem 12.8 then follow from the commutativity and associativity of the conjunction and disjunction, respectively. The equivalences in Problem 12.9 do not hold, since conjunction and disjunction cannot, in general, be interchanged.

Below, it is essential that the equivalences spoken about in Problem 12.9 hold if one of the predicates (A or B) is independent of x.

12.10. Prove the absolute equivalences

$$\forall x(A(x) \vee B) = \forall x A(x) \vee B$$

$$\exists x(A(x) \;\&\; B) = \exists x A(x) \;\&\; B,$$

where B is independent of x (x is not a free variable of B).

Now let us consider the commutativity of the quantifiers.

12.11. Prove that predicates of the same type may be interchanged, i.e.,

$$\forall x \forall y A(x, y) = \forall y \forall x A(x, y),$$

$$\exists x \exists y A(x, y) = \exists y \exists x A(x, y).$$

12.12. Prove that quantifiers of different types cannot, in general, be interchanged.

Note that in the case of a finite domain the commutativity of quantifiers of the same type follows from the commutativity of the disjunction and conjunction.

12.13. Explain the geometric meaning of the propositions $\exists x \forall y A(x, y)$ and $\forall y \exists x A(x, y)$ in the case where the atomic domain consists of a set of points on the line.

By analogy with the duality of the disjunction and conjunction, there is a duality between the quantifiers.

12.14. Prove that

$$\overline{\forall x A(x)} = \exists x \overline{A(x)};$$

$$\overline{\exists x A(x)} = \forall x \overline{A(x)}.$$

The equivalences of Problem 12.14 should be compared to de Morgan's laws (2.8) and (2.9). By means of these equivalences and the law of duality in algebraic logic, it is possible to transform a formula of predicate logic into an equivalent formula in which only elementary predicates are negated. A rigorous proof that such a reduction is possible may be performed by induction on the construction of the formula. (We suggest the reader carry out the construction.) The resulting formula is called an *almost normal form* of the initial formula.

Now we can express the wish usually stipulated when constructing definitions of negative concepts in the language of predicate logic. That is, *every definition written as a formula of predicate logic must be in almost normal form.* In this book, we have tried repeatedly to adhere to this constraint in the construction of the definitions of negative concepts (cf. Problems 1.10, 3.6, 5.1, etc.).

12.15. Reduce the following formulas to almost normal form:

(a) $\forall x \overline{\exists y(P(x) \to Q(x))}$;

(b) $\exists x\overline{(\forall y P(x,y,z) \to \exists u Q(x,u)})\ \&\ \forall\overline{(\forall v(A(t) \vee B(v))})$.

12.16. Solve Problems 3.6 and 5.1 using the symbolism of predicate logic.

12.17. Give a definition of a sequence which does not have a finite limit.

Formulas in almost normal form may be transformed further. That is, since we have agreed to denote all bound variables by distinct letters, by Problems 12.8 and 12.10 we may extract the quantifiers from a formula so that it assumes the form of a quantifier-free formula to which quantifiers are sucessively applied. The resulting formula is called the *normal form* of the initial formula.*

12.18. Reduce the formulas of Problem 12.15 to normal form.

It is usually not necessary for a definition to be stated as a formula in normal form, though as a rule, it must be in almost normal form.

6. *Examples of Assertions Written as Formulas of Predicate Logic.* Let us make a number of remarks on the notation used

*Note that the normal and almost normal form are not unique.

to write out assertions and definitions in the form of formulas. It is first necessary to carefully make certain that there are no redundant free variables in the resulting formulas (only those atomic variables asserted to be free may be free variables).

Second, it is often necessary to define the quantifiers over some sub-domain $\mathfrak{M}'$ of the entire atomic domain $\mathfrak{M}$, rather than over all of $\mathfrak{M}$. For example, a quantifier may be defined over the positive numbers, rather than over all the real numbers; over the lines that pass through a given point, rather than over all lines. We thus obtain what are known as *restricted quantifiers*, denoted $\forall_{\mathfrak{M}'} x$ and $\exists_{\mathfrak{M}'} x$, with the former standing for the phrase, "for all x belonging to $\mathfrak{M}'$," and the latter, "there exists an element x belonging to $\mathfrak{M}'$." However, restricted quantifiers may be expressed in terms of the ordinary quantifiers, and thus, we will not make further use of the symbols for the restricted quantifiers.

12.19. Express the restricted quantifiers in terms of formulas that contain only quantifiers defined over the entire predicate domain.

Here are several more exercises on writing out definitions in the form of formulas of predicate logic and on the construction of the negations of these formulas.

12.20.* Write out the following definitions as formulas of predicate logic:

(a) a function $f(x)$ continuous on (0, 1);
(b) a function $f(x)$ not discontinuous on (0, 1);
(c) a function uniformly continuous on (0, 1);
(d) a function continuous, but not uniformly continuous on (0, 1);
(e) a sequence of functions $f_n(x)$ convergent on (0, 1);
(f) a sequence of functions $f_n(x)$ convergent uniformly on (0, 1);
(g) a sequence of functions $f_n(x)$ convergent on (0, 1).

7. *Quantifiers Defined on Predicate Variables.* Let us consider the simplest cases in which it is natural to resort to

*We agree to consider only functions defined over the entire open interval (0, 1).

the use of quantifiers on predicates. One such case occurs when it is necessary to describe some individual predicate. The identity equality predicate $x = y$ is defined uniquely by the conditions

$$\forall x(x = x);$$

$$\forall A \forall x \forall y(x = y \to (A(x) \to A(y))).$$

12.21. Prove that the single predicate $x = y$ that satisfies the above two conditions is the identity equality predicate on every atomic domain $\mathfrak{M}$.

It is also possible to describe a designated set of atomic domains by means of formulas with quantifiers defined on atomic variables. We present a very simple example.

12.22. Prove that the proposition

$$\forall A \forall x \forall y(A(x) \vee \overline{A(y)})$$

is true for domains consisting of a single element, and only for such domains.

In general, formulas consisting of quantifiers defined on both atomic as well as predicate variables are used to describe an arbitrary set of atomic domains with designated individual predicates defined on them (signature). A system of such formulas is called a *system of axioms* and sets with individual predicates that satisfy these axioms are called *interpretations* of the system of axioms.

12.23. Describe by means of the axioms domains consisting of (a) more than one element; (b) at most two elements; (c) two elements; (d) finite domains; (e) infinite domains.

12.24. Prove that the set of all finite domains (as well as all infinite domains) cannot be described by means of axioms that contain only one-place predicates.

As an example, we present a system of axioms for the natural number series. An atomic domain $\mathfrak{M}$ equipped with the individual object 0 and individual predicates $x = y$ and

$x < y$ that satisfy Axioms (1) – (6) below is called a *natural number series.* The following notation is used for the predicates occurring in the axioms:

$$x \leqslant y = x < y \vee x = y;$$

$$\sigma(x, y) = x < y \ \& \ \forall z(z \leqslant x \vee y \leqslant z);$$

$\sigma(x, y)$ is the predicate, "y directly follows x."

In the following formulas, all variables not bound by any quantifier, whether they are atomic or predicate variables, are assumed to be bound by the universal quantifiers with which these formulas begin.

Axioms of the Natural Number Series

(1) $x = x$;
(2) $x = y \to (A(x) \to A(y))$;
(3) $\overline{x < x}$;
(4) $x < y \to (y < z \to x < z)$;
(5) $\exists y\sigma(x, y) \ \& \ \forall z(\sigma(x, z) \to z = y)$;
(6) $A(0) \ \& \ (A(x) \ \& \ \sigma(x, y) \to A(y)) \to A(z)$.

Axioms (1) and (2) define the identity equality predicate, which we have already discussed; Axioms (3) and (4) assert that the predicate $x < y$ introduces an order relation in $\mathfrak{M}$; Axiom (5) asserts that there exists a unique, immediate successor element; and Axiom (6) is what is known as the axiom of complete induction, according to which if some assertion is true for 0 and if it is true for the element that is the immediate successor of some element x as a consequence of its truth for x, then the assertion is true for all the elements of $\mathfrak{M}$. It can be proved that all interpretations of Axioms (1) – (6) are isomorphic in some sense, i.e., the natural number series is unique to within isomorphism. Further information on the axiomatic structure of the natural number series may be found in Refs. 1 and 2.

Hints

12.1. 2^{n^k}.

12.2. This formula is not equivalent to 1 even on domains consisting of two variables.

12.3. (a) Introduce the following predicates on the natural number series:

$A(x)$: x is divisible by 12 (i.e., $A(x) = 1$ if and only if x is divisible by 12);
$B(x)$: x is divisible by 2;
$C(x)$: x is divisible by 4;
$D(x)$: x is divisible by 6;

(b) Introduce the following predicates on the set of human beings:

$A(x)$: x lives in Switzerland;
$B(x)$: x speaks French;
$C(x)$: x speaks Italian;
$D(x)$: x speaks German;

(c) Introduce the following predicates on the set of functions:

$A(f)$: f is a continuous function (on some closed interval);
$B(f)$: f preserves the sign;
$C(f)$: f vanishes.

Besides $A(f)$, we could also introduce predicates on the field of real numbers, for example: $x \in [0,1]$: x belongs to the closed interval $[0,1]$; $x \geqslant 0$, $x = 0$.

12.4. Whenever possible, use the ordinary notation for the elementary predicates. Introduce the following predicates:
(a) the two-place predicate $f(\alpha) = 0$, where f belongs to the atomic domain of polynomials of one variable with real coefficients and α is the atomic domain of complex numbers;
(b) the three-place predicate $A(x,y,z)$ which asserts that the point z lies between the points x and y; the two-place predicate $x \in \ell$: the point x belongs to the line ℓ; the two-place precicate $x = y$; the points x and y coincide.
(c) $A(x, y)$: the student x has completed laboratory project y;
(d) $x|y$: the natural number y is divisible by x;

$A(x)$: x is a prime number (incidentally, this predicate may be expressed in terms of $x|y$ and $x = y$).

12.5. Prove that the result of applying a quantifier to a quantifier-free formula with individual one-place predicates may be written in terms of the same predicates in quantifier-free form. The rest of the proof is performed by induction.

12.6. Introduce one-place predicates equal to 1 on a unique element.

12.7. A necessary and sufficient condition under which the representation is possible is that the truth set (sequences of values of the arguments under which the predicate is equal to 1) is the union of a finite number of direct products of subdomains (subsets) of the initial domain. In particular, the two-place identity (coincidence) equality predicate $x = y$ cannot be represented in this form on any infinite domain.

12.8. Verify directly from the definition.

12.9. The implication $\forall x(A(x) \vee B(x)) \rightarrow \forall yA(y) \vee \forall zB(z)$ may be false.

12.10. Verify directly.

12.11. The truth of the proposition $\forall x \forall y A(x, y)$ denotes that the predicate $A(x, y)$ is true for all pairs (x, y), while the truth of $\exists x \exists y A(x, y)$ denotes that $A(x, y)$ is true for at least one pair (x, y).

12.12. Consider, for example, the predicate $x < y$ on the natural number series.

12.13. Note that the truth set of the predicate $A(x, y)$ is a set in the plane (x, y).

12.14. Let us list the propositions whose truth must be established:

"$A(x)$ does not hold for all x if and only if there exists an x such that $A(x)$ does not hold."

"There does not exist an x such that $A(x)$ holds if and only if $A(x)$ does not hold for all x."

It is self-evident that these assertions are true, though this does not prevent mistakes when interpreting negations of quantifiers. Often the same quantifier is retained after a negation has been applied. ("It is not the case that all cats are gray" – "All cats are gray.").

12.15. (a) $\overline{\forall x \forall y (P(x) \& \overline{Q(y)})}$; (b) $\exists x \exists y (P(x, y, z) \& \overline{Q(x, y)}) \& \forall t \exists v (A(t) \& \overline{B(v)})$.

12.16. Write down the definitions of self-dual and monotone functions in the form of formulas of predicate logic and reduce the negations of these formulas to almost normal form.

12.17. Recall that the number A is the limit of the sequence $a_1, a_2, \dots, a_n, \dots$ ($\lim_{n\to\infty} a_n = A$) if for every $\varepsilon > 0$, there exists a natural number N such that $|a_n - A| < \varepsilon$ if $n > N$. First write out the proposition that the sequence $\{a_n\}$ has a limit in the form of a formula of predicate logic.

12.18. (b) $\exists x \forall y \forall t \exists u (P(x,y, z) \& \overline{Q(x, y)} \& \overline{A(t)} \& \overline{B(u)})$.

12.19. $\forall_{\mathfrak{M}'} xA(x) = \forall x(x \in \mathfrak{M}' \to A(x)) = \forall x(x \notin \mathfrak{M}' \vee A(x))$; $\exists_{\mathfrak{M}'} xA(x) = \exists x(x \in \mathfrak{M}' \& A(x))$.

These equivalences may be verified directly.

12.20. (a) Recall that $f(x)$ is continuous at the point x_0 if $\lim_{x\to x_0} f(x)$ exists and is equal to $f(x_0)$, i.e., for every $\varepsilon > 0$,

there exists a number $\delta > 0$ such that $|f(x) - f(x_0)| < \varepsilon$ if $|x-x_0| < \delta$. A function is continuous on the interval (0, 1) if it is continuous at every point of this interval.

(b) Construct the negation of the preceding formula.

(c) A function is uniformly continuous on (0, 1) if the number $\delta > 0$ occurring in the definition of continuity can be one and the same for each $\varepsilon > 0$ at all points in the interval;

(d) The sequence of values of the functions converges at every point $x \in (0, 1)$.

12.23. Formulas that contain an individual identity equality predicate defined on an arbitrary domain can be easily constructed. Try to also derive formulas without any individual predicates.

(d) Consider the formula

$$\forall A \exists x \exists y \exists z \forall u (A(x,y) \vee A(x,y)A(y,z)\overline{A(x,z)} \vee \overline{A(x,u)}).$$

12.24. It may be assumed that the axioms contain only bound variables (predicate and atomic), i.e., are propositions on each domain. Then considering a conjunction of the axioms, we may assume that we have a single axiom. Prove that if some formula without any free variables and containing only one-place predicates is true on some domain $\mathfrak{M}$, it is true on some finite domain $\mathfrak{M}$.

Solutions

12.1. The number of one-place predicates on some set $\mathfrak{M}$ is equal to the number of subsets of this set (with each predicate we associate the set of elements on which it is equal to 1); i.e., on a set of n elements there are 2^n one-place predicates. We can then use the fact that the k-place predicates on $\mathfrak{M}$ can be considered one-place predicates on a k-fold direct product of sets $\mathfrak{M}$, i.e., on a domain of n^k elements.

12.2. Let $\mathfrak{M}$ consist of the elements a and b and let $A(a) = 0$ and $A(b) = 1$.

12.3. (a) $\forall x(A(x) \to B(x)C(x)D(x))$;

(b) $\forall x(A(x) \to B(x) \vee C(x) \vee D(x))$;

(c) $\forall f(A(f) \to B(f) \vee C(f))$.

Here is another solution:

$$\forall f(A(f) \to \forall x \forall y(x \in [0,1] \ \& \ y \in [0,1] \to f(x)f(y) \geqslant 0) \vee \exists z(f(z) = 0)).$$

12.4. (a) $\forall f \forall \alpha(f(\alpha) = 0 \to f(\overline{\alpha}) = 0)$;

(b) $\forall x \forall y \forall \ell(x \in \ell \ \& \ y \in \ell \ \& \ \overline{x=y} \to \exists z(z \in \ell \ \& \ A(x, y, z) \ \& \ \overline{x = z} \ \& \ \overline{y = z}))$;

(c) $\forall x \forall y \overline{(x = y} \to \exists P(x \in P \;\&\; y \in P \;\&\; \forall Q(x \in Q \;\&\; y \in Q \to P = Q)))$;

(d) $\forall x \exists y A(x, y)$;

(e) $\forall x \forall y \forall z(z|xy \;\&\; A(z) \to z|x \vee z|y)$.

We have $A(z) = \underline{\forall x(x|z \to x = z \;\vee x = 1)}$,

(f) $\forall x \forall y \forall z(\exists \ell(x \in \ell \;\&\; y \in \ell \;\&\; z \in \ell) \to \exists U(x \in U \;\&\; y \in U \;\&\; z \in U \;\&\; \forall V(x \in V \;\&\; y \in V \;\&\; z \in V \to U = V)))$; x, y, and z are points, ℓ is a line, and U and V are planes.

Note that in all these examples we are dealing with propositions, so that all the atomic variables in the resulting formulas are bound.

12.5. Suppose that $\mathfrak{B}$ is a quantifier-free formula containing the individual one-place predicates $A_1, \ldots, A_m$ and the atomic variables $x_1, \ldots, x_m$. Then there exists a logical function* $\omega(X_{ij})$ $(i = 1, \ldots, m;\ j = 1, \ldots, n)$ such that $\mathfrak{B}(x_1, \ldots, x_n) = \omega(A_1(x_1), \ldots, A_m(x_1); A_1(x_n), \ldots, A_m(x_n))$. The atomic constants may be replaced by the logical constants (0 or 1). Formally, we assume that all the predicates $A_i(x_j)$ occur in $\mathfrak{B}$. Let us consider

$$\tilde{\mathfrak{B}}(x_1; X_{12}, \ldots, X_{m2}; \ldots; X_{1n}, \ldots, X_{mn})$$
$$= \omega(A_1(x_1), \ldots, A_m(x_1); X_{12}, \ldots, X_{m2}; \ldots;$$
$$X_{1n}, \ldots, X_{mn}).$$

Then** $Qx_1\tilde{\mathfrak{B}}$ is the logical function $\psi(X_{ij})$ $(i = 1, \ldots, m;\ j = 1, \ldots, n)$ and

$$Qx_1\mathfrak{B}(x_1, \ldots, x_n) = \psi(A_1(x_2), \ldots, A_m(x_2);$$
$$\ldots; A_1(x_n), \ldots, A_m(x_n)).$$

On the right side is a quantifier-free formula. The case of several quantifiers may be treated by induction.

---

*In this section we are denoting the logical variables (which take the values 0 and 1) by capital Latin letters.

**We are using the notation Qx whenever it is of no importance which quantifier occurs in the formula.

12.6. A predicate is determined by its truth set (set of sequences of values of the arguments on which the predicate is equal to 1). It is convenient to introduce the predicates

$$\delta_a(x) = \begin{cases} 1 & \text{if } x = a \\ 0 & \text{otherwise} . \end{cases}$$

The truth set of this predicate consists of the single element a. Since the predicate $\delta_{a_1}(x_1) \dots \delta_{a_n}(x_n)$ is equal to 1 on the single sequence $(a_1, \dots , a_n)$, and since we may associate with a disjunction of predicates the union of their truth sets, every predicate may be found by taking a disjunction of predicate of the form $\delta_{a_1}(x_1), \dots , \delta_{a_n}(x_n)$.

12.7. Let us prove the assertion given in the hint.

Necessity. By Problem 12.5, we may limit the discussion to quantifier-free formulas. A formula is a logical function of individual one-place predicates. Let us represent it in DNF. Introducing, if necessary, new one-place predicates, we may assume that every term of the disjunction is a conjunction of one-place predicates that depend on distinct atomic variables. (The new predicates are conjunctions of predicates occurring in some elementary conjunction and thus all depend on the same variable; we need only conisder all the elementary, though not necessarily complete, conjunctions of predicates occurring in a formula.) It is clear that the truth set of this conjunction is the direct product of the truth sets of the predicates occurring in the conjunction. The union of these direct products corresponds to the entire formula.

To prove sufficiency, note that the correspondence between conjunctions of one-place predicates of distinct variables and direct products of subsets of an atomic domain is one-to-one. A disjunction of conjunctions corresponds to a union of direct products.

The solution of Problem 12.6 essentially asserts that in the case of a finite domain, a truth set may be partitioned into a union of one-element sets, each of which, of course, is a direct product.

The truth set of the predicate $x = y$ consists of the "diagonal of a direct product," or set of pairs (a,a) (for example, the line $x = y$ in the plane (x, y)). The only partitioning of this set into a union of direct products is a partitioning into sets each consisting of a single point. Therefore, there is no finite partitioning of an infinite set.

12.9. We need only present some examples.

1. Suppose that $A(x)$ is the predicate defined on the natural number series which asserts that "x is an even number" and $B(x)$ is likewise a predicate defined on the natural series which asserts that "x is an odd number." Then the proposition $\forall x(A(x) \vee B(x))$, which asserts that "every natural number is odd or even," is true. However, the proposition $\forall xA(x) \vee \forall yB(y)$, which asserts that "every natural number is even or every natural number is odd," is false (every term of the disjunction is false).

Similarly, from the assertion that "all the students went to the movies or to the theater" it does not, in general, follow that "all the students went to the movies or all the students went to the theater." (They may have gone to different places, some to the movies, and others to the theater.)

As can be easily verified, the implication

$$\forall xA(x) \vee \forall yB(y) \rightarrow \forall z(A(z) \vee B(z)).$$

is, in general, always true, though the implication

$$\forall x(A(x) \vee B(x)) \rightarrow \forall yA(y) \vee zB(z)$$

can be false.

2. We may similarly consider the other distributive law. From the assertion that "there is a boy with azure eyes, and there is a boy with hazel eyes," it does not, of course, follow that "there is a boy who has both azure and hazel eyes." Thus, the implication

$$\exists xA(x) \ \& \ \exists yB(y) \rightarrow \exists z(A(z) \ \& \ B(z))$$

may be false, whereas the implication

$$\exists x(A(x) \ \& \ B(x)) \rightarrow \exists yA(y) \ \& \ \exists zB(z),$$

is, of course, always true.

12.12. The proposition $\forall x \exists y\ (x > y)$ ("for every natural number, there exists a larger number") is true on the natural number series, while the proposition $\exists y \forall x\ (x < y)$ is false, since there is no greatest natural number. It is similarly true that "every book has been read by some person" (the author, for example), though it is not true that "there is a person who had read every book."

The truth of the implication*

$$\exists x \forall y A(x, y) \rightarrow \forall y \exists x A(x,y)$$

can be verified, whereas the converse implication can be false.

12.13. 1. The predicate $\forall y A(x, y)$ is equal to 1 for those x_0 for which the vertical line $x = x_0$ is in the set $A(x, y)$. The proposition $\exists x \forall y A(x, y)$ is true if the truth set of $A(x, y)$ contains a vertical line.

2. The predicate $\exists x A(x, y)$ is equal to 1 for those y contained in the projection of the truth set of $A(x, y)$ on the y-axis. Therefore, the proposition $\forall y \exists x A(x,y)$ is true if the projection of the truth set of the predicate $A(x,y)$ on the y-axis coincides with the entire axis.

These examples illustrate the fact that the truth of $\exists x \forall y A(x, y)$ implies the truth of $\forall y \exists x A(x, y)$, though the converse is, in general, false. (We could consider the equality predicate, with which we associate the line $x = y$.)

12.15. (a)

$$\forall x \forall y \overline{(P(x) \rightarrow Q(y))} = \forall x \forall y \overline{(\overline{P(x)} \vee Q(y))}$$

$$= \forall x \forall y (P(x) \,\&\, \overline{Q(y)});$$

(b)

$$\exists x(\forall y(P(x,y,z) \,\&\, \overline{\exists u Q(x,u)})) \,\&\, \forall t \exists v \overline{(A(t) \vee B(v))}$$

$$= \exists x(\forall y(P(x,y,z) \,\&\, \forall u \overline{Q(x,u)}))$$

*Note that the sentence corresponding to the proposition $\exists x \forall y(A(x,y)$ is read thus: "There exists an x such that for every y, A(x,y) is true," while the proposition $\forall y \exists x A(x,y)$ is read "For every y, an x may be found such that A(x,y) is true."

$$\& \ \forall t \exists v(\overline{A(t)} \ \& \ \overline{B(v)})$$

$$= \exists x \forall y(P(x,y,z) \ \& \ \overline{Q(x,y)}) \ \& \ \forall t \exists v(\overline{A(t)} \ \& \ \overline{B(v)}).$$

On the last step, we have used the distributive property of the universal quantifier. (However, we did not have to do this, since an almost normal form has already been obtained on the preceding step.)

12.16. (1) (Problem 3.6) We write out the definition of a self-dual function. A function f is self-dual if

$$\forall \alpha(f(\alpha) = \overline{f(\bar{\alpha})}).$$

The quantifier is defined on a set of binary sequences α of suitable length and $\bar{\alpha}$ is the opposite sequence. Then f is a non-self-dual function if

$$\overline{\forall \alpha(f(\alpha) = \overline{f(\bar{\alpha})})},$$

i.e., if

$$\exists \alpha\overline{(f(\alpha) = \overline{f(\bar{\alpha})})}$$

or, what is the same thing, $\exists \alpha(f(\alpha) = f(\bar{\alpha}))$.

(2) (Problem 5.1.) We give the definition of a monotone function:

$$\forall \alpha \forall \beta(\alpha \preccurlyeq \beta \rightarrow f(\alpha) \leqslant f(\beta)).$$

Therefore f is a non-monotone function if

$$\overline{\forall \alpha \forall \beta(\alpha \preccurlyeq \beta \rightarrow f(\alpha) \leqslant f(\beta))}$$

$$= \exists \alpha \exists \beta\overline{(\alpha \preccurlyeq \beta \rightarrow f(\alpha) \leqslant f(\beta))}$$

$$= \exists \alpha \exists \beta(\alpha \preccurlyeq \beta \ \& \ \overline{f(\alpha) \leqslant f(\beta)})$$

$$= \exists \alpha \exists \beta(\alpha \preccurlyeq \beta \ \& \ f(\alpha) > f(\beta)).$$

12.17. The sequence $\{a_n\}$ has finite limit if

$$\exists A \forall \varepsilon(\varepsilon > 0 \rightarrow \exists N \forall n(n > N \rightarrow |a_n - A| < \varepsilon)).$$

We will not enumerate the sets over which the quantifiers are defined. Applying negation and reducing the result to almost normal form, we find that the sequence $\{a_n\}$ does not have a finite limit if

$$\forall A \exists \varepsilon(\varepsilon > 0 \;\&\; \forall N \exists n(n > N \;\&\; |a_n - A| \geqslant \varepsilon)),$$

i.e., the sequence $\{a_n\}$ does not have a finite limit if for every A, a number $\varepsilon > 0$ can be found such that for every number n there exists a number $n > N$ such that $|a_n - A| \geqslant \varepsilon$.

12.18. (a) An almost normal form is already in normal form.

(b) In the formula already obtained, it is only necessary to extract the quantifiers:

$$\exists x \forall y(P(x,y,z) \;\&\; \overline{Q(x, y)}) \;\;\&\;\; \forall t \exists v(\overline{A(t)} \;\&\; \overline{B(v)})$$
$$= \exists x \forall y \forall t \exists v(P(x,y,z) \;\&\; \overline{Q(x, y)} \;\&\; \overline{A(t)} \;\&\; \overline{B(v)}).$$

12.20. (a)

$$\forall x_1(x_1 \in (0,1) \to \forall \varepsilon(\varepsilon > 0 \to \exists \delta\, (\delta > 0 \;\&\; \forall x_2(|x_2 - x_1| < \delta \to |f(x_2) - f(x_1)| < \varepsilon)))).$$

In this formula, the only free atomic variable is the function f. A common error is to forget to relate the variables x_1 and x_2 in writing out the definition. In general, it is often a simple matter to omit some universal quantifier. Note that all four of the quantifiers are restricted quantifiers.

(b) We at once reduce the negation of the preceding formula to almost normal form, noting that $\overline{x \to y} = x \;\&\; \overline{y}$. We find that

$$\exists x_1(x_1 \in (0,1) \;\&\; \exists \varepsilon(\varepsilon > 0 \;\&\; \forall \delta\, (\delta > 0 \to \exists x_2(|x_2 - x_1| < \delta \;\&\; |f(x_2) - f(x_1)| \geqslant \varepsilon)))).$$

We advise the reader to find out how the definition of discontinuity would be affected is some of the quantifiers in the definition of continuity were omitted.

(c) $$\forall \varepsilon (\varepsilon > 0 \rightarrow \exists \delta (\delta > 0 \ \& \ \forall x_1 (x_1 \in (0,1) \rightarrow \forall x_2 (|x_2 - x_1| < \delta \rightarrow |f(x_2) - f(x_1)| < \varepsilon)))).$$

The definition of uniform continuity may be obtained from the definition of continuity by interchanging the quantifiers. Interchanging quantifiers of the same type (e.g., $\forall x_1$ and $\forall \varepsilon$), does not alter the meaning of a proposition, as is well known, though interchanging the quantifiers $\forall x_1$ and $\exists \delta$ does, as is shown by the examples constructed in our analysis of continuous, but not uniformly continuous functions. Note that continuity follows from uniform continuity, which is in agreement with the general remark made in discussing the possibility of interchanging quantifiers (cf. solution of Problem 12.12).

(d) We must consider the conjunction of the definition of continuity and the following proposition, which is the negation of the definition of uniform continuity:

$$\exists \varepsilon (\varepsilon > 0 \ \& \ \forall \delta (\delta > 0 \rightarrow \exists x_1 (x_1 \in (0,1) \ \& \ \exists x_2 (|x_2 - x_1| < \delta \ \& \ |f(x_2) - f(x_1)| \geqslant \varepsilon)))).$$

(e)

$$\forall x (x \in (0,1) \rightarrow \exists A \forall \varepsilon (\varepsilon > 0 \rightarrow \exists N \forall n (n > N \rightarrow |A - f_n(x)| < \varepsilon))),$$

or what is the same thing,

$$\exists f \forall x (x \in (0,1) \rightarrow \forall \varepsilon (\varepsilon > 0 \rightarrow \exists N \forall n (n > N \rightarrow |f(x) - f_n(x)| < \varepsilon)));$$

In other words, the existence of a limiting domain is asserted.

(f)

$$\exists f \forall \varepsilon (\varepsilon > 0 \rightarrow \exists N \forall x \forall n (x \in (0,1) \ \& \ n > N \rightarrow |f(x) - f_n(x)| < \varepsilon)),$$

i.e., by comparison with the second formula in (e), in (f) the quantifiers $\forall x$ and $\exists N$ have been interchanged (it is not important the $\forall x$ and $\forall \varepsilon$ were interchanged).

(g) It is necessary to consider the conjunction of the formula in (e) and the negation of the formula in (f), i.e., the formula

$$\forall f\, \exists \varepsilon\, (\varepsilon > 0\ \&\ \forall N \exists x\, \exists n (x \in (0,1)\ \&$$

$$n > N\ \&\ |f(x) - f_n(x)| \geqslant \varepsilon)).$$

The resulting formula may be simplified somewhat. By virtue of the uniqueness of the limit function, the quantifier $\forall f$ may be omitted from the second term of the conjunction and the quantifier $\exists f$ extended to both terms.

In constructing the negations of formulas, note that the restricted quantifiers obey duality relations.

12.21. Suppose that a predicate $(x = y)'$ (not the identity equality) is defined on some set and satisfies the two conditions stated. Further, suppose that a and b are two distinct elements such that the predicate $(a = b)'$ is equal to 1 (by the first condition $(a = a)'$ is always equal to 1). If we consider a predicate $A(x)$ such that $A(a) = 1$ and $A(b) = 0$, we find that the proposition $(a = b)' \to (A(a) \to A(b))$ is false, and we have arrived at a contradiction.

12.22. That the formula is true for domains consisting of a single element is obvious and has already been noted. If there are two distinct elements a and b in the domain, then, by considering a predicate $A(x)$ such that $A(a) = 0$ and $A(b) = 1$, we find that the formula is false (cf. solution of Problem 12.2).

12.23. (a) $\exists x \exists y (\overline{x = y})$; the negation of the formula from Problem 12.22 ($\exists x \exists y\, A(\overline{A(x)}\ \&\ A(y))$ could also be constructed.

(b) $\forall x \forall y \forall z \forall A (A(x) \vee A(y) \vee \overline{A(z)})$. It is clear that, in general, domains containing at most k elements are described by the formula

$$\forall x_1 \ldots \forall x_k \forall y \forall A (A(x_1) \vee \ldots \vee A(x_k) \vee \overline{A(y)}).$$

The proof is analogous to the solution of Problem 12.22.

(c) It is necessary to consider the conjunction of the formulas in (a) and (b):

$$\exists x\exists y\exists A(\overline{A(x)}\ \&\ A(y)\ \&\ \forall z\forall t\forall u\forall B(B(z)\vee B(t)\vee\overline{B(u)})).$$

All the quantifiers may be extracted. We may analogously describe domains containing precisely k elements.

(d) Let us consider the formula

$$A(x,x)\vee A(x,y)\,A(y,z)\overline{A(x,z)}\vee\overline{A(x,u)}$$

for a predicate A on a finite domain $\mathfrak{M}$. Let us suppose that the propositions $\forall x\overline{A(x,x)}$, $\forall x\exists uA(x,u)$, and $\forall x\forall y\forall z(\overline{A(x,y)}\vee\overline{A(y,z)}\vee A(x,z))$ are true. Suppose that $x_1\in\mathfrak{M}$ and let x_2 be an element of $\mathfrak{M}$ such that $A(x_1,x_2)=1$ (x_2 exists by virtue of the truth of the second of the above propositions); in general, x_{i+1} is selected from the condition $A(x_i,x_{i+1})=1$. We obtain the sequence $x_1,x_2,\ldots,x_i,\ldots$. By induction, we prove that, because the third of the above formulas is true, $A(x_i,x_j)=1$ if $i<j$, whence by the first formula $x_1,x_2,\ldots,x_i,\ldots$ are all distinct. We have arrived at a contradiction, since $\mathfrak{M}$ is a finite domain. Thus, the formula presented in the hint is true for any finite domain.

Now suppose that $\mathfrak{M}$ is an infinite domain. We partition all its elements into non-empty non-intersecting sets $\mathfrak{M}_0,\mathfrak{M}_1,\ldots,\mathfrak{M}_n,\ldots$. For this purpose, we may consider any countable subset $x_1,x_2,\ldots,x_n,\ldots$ in $\mathfrak{M}$ (which exists in any infinite domain) and consider a set $\mathfrak{M}_i$ ($i>0$) consisting of a single element x_i, all the other elements of $\mathfrak{M}$ having been placed in $\mathfrak{M}_0$. We let the predicate $A(x,y)$ equal 1 if $x\in\mathfrak{M}_i$ and $y\in\mathfrak{M}_j$, where $i<j$, and let it equal 0 for other pairs. The proposition

$$\exists x\exists y\exists z\forall u(A(x,x)\vee A(x,y)A(y,z)\overline{A(x,z)}\vee\overline{A(x,u)}),$$

is true for this predicate, so that the formula in the hint is false for this problem.

(e) A formula that describes all infinite domains may be obtained if the formula in (d) is negated.

12.24. Suppose that the formula $\mathfrak{U}$ without free variables and with one-place predicates is true in the domain $\mathfrak{M}$, and let $A_1,\ldots,A_n$ be predicate variables bound in $\mathfrak{U}$ by existential quantifiers, $A_1^0,\ldots,A_n^0$ predicates on $\mathfrak{M}$ which, upon replacing

$A_1, \dots, A_n$, yield a true formula for any substitution of the other predicate variables. We partition $\mathfrak{M}$ into disjoint subsets $_i$, including in a single subset all elements from $\mathfrak{M}$ on which each of the predicates $A_1^0, \dots, A_n^0$ takes the same values. The set of these subsets $\mathfrak{M}$ is also a finite domain on which $\mathfrak{U}$ is true. We need only note that the one-place predicates $\tilde{A}$ on $\tilde{\mathfrak{M}}$ are in one-to-one correspondence with the one-place predicates A on $\mathfrak{M}$ that are constant on all the $\mathfrak{M}_i$ (the value of A on $\mathfrak{M}_i \in \mathfrak{M}$ coincides with the value of A on elements of $\mathfrak{M}_i$). A_j^0 is one of the latter predicates. We will not present the detailed reasoning.

Hence, it follows that no system of axioms containing only one-place predicates can describe the set of all infinite domains, thus neither the set of all finite domains, since one axiomatic structure may be obtained from another axiomatic structure by introducing the disjunction of the negations of the axioms.

APPENDIX

1. *The Basic Logical Operations.*

x	y	$\bar{x}$	$x \& y$	$x \vee y$	$x \to y$	$x \sim y$
0	0	1	0	0	1	1
0	1	1	0	1	1	0
1	0	0	0	1	0	0
1	1	0	1	1	1	1

2. *The Basic Logical Equivalences in Algebraic Logic.*

$$\bar{\bar{x}} = x; \tag{2.1}$$

$$xy = yx; \tag{2.2}$$

$$(xy)z = x(yz); \tag{2.3}$$

$$x \vee y = y \vee x; \tag{2.4}$$

$$(x \vee y) \vee z = x \vee (y \vee z); \tag{2.5}$$

$$x(y \vee z) = xy \vee xz; \tag{2.6}$$

$$x \vee yz = (x \vee y)(x \vee z); \tag{2.7}$$

$$\overline{x \vee y} = \bar{x}\bar{y}; \tag{2.8}$$

$$\overline{xy} = \overline{x} \vee \overline{y}; \tag{2.9}$$

$$x \vee x = x; \tag{2.10}$$

$$xx = x; \tag{2.11}$$

$$1x = x; \tag{2.12}$$

$$0 \vee x = x; \tag{2.13}$$

$$f(x_1, \dots, x_n) = \bigvee_{f(\sigma_1, \dots, \sigma_n) = 1} x_1^{\sigma_1} \dots x_n^{\sigma_n}; \tag{2.14}$$

$$f(x_1, \dots, x_n) = \bigvee_{(\sigma_1, \dots, \sigma_n)} f(\sigma_1, \dots, \sigma_n) x_1^{\sigma_1} \dots x_n^{\sigma_n}; \tag{2.15}$$

$$f(x_1, \dots, x_n; y_1, \dots, y_m) = \bigvee_{(\sigma_1, \dots, \sigma_n)} f(\sigma_1, \dots, \sigma_n; y_1, \dots, y_m) x_1^{\sigma_1} \dots x_n^{\sigma_n}; \tag{2.16}$$

$$f(x_1, \dots, x_n) = \prod_{f(\overline{\sigma}_1, \dots, \overline{\sigma}_n) = 0} x_1^{\sigma_1} \quad \dots \quad x_n^{\sigma_n}; \tag{2.17}$$

$$f(x_1, \dots, x_n) = \prod_{(\sigma_1, \dots, \sigma_n)} f(\overline{\sigma}_1, \dots, \overline{\sigma}_n) \; x_1^{\sigma_1} \vee \dots \vee x_n^{\sigma_n}; \tag{2.18}$$

$$f(x_1, \dots, x_n; y_1, \dots, y_m) = \prod_{(\sigma_1, \dots, \sigma_n)} f(\overline{\sigma}_1, \dots, \overline{\sigma}_n; y_1, \dots, y_m) \vee x_1^{\sigma_1} \vee \dots \vee x_n^{\sigma_n}; \tag{2.19}$$

$$x \vee xy = x; \tag{2.20}$$

$$x(x \vee y) = x; \tag{2.21}$$

$$x \vee \bar{x}y = x \vee y; \tag{2.22}$$

$$\bar{x} \vee xy = \bar{x} \vee y; \tag{2.23}$$

$$x(\bar{x} \vee y) = xy; \tag{2.24}$$

$$\bar{x}(x \vee y) = \bar{x}y. \tag{2.25}$$

3. *Description of the Functionally Closed Classes.* General notation stipulations. If Q is a functionally closed class, Q_0 is the class consisting of the elements of Q that preserve the additive identity (zero) ($Q_0 = Q \cap \mathbf{P}_0$); Q_1 consists of the elements of Q that preserve the multiplicative identity (unit); $Q_{01} = Q_0 \cap Q_1$ is the set of elements of Q that preserve the additive and multiplicative identities. We let Q^0, Q^1, and Q^{01} denote the functionally closed classes obtained from Q by an extension by means of 0, 1, and 0 and 1, respectively. By an *extension*, we understand the smallest functionally class containing Q and the corresponding constants; this class does not, in general, coincide with the union of Q and these constants. (For example, as follows from Post's theorem, $\mathbf{P}_0^1 = \mathbf{P}$.) However, in those rare cases when we do make use of this notation, Q^0, Q^1, and Q^{01} may be obtained from Q by the simple adjunction of the corresponding constants. Finally, if Q and R are functionally closed classes, QR is the functionally closed class which is the intersection of Q and R. With this notation, one and the same class may be denoted in different ways.

Let us now describe the classes:

$\mathbf{P}$: The class of all the logical functions.

$\mathbf{P}_0$: The class of all the functions which preserve the additive identity ($f \in \mathbf{P}_0 \Leftrightarrow f(0,0, \ldots, 0) = 0$).

$\mathbf{P}_1$: The class of all the functions which preserve the multiplicative identity (P_1 is dual to P_0).

$\mathbf{P}_{01} = \mathbf{P}_0\mathbf{P}_1$: This class is self-dual, i.e., a function dual to a function in the class is also in the class (that a class is self-dual does not, of course, mean that its elements are self-dual functions).

$\mathbf{M}$: The class of monotone functions (Definition 3.2); this class is self-dual.

$\mathbf{M}_1$: The class of monotone functions which preserve the multiplicative identity (that is, all the monotone functions, other than the additive identity), i.e., $\mathbf{M}_1$ is obtained from $\mathbf{M}$ by discarding zero.

$\mathbf{M}_0$: The class of monotone functions which preserve the additive identity; it is obtained from $\mathbf{M}$ discarding unit; $\mathbf{M}_0$ and $\mathbf{M}_1$ are dual.

The set of all monotone functions which are not constants coincides with the class $\mathbf{M}_{01}$ of all monotone functions which preserve the additive and multiplicative identities; this class is self-dual.

S: The class of self-dual functions ($f \in \mathbf{S} \Leftrightarrow f(x_1, \dots, x_n) = \overline{f(\overline{x_1}, \dots, \overline{x_n})}$); this class and all its subclasses are self-dual.

$\mathbf{S}_{01}$: The class of self-dual functions which preserve the additive and multiplicative identities ; $\mathbf{S}_{01} = \mathbf{S}_0 = \mathbf{S}_1$, i.e., from the fact that some self-dual function preserves the additive identity, it follows that it preserves the multiplicative identity, and conversely.

SM: The class of self-dual monotone functions.

L: The class of linear functions ($f \in \mathbf{L} \Leftrightarrow f(x_1, \dots, x_n) = x_1 + \dots + x_n + a$); **L** is a self-dual class.

$\mathbf{L}_0$: The class of linear functions which preserve the multiplicative identity; i.e., functions of the form $x_1 + \dots + x_n$.

$\mathbf{L}_1$: The class of linear functions which preserve the multiplicative identity, i.e., functions of the form $x_1 + \dots + x_{2k} + 1$ or $x_1 + \dots + x_{2k+1}$. The classes $\mathbf{L}_0$ and $\mathbf{L}_1$ are dual.

LS: The class of self-dual linear functions; it coincides with the set of functions of the form $x_1 + \dots + x_{2k+1} + a$.

$\mathbf{L}_{01} = \mathbf{L}_0\mathbf{L}_1$: The self-dual class consisting of functions of the form $x_1 + \dots + x_{2k+1}$; $\mathbf{L}_{01} = \mathbf{LS}_0 = \mathbf{LS}_1$.

0 = The class of functions of one variable; it consists of the functions 0, 1, x, and $\overline{x}$.

The set $\{0, 1, x\}$ coincides with both the class **QM** of monotone functions of one variable and the class **ML** of monotone linear functions; it is a self-dual class.

The class of two functions* $\{x, \overline{x}\}$ may also be described as a set of self-dual functions of one variable.

The pairs of functions* $\{0, x\}$ and $\{1, x\}$ are classes of functions of one variable which preserve the additive and multiplicative identities, respectively ($\mathbf{0}_0$, $\mathbf{0}_1$); they are dual to each other.

*To within a renaming of variables.

D(K): The class consisting of the disjunction $x_1 \vee \dots \vee x_k$, $k \geqslant 1$ (correspondingly, the conjunction $x_1 \dots x_k$, $k \geqslant 1$). The classes $\mathbf{D}^0$, $\mathbf{D}^1$, $\mathbf{D}^{01}$, $\mathbf{K}^0$, $\mathbf{K}^1$, and $\mathbf{K}^{01}$ are obtained from **D** and **K** by adding the corresponding constants. The classes **D**, $\mathbf{D}^0$, $\mathbf{D}^1$, and $\mathbf{D}^{01}$ are dual to the classes **K**, $\mathbf{K}^0$, $\mathbf{K}^1$, and $\mathbf{K}^{01}$ respectively.

The self-dual class* $\{x\}$ coincides with $\mathbf{KD} = \mathbf{OS}_0 = \mathbf{OS}_1 = \boldsymbol{OS}_{01}$.

A function f belongs to the class $\mathbf{F}^{(k)}$, $k \geqslant 2$, if any k sequences on which it is equal to 0 share a common additive identity at some place. Correspondingly, f belongs to the class $\mathbf{G}^{(k)}$, $k \geqslant 2$, if any k sequences on which it is equal to 1 share a common multiplicative identity.

$F^{(\infty)}$: The class of functions f such that all sequences on which f is equal to 0 share a common additive identity.

$G^{(\infty)}$: The class of functions f such that all sequences on which f is equal to 1 share a common multiplicative identity.

$\mathbf{MF}^{(k)}$: The class of monotone functions in $\mathbf{F}^{(k)}$, $k \leqslant \infty$.

$\mathbf{MG}^{(k)}$: The class of monotone functions in $\mathbf{G}^{(k)}$, $k \leqslant \infty$.

$\mathbf{F}_0^{(k)}$: The class of functions in $\mathbf{F}^{(k)}$ which preserve the additive identity.

$\mathbf{G}_1^{(k)}$: The class of functions in $\mathbf{G}^{(k)}$ which preserve the multiplicative identity. The classes $\mathbf{F}^{(k)}$, $\mathbf{MF}^{(k)}$, and $\mathbf{F}_0^{(k)}$ are dual to the classes $\mathbf{G}^{(k)}$, $\mathbf{MG}^{(k)}$, and $\mathbf{G}^{(k)}$, $(2 \leqslant k \leqslant \infty)$, respectively.

There are three other classes consisting of the constants: $\{0, 1\}$, $\{0\}$, and $\{1\}$.

In problems involving extended composition, keep in mind that only seven classes contain both the constants: **P**, **L**, **M**, **O**, $\mathbf{D}^{01}$, $\mathbf{K}^{01}$, and **OM**.

---

*To within a renaming of variables.

LITERATURE

Chapter 1*

1. J. T. Culbertson. *The Mathematics and Logic of Digital Computers* [sic].

2. L. A. Kaluzhnin. *What Is Mathematical Logic* [in Russian]. Nauka, Moscow, 1964.

3. J. Kemeny, J. Snell, and J. Thompson. *Introduction to Finite Mathematics.* Englewood Cliffs, Prentice Hall, 1974.

4. R. Stoll. *Sets, Logic, and Axiomatic Theories.* San Francisco, W. H. Freeman, 1974.

5. I. M. Yaglom. *Unusual Algebras.* Moscow, Mir, 1978 (in English).

Chapter 5

1. V. K. Korobkov. "How many monotone logical functions are there?" *Diskretnyi Analiz*, No. 1, Novosibirsk, 1963.

Chapter 6

1. G. A. Shestopal. "How many simple bases of Boolean functions are there?" *Dokl. Akad. Nauk SSSR*, **140**, No. 2, pp. 314-317, 1961. [Translated in *Soviet Mathematics.*]

*The list of literature for Chapter 1 includes books which may be useful to read prior to reading the current book.

Chapter 7

1. E. Post, The Two-Valued Iterative Systems of Mathematical Logic. Princeton, Princeton University Press, 1941.

2. S. V. Yablonskii, G. P. Gavrilov, and V. B. Kudyavtsev. *The Logical Functions and Post Classes* [in Russian]. Moscow, Nauka, 1966.

3. S. G. Gindikin and A. A. Muchnik. "Solving the completeness problem for systems of logical functions with unreliable realization." *Problemy Kibernetiki,* No. 15, pp. 65-84, 1965. [Translated in *Problems of Cybernetics.*]

Chapter 8

1. V. B. Kudryavtsev. "Completeness Theorems for a class of automata without feedback." *Problemy Kibernetiki,* No. 8, pp. 91-115, 1962 [Translated in *Problems of Cybernetics.*]

2. M. I. Kratko. "The algorithmic unsolvability of the problem of recognizing completeness for finite automata." *Dokl. Akad. Nauk SSSR,* **155**, No. 1, pp. 35-37 (1964) [Translated in *Soviet Mathematics.*]

3. (n.a.) *Automata* [in Russian]. Moscow, IL, 1956.

4. N. E. Kobrinskii and B. A. Trakhtenbrot. *Introduction to the Theory of Finite Automata* (Translation edited by J. C. Shepherdson). Amsterdam, North-Holland, 1965.

5. M. L. Tsetlin. *Studies on Automata Theory and the Modeling of Biological Systems* [in Russian]. Moscow, Nauka, 1969.

6. J. von Neumann. "Probabilistic logic and the synthesis of reliable organisms from unreliable components." In: A. H. Taub, *Collected Works of John von Neumann*, New York, Pergamon Press, Vol. 5, pp. 329-337, 1961-63.

Chapter 9

1. O. B. Lupanov. "Design of contact networks." *Dokl. Akad. SSSR*, **119**, No. 1, pp. 23-26, 1958. [Translated in *Soviet Mathematics.*]

2. E. I. Nechiporuk. "A Boolean function." *Dokl. Akad. Nauk SSSR*, **169**, No. 4, pp. 765-766, 1966. [Translated in *Soviet Mathematics.*]
3. S. V. Yablonskii. "Algorithmic difficulties in the design of minimal contact networks." *Problemy Kibernetiki*, no. 2, pp. 75-121, 1959. [Translated in *Problems of Cybernetics.*]
4. O. B. Lupanov. "Design of certain classes of control functions." *Problemy Kibernetiki*, No. 10, pp. 63-98, 1963. [Translated in *Problems in Cybernetics.*]
5. S. V. Yablonskii. "Realization of a linear function in the class of pi-networks." *Dokl. Akad. Nauk SSSR*, **94**, No. 5, pp. 805-806, 1954. [Translated in *Soviet Mathematics.*]
6. B. A. Subbotovskaya. "Realization of linear functions by means of formulas in the basis &, , and . " *Dokl. Akad. Nauk SSSR*, **136**, No. 3, pp. 553-555, 1961. [Translated in *Soviet Mathematics.*]
7. O. B. Lupanov. "Complexity of the realization of logical functions by means of formulas." *Problemy Kibernetiki*, No. 3, pp. 61-80, 1980. [Translated in *Problems of Cybernetics.*]
8. A. Karatsuba and Yu. Ofman. "Multiplication of multi-digit numbers on automata." *Dokl. Akad. Nauk SSSR*, **145**, No. 2, pp. 293-294, 1962. [Translated in *Soviet Mathematics.*]
9. A. L. Toom. "Complexity of a circuit of functional elements that realizes integer multiplication." *Dokl. Akad. Nauk SSSR*, **150**, No. 3, pp. 496-498, 1963. [Translated in *Soviet Mathematics.*]

Chapter 10

1. A. M. Yaglom and I. M. Yaglom. *Probability and Information* (translated by V. K. Jain), Boston, Kluwer Academic, 1983.
2. W. Feller. *An Introduction to Probability Theory and Its Applications*, Vol. 1. New York, Wiley, 1968.
3. S. N. Bernshtein. *Probability Theory.* Moscow, 1946.
4. J. Kemeny, J. Snell, and J. Thompson (*op. cit.*).
5. D. Polya. *Mathematics and Plausible Reasoning.* Princeton, Princeton University Press, 1969.
6. M. L. Tsetlin (*op. cit.*).
7. E. Moore and C. Shannon. "Reliable networks from unreliable relays." In: A. H. Taub (*op. cit.*).

8. S. G. Gindikin. "On Bernshtein polynomials associated with the logical functions." In: (n.a.) *Studies in Modern Problems of the Constructive Theory of Functions.* Baku, pp. 590-595, 1965.
9. S. G. Gindikin and A. A. Muchnik (*op. cit.*)

Chapter 11

1. S. V. Yablonskii. "Functional constructions in k-valued logic." *Trudy Matem. In-ta Akad. Nauk SSR im. Steklova,* **51**, pp. 5-142, 1958.
2. I. M. Vinogradov. *Elements of Number Theory* (translated by Sol Kravetz). New York, Dover, 1954.
3. A. Salomaa. "Certain completeness criteria for sets of multi-valued logical functions." *Kiberneticheskii Sbornik,* No. 8, pp. 7-32 (Mir), 1964.
4. V. V. Martynyuk. "Certain classes of functions in multi-valued logics." *Problemy Kibernetiki,* No. 3, pp. 49-60, 1960. [Translated in *Problems of Cybernetics.*]
5. I. Rosenberg. "Structure of functions of many variables on a finite set." *Comptes Rendues Acad. Sci. Paris,* **260**, Gr. 1, pp. 3817-3819, 1965.
6. E. Yu. Zakharov, V. B. Kudryavtsev, and S. v. Yablonskii. "Pre-complete classes in k-valued logics." *Dokl. Akad. Nauk SSSR,* **186**, No. 3, pp. 509-512, 1969. [Translated in *Soviet Mathematics.*]
7. Yu. I. Yanov and A. A. Muchnik. "On the existence of k-valued closed classes without a finite basis." *Dokl. Akad. Nauk SSSR,* 127, No. 1, 1959. [Translated in *Soviet Mathematics.*]

Chapter 12

1. E. Landau. *Foundations of Analysis* (translated from German by F. Steinhardt). New York, Chelsea, 1951.
2. C. Feferman. *The Number System. Foundations of Algebra and Analysis.* Reading, Mass. Addison-Wesley, 1964.

INDEX